T0312876

Forest Value Chain Optimization and Sustainability

Forest Value Chain Optimization and Sustainability

Sophie D'Amours
Mustapha Ouhimmou
Jean-François Audy
Yan Feng

CRC Press
Taylor & Francis Group
Boca Raton London New York

CRC Press is an imprint of the
Taylor & Francis Group, an **informa** business

CRC Press
Taylor & Francis Group
6000 Broken Sound Parkway NW, Suite 300
Boca Raton, FL 33487-2742

Printed on acid-free paper
Version Date: 20160705

International Standard Book Number-13: 978-1-4987-0486-1 (Hardback)

Library of Congress Cataloging-in-Publication Data

Names: D'Amours, Sophie, editor.
Title: Forest value chain optimization and sustainability / [edited by] Sophie D'Amours, Mustapha Ouhimmou, Jean-Franðcois Audy, and Yan Feng.
Description: Boca Raton, FL : CRC Press, 2017. | Includes bibliographical references and index.
Identifiers: LCCN 2016009689 | ISBN 9781498704861 (alk. paper)
Subjects: LCSH: Forest products industry--Canada.
Classification: LCC HD9764.C3 F68 2017 | DDC 338.1/74980971--dc23
LC record available at https://lccn.loc.gov/2016009689

Visit the Taylor & Francis Web site at
http://www.taylorandfrancis.com

and the CRC Press Web site at
http://www.crcpress.com

Dedication

This book is dedicated to Professor Eldon Gunn for his distinct and inspiring contributions to the research program in the field of forest value chain optimization and sustainability and his lifelong services to the forest research community at large.

Contents

SECTION III   *Forest Value Chain Tactical Planning and Wood Flows*

Preface

This book comes as a result of 6 years of a national collaborative research effort engaged in enhancing knowledge and innovation in the field of value chain optimization in the forest sector. Under the leadership of high-profile Canadian researchers, the guidance of a business-led board of directors, and an incredible team of employees, the Natural Sciences and Engineering Research Council (NSERC) national strategic Value Chain Optimization (VCO) Network brought together 17 universities, 60 professors and researchers, and more than 55 graduate students, and supported the realization of more than 93 research projects with the support of FPInnovations and more than 74 public and private partners.

This fantastic journey was launched in 2009 with a successful grant application to the NSERC strategic network program. The scientific leaders—Eldon Gunn, David Martell, Paul Stuart, Mustapha Ouhimmou, Mikael Rönnqvist, Bernard Gendron, Reino Pulkki, Robert Kozak, and myself—have established and developed a unique Canadian team to foster the benefit of a value chain approach to support decision-making in the forest sector. The unique mix of expertise of the scientific leaders served to deploy a challenging research program and a multidisciplinary scientific leadership.

The strong partnership with FPInnovations and its many contributions helped in conducting the real case-based research and, in particular, in translating the research results to innovative solutions for government and industry. This has permitted the implementation of one of the first national collaborative research and innovation models. We are grateful that FPInnovations shared the vision with us that a stronger forest sector resided in our joint ability to grasp the whole value chain perspective to improve decision-making.

The scientific community that was built through this initiative is fully engaged in contributing to the improvement of the competitiveness and sustainability of the forest sector. Broad knowledge has been developed through university, industry, and government collaborative research projects and disseminated through workshops, summer schools, seminars, and webinars. It became clear that all the knowledge created within the community and perspectives that have driven the thinking process of the VCO Network should be shared. This book is an ultimate effort in assembling the knowledge that has developed in this first phase of the VCO Network research. We expect that it will serve as a knowledge foundation for the future research in this field.

This being said, the outcomes of the VCO Network have gone far beyond the results of the research program, as long-lasting friendships have resulted from the collaborative efforts. The VCO Summer Schools in Université Laval, Lakehead University, FPInnovations, Dalhousie University, and École Polytechnique de Montréal gave great opportunities for many fun-filled and inspiring gatherings. There have also been special moments to appreciate the impressiveness of the upcoming generation of scientists and the wisdom of our pioneers. In that regard, I recall a very special moment when we could thank and acknowledge the incredible pioneering contributions of professor Eldon Gunn, Dalhousie University. I have had the privilege to be the first director of the VCO Network and to be mentored by Professor Gunn, an exceptional and generous colleague.

Let me conclude by acknowledging the steady and engaged support of the board members of the VCO Network. They have supported the VCO research community all the way, believing in our capacity in having a significant impact and providing their time and advice generously for the last 6 years. This book is a way to thank them sincerely for their invaluable contributions. Jason Linkewich has served as the President of the board for the last 4 years; and his vision and leadership has stimulated our community. He was supported by Winnifred Hays-Byl and David Chamberlain, who have also guided the VCO initiative over the last 6 years, offering their extensive experience within the forest sector. Over its first phase of development, the VCO Network also counted on Paule Têtu (past president), Yves Bergeron, Catherine Cobden, Peter Lister, Jean-Claude Savoie, Darcie Booth, Jean-François Levasseur, Trevor Stuthridge, Robert Beauregard, Reginald Theriault, Samir Boughaba, and Driss Haboudane. Finally, the VCO Network would not have been capable of delivering its full value without the dedication of past and current employees Mathieu Bouchard, Catherine Lévesque, Mustapha Ouhimmou, Jean-François Audy, Yan Feng, Clémence Gaborieau, Catherine Savard, Julie Richard, and Barbara McKenzie.

I hope this book will contribute to the development of competitive forest companies and sustainable forest sectors. I believe it is a foundation to build on and will serve further developments. This book has truly been a joint effort. I would like to thank all the contributors. It is their hard work, passion, and dedication that made this book possible. I want to thank all the reviewers for their valuable comments as they have contributed significantly to the high quality of the book. I would like to thank the great efforts made in gathering chapters written by many teams of the VCO Network. My colleague editors, Yan Feng, Mustapha Ouhimmou, and Jean-François Audy, have invested so many hours to make it a success, facilitating the planning of the book and the review process of each of its chapters. I would like to especially thank Lucie Verret for her expert advice on various matters in contracting and book publication. I would like to thank Mathieu Bouchard, Catherine Lévesque, Catherine Savard, and Clémence Gaborieau for their administrative support. I would like to extend our thanks to the team of Taylor & Francis, especially Joseph Clements, Andrea Dale, Kathryn Everett, and colleagues for their friendly support and patience.

Sophie D'Amours
On behalf of the VCO Community

Editors

Sophie D'Amours, PhD, is a professor of industrial engineering at Université Laval, Québec, Québec, Canada. She holds a Canada Research Chair (CRC) on planning sustainable forest networks. Her research interests cover a wide array of challenging issues related to value chain management and optimization in the forest sector. She was the founding director of the Forest-to-Customer Research Consortium (FORAC) and led the launching of the National Strategic NSERC Network on Value Chain Optimization in Canada.

Mustapha Ouhimmou, PhD, is an associate professor at the Automated Manufacturing Department at École de technologie supérieure, Université du Québec, Montréal, Québec, Canada. He holds a PhD degree in industrial engineering from Université Laval, Québec, Québec, Canada. He has industrial experience in the pulp and paper and wood furniture industries. His primary research interests are in value chain optimization in the forest products industry. He is the scientific codirector of the NSERC strategic network on value chain optimization (VCO Network).

Jean-François Audy, PhD, is a professor of operations management and logistics in the Department of Management, Business School at Université du Québec à Trois-Rivières, Québec, Canada, and member of the Forest-to-Customer Research Consortium (FORAC). His primary research interests are in the development of decision support tools and interfirm collaboration within and across value chains as a vector for competitive and sustainable networks.

Yan Feng, PhD, is a modeling and technical advisor in the NSERC Strategic Research Network on Value Chain Optimization at Université Laval, Québec, Québec, Canada. She has university teaching, research, and industrial experience. Her research interests are in the areas of value chain sustainability and integration, operations research, and robust decision-making methodologies in the forest products industry.

Contributors

Jean-François Audy
Département de management
École de Gestion
Université du Québec à Trois-Rivières
Trois-Rivières, Québec, Canada

Riadh Azouzi
Département de génie mécanique
Université Laval
Québec, Québec, Canada

Justin G. Bull
Department of Wood Science
Faculty of Forestry
University of British Columbia
Vancouver, British Columbia, Canada

Marc-André Carle
Département d'opérations et systèmes
    de décision
Université Laval
Québec, Québec, Canada

Virginie Chambost
EnVertis Consulting Inc.
Montreal, Quebec, Canada

Denis Cormier
Fibre Supply
FPInnovations
Pointe-Claire, Québec, Canada

Sophie D'Amours
Département de génie mécanique
Université Laval
Québec, Québec, Canada

Cédric Diffo Téguia
Department of Chemical
    Engineering
École Polytechnique de Montréal
Montréal, Québec, Canada

Jean Favreau
Modeling and Decision Support
FPInnovations
Pointe-Claire, Québec, Canada

Yan Feng
Département de génie mécanique
Université Laval
Québec, Québec, Canada

Jean-Marc Frayret
Département de mathématiques et de
    génie industriel
École Polytechnique de Montréal
Montréal, Québec, Canada

Christopher Gaston
Department of Wood Science
Faculty of Forestry
University of British Columbia
Vancouver, British Columbia, Canada

Bernard Gendron
Département d'informatique et de
    recherche opérationnelle
Université de Montréal
CIRRELT (Interuniversity
    Research Centre on Enterprise
    Networks, Logistics and
    Transportation)
Montréal, Québec, Canada

Eldon A. Gunn
Department of Industrial
    Engineering
Dalhousie University
Halifax, Nova Scotia, Canada

Robert A. Kozak
Department of Wood Science
Faculty of Forestry
University of British Columbia
Vancouver, British Columbia, Canada

Luc LeBel
Faculté de foresterie, géographie et
    géomatique
Université Laval
Québec, Québec, Canada

Alexandra F. Marques
Centre for Enterprise Systems
    Engineering
INESC TEC
Porto, Portugal

David L. Martell
Faculty of Forestry
University of Toronto
Toronto, Ontario, Canada

Azadeh Mobtaker
Département de génie de la production
    automatisée
École de Technologie Supérieure
Montréal, Québec, Canada

Mustapha Ouhimmou
Département de génie de la production
    automatisée
École de Technologie Supérieure
Montréal, Québec, Canada

Nathalie Perrier
Département de mathématiques et de
    génie industriel
École Polytechnique de Montréal
Montréal, Québec, Canada

Marius Posta
Département d'informatique et de
    recherché opérationnelle
Université de Montréal
Montréal, Québec, Canada

Reino Pulkki
Faculty of Natural Resources
    Management
Lakehead University
Thunder Bay, Ontario, Canada

Catalin Ristea
Modeling and Decision Support
FPInnovations
Vancouver, British Columbia, Canada

François Robichaud
Business Analysis
FPInnovations
Québec, Québec, Canada

Mikael Rönnqvist
Département de génie
    mécanique
Université Laval
Québec, Québec, Canada

Shabnam Sanaei
Department of Chemical
    Engineering
École Polytechnique de Montréal
Montréal, Québec, Canada

Paul R. Stuart
Department of Chemical
    Engineering
École Polytechnique de Montréal
Montréal, Québec, Canada

1 Introduction

Yan Feng, Jean-François Audy, Mustapha
Ouhimmou, and Sophie D'Amours

CONTENT

The forest industry is gaining an increasing importance worldwide owing to its significant contributions to our society by creating jobs, creating economic value, improving our environment, generating renewable energy with remarkable potential for substituting fossil fuels and chemicals, and mitigating climate change. With the third-largest forest area in the world and its abundant forest resources, Canada has an important role to play in both the national and international landscapes to transform the forest industry for greater value creation. This is particularly meaningful to forest-dependent communities and First Nations in Canada as it supports more than 200 communities, creates approximately 235,000 direct jobs, and accounts for 2% gross domestic product (GDP) nationwide (Lindsay, 2014). Nevertheless, the Canadian forest industry is facing great challenges. Markets for traditional forest commodities, such as commercial printing papers, lumber, and wood products, are becoming increasingly competitive because of market globalization, volatile commodity prices, shrinking demand, and increased raw material and energy costs. Meanwhile, the emerging modern bioeconomy is characterized by rapid advancement in biotechnologies, commercialization, and the expanding market for value-added bioproducts, bioenergy, and biofuels. This business environment has forced Canadian forest companies to reevaluate their business strategies, develop new business models, and seek new opportunities for business transformation.

Historically, the Canadian forest industry has relied on its massive harvesting operations that "push" resources to the processing mills and market in a traditional "supply push" business model. The premises of this model have been cheap fiber, energy, and labor, combined with a very large, nearby market such as the United States for traditional products such as paper, lumber, and wood products. These competitive factors are vanishing due to global economic volatility and worldwide competition. It has become clear that this low-value commodity-focused business model will not be economically competitive and it will not be sustainable for the industry and forest-dependent communities. This raises several questions to forest industry companies, government, and researchers. What are the weaknesses and consequences of the current state of forest management and forest industry operations? What new business approaches and strategies should the Canadian forest industry consider? How do we ascertain that a proposed business strategy is the best solution for a company? What are the potential risks and how can they be mitigated? How do we improve the global competitiveness of the Canadian forest industry? These questions challenge us to think innovatively beyond our traditional thinking frame to explore new knowledge and methodologies for reengineering the forest sector. Significant research and development, funded by the federal and provincial governments, has taken place, aiming at improving the competitiveness of the Canadian forest products industry by diversifying the revenue streams from the forest resources and transforming the forest industry toward an integrated forest bioeconomy.

The objectives of this book are to assemble the recent research and development in forest value chain transformation and optimization. It is aimed at being a reference book for researchers and practitioners within the field. Thus, the book does not provide new methods or results but rather presents literature reviews, strategic research orientations, assessment of current key issues, and state-of-the-art methodologies of decision support for forest value chain optimization.

This book addresses the concept of value chain in the forest industry. The value chain recognizes that business activities in the forest industry including forest resource management, harvesting operations, logistics, manufacturing, product distribution, and sales and marketing performed by a single company or several companies are not isolated. They are interdependent and therefore a decision made in one area of business has subsequent effects on the others. This chain of business activities needs to be managed and coordinated so that greater values can be created throughout the transformation process from raw materials to the products that meet the needs of the consumers in the market. Uncoordinated business decisions can result in significant economic consequences as a result of mismatched supply and demand, excessive inventories, waste of resources, long lead times, uncertainties, and customer dissatisfaction. This value chain–based thinking, thus, is fundamental for helping the decision-makers to focus on the entire business processes and coordinations to reduce cost and maximize values.

Emphasis is placed on the optimization of the value chain in the Canadian forest products industry. Carle et al. (Chapter 2) and Audy et al. (Chapter 10) have pointed out that forest industry consists of several value chains with different core businesses that are complex, nonlinear, and interconnected. Figure 1.1 shows the flows of raw materials, products, by-products, and energy within and among the key forest industry value chains of sawmills, pulp and paper mills, wood panel mills, and the integrated biorefineries and biofuel plants. These interconnected value chains result in a complex network connecting market to forest with many divergent processes performed by different business units and business sectors. Although the companies within the value chain network are typically independent and autonomic entities with different business goals and strategies, they are interdependent in many ways through business transactions and customer–supplier relationships. This interdependency raises many collaboration and value creation opportunities when a systematic and holistic approach and value chain optimization methodology are adopted to view this industry as a whole. To optimize such a complex value chain network, a full understanding of forest industry interrelationships is needed including the integrated forest bioeconomy of Canada representing the new visions and the biopathways proposed by Natural Resources Canada, FPAC, and FPInnovations. As the modern forest industry value chain extends from its traditional core businesses to several

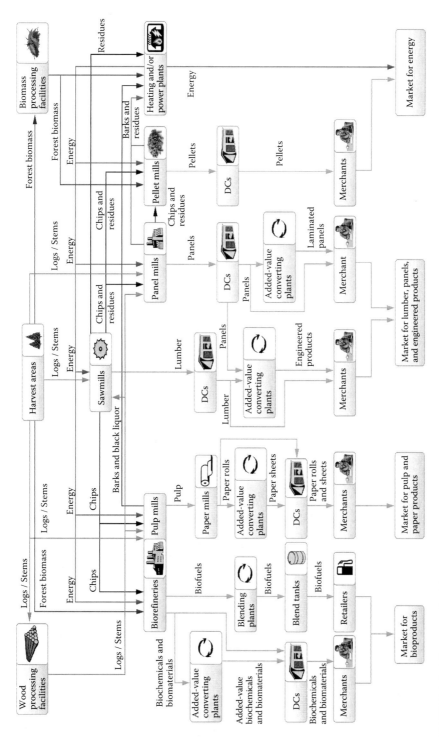

FIGURE 1.1 The network of forest bioeconomy value chain.

value-added businesses in biochemicals, biofuels, and bioenergy, it will have the potential to increase not only the economic value but also the social and environmental benefits nationwide.

Emphasis is also given to forest industry sustainability, which is an important value from the very nature of the Canadian forest society and multiple stakeholders' point of view. Forest industry value chains incorporate many actors including the forest industry companies, the federal and provincial governments, landowners, nonprofit organizations, and local communities. These actors often have very different preferences and priorities in regard to forest management and usage. Government policies and business decisions must be made with respect to these values and to the very nature of the Canadian forest (e.g., varying species and sustainable forest management) to create the maximum value with market opportunities. This includes business reengineering and value chain redesign taking into account the unique advantages of the forest industry in developing and advancing the technologies and capacities for renewable energy while mitigating the carbon footprint throughout forest operations, manufacturing, and logistics value chains. The decisions must also respect the social, environmental, and ecological interests of the stakeholders to build long-term sustainability and prosperity for the forest industry and society as a whole.

This book consists of 12 chapters covering a wide range of subjects with comprehensive literature surveys and insightful discussions. The knowledge covers from strategic forest bioeconomy transformation to tactical forest industry value chain planning; from forest management to manufacturing, logistics, and markets; from lumber and pulp and paper industry to biomass and biorefinary value chain diversifications; from decision support to collaborations and information sharing; and from government policies and national research to industry analysis. Each chapter is unique and the depth of knowledge is profound.

We organized the book into three sections. Section I, "Forest Biorefinery Transformation and Strategic Planning," consists of Chapters 2 through 5 and addresses the forest bioeconomy transformation and strategic planning. Chapter 2 focuses on the strategic research framework built on the forest value chain fundamentals for reengineering forest value chains to capture the broad benefits as forest industry developing the bioeconomy capacities. The concept of the value chain and its potential benefits in the Canadian forest industry to improve economic, social, and environmental values are illustrated. The challenges of decision-making in the multistakeholder business environment with a dynamic market and uncertainties are discussed. Chapter 3 emphasizes strategic planning, exposing the need for systematic approaches and management for forest biorefinery transformations. It addresses many challenges that forest product companies are facing when engaging in a business transformation and product diversification under market volatilities, technology, and financial risks. The chapter consists of two parts. Part I reviews several key concepts and tools for the assessment and evaluation of biorefinery business models. Part II proposes a systematic approach for the identification of promising transformative models from business strategies to the design of product portfolios, processes, and value chains. Two case studies in the pulp and paper sector were provided for illustrations. Chapter 4 discusses the roles of FPInnovations (the national research institution), governments, and industry in Canada's nationwide forest industry's transformation through collaborative research, knowledge dissemination, and technology transfer. Government initiatives and policies supporting the forest industry transformation over the last decade are reviewed. The need for a value chain optimization approach and its challenges in the forest industry are discussed. Chapter 5 addresses strategic forest management planning and industrial capacity planning from a value chain perspective. Knowledge about forest management and various challenges faced in the forest sector are discussed including sustainability, environmental strategy, unexpected events, and spatial and temporal issues in a complex multistakeholder environment. The need for a decision support system that supports strategic sustainable forest management planning considering the downstream industry value chain capacities is discussed.

Section II, "Forest Value Chain Sustainability, Market Research, Collaboration, and Agility," is composed of Chapters 6 through 9. It explores important knowledge on forest value chain sustainability, market research, collaboration, and agility. As the focus on sustainability has become widespread, questions arise concerning what it means and why it concerns forest value chains. Chapter 6 explores the meanings of sustainability, the wide spectrum of drivers for sustainability from regulatory initiatives of governments to advocacy campaigns, and the responses to sustainability from the private and public sectors. Two case studies are presented with implications for the value chain in the Canadian forest product sector. Chapter 7 addresses the importance of connecting market needs with value chain strategies. Emphasis has been given to market research, distribution channels, and the strategic changes experienced in the distribution networks and market, as well as the implications in the Canadian wood products industry value chain management. This is highly relevant as market research in value chains permits a better assessment of the multiple values of alternative forest and business strategies. Creation of forest products and their delivery to the end user markets involve many business activities that require collaborations between actors to coordinate these complex business activities. Chapter 8 discusses the importance of value chain collaboration and the need for decision support tools such as collaborative modeling and simulation tools for the forest products value chain. Surveys of literature on agent-based modeling and standardization of information and knowledge sharing are presented in the context of the forest products value chain.

Chapter 9 introduces the concepts of agility and flexibility in complex manufacturing and value chain systems. Their important roles in helping companies to achieve quick responses to the volatile business environment are discussed. Agility is analyzed from different dimensions including information, organization and operation requirements, system design and flexibility, decision support system for real-time decision-making, and collaboration to better understand how a value chain can adapt itself to become agile. A softwood lumber value chain case is included for illustration.

Section III, "Forest Value Chain Tactical Planning and Wood Flows," includes Chapters 10 through 12. It provides comprehensive knowledge on tactical planning and wood flows in the forest industry value chain. Chapter 10 is focused on tactical planning and decision support tools and systems in the forest-based value creation network. A broad review of literature on decision support models, solution methodologies, and decision-making frameworks is discussed, highlighting the most successful decision support system development worldwide. Significant applications and benefits and implementation challenges are reported. A synthesis of the trends in planning methods and decision support systems and future research directions is presented. In Canada, forest operations generate a considerable amount of forest residues. Chapter 11 focuses on the forest biomass and specifically forest logging residues and inferior-quality timbers, which are not suitable for producing higher value forest products, but do have economic and environmental potential through production of renewable energy and biofuels. The value chain of forest biomass, sustainable network design, management and utilization of bioenergy, and biorefining transformations in Canada are discussed. Chapter 12 presents a wide range of knowledge on wood transportation, one of the most challenging research areas in the Canadian forest value chain business network. Multimodal wood transportation, strategic, tactical, and operational planning problems, and operations research methodologies are discussed.

The intended readers of this book are researchers, engineers, managers, planners, government agents, policy-makers, university graduate students, and faculty members working for academia, industry, and government. It will benefit more significantly those who work in the fields of forest management; regional sustainable development; biomass utilization and biorefinery/bioenergy strategies; forest value chain optimization; decision support; manufacturing agility, flexibility, and efficiency; logistics and distribution systems; and marketing and sales in the broad forest industry sector. However, the knowledge can be applied in the forest products industry in other countries.

Our objectives are to provide as much as possible of the state-of-the-art knowledge and methodologies from the latest research addressing the many key issues, concepts, and knowledge development in the field of forest value chain management. We hope that readers find this book valuable and that it serves well as a useful reference book to inspire new innovations in this important and transforming industry.

REFERENCE

Lindsay, D., 2014. Vision 2020 report card: 2010 to 2012: Pathways to Prosperity for Canada's forest products sector. FPAC/APFC.

Section I

Forest Biorefinery
Transformation and
Strategic Planning

2 Framework for Value Chain Optimization of Forest Networks in the Emerging Canadian Bioeconomy

Marc-André Carle, Sophie D'Amours,
Eldon A. Gunn, and Reino Pulkki

CONTENTS

2.1 INTRODUCTION

Natural resources are an inseparable part of human history, in particular forest resources, which continue to play a major role in national prosperity. For example in Canada, forests are a key contributor to society, particularly for First Nations and many forest-dependent communities. As we undergo a period of massive change in terms of product markets, availability of fossil fuel and other nonrenewable resources, and the ability of the natural environment to absorb the emissions of greenhouse gases and other pollutants, it is important to once again look to the forests to provide innovative, renewable sources of materials and energy that can be a fundamental strength of the modern bioeconomy.

The worldwide forest products industry is at a crossroads. Its business environment has drastically changed in a very short period forcing companies to improve their value propositions. International and national factors such as increased overseas competition, fluctuation of currency values, new production technologies, increasing cost of energy, trade disputes, and reduced access to capital due to poor past returns have all negatively impacted the competitive advantage of the traditional forest products industry.

The Canadian forest industry has always relied on its massive fiber supply as well as on a "push to market" business model. However, it has become clear that this low-cost commodity-focused business model will not be economically, environmentally, or socially sustainable for the industry or the forest-dependent communities. The premises of this model have been cheap fiber, energy, and labor, combined with a very large, close-by market such as the United States for both paper and lumber products. These competitive factors are vanishing, raising the need for new premises and fundamentals for reengineering the forest sector.

The value chain concept is providing new grounds to support the decision-making process. The objective of this chapter is threefold: (1) to illustrate how value chain concepts can be applied to the Canadian forest sector; (2) to propose an integrated decision-making framework, which supports multiple actors, business models, and value chains; and (3) to highlight how the implementation of the framework can support the reengineering of the Canadian forest value chains. Some definitions of value chain fundamentals and also the value chains found in the Canadian forest sector are discussed in this chapter. The proposed framework is then presented and discussed. Each of its main components is discussed in a separate section.

2.2 FORESTS AND VALUE CHAINS

2.2.1 Value Chain Concepts

Simply put, a value chain is a sequence of business activities performed to deliver a valuable product or service for a given market (Porter, 1985). It typically involves several activities such as raw materials sourcing, several processing or manufacturing steps, as well as logistics and sales. These activities can be performed by a single company or divided between several actors. Value chain–based thinking is important because it forces companies to focus on the whole sequence of processes that add value to the product up to the final consumer, rather than just the steps that are under their direct control. In particular, value chain concepts can help the industrial sector to have the capacity to effectively manage "change" and lower resistance to change.

The forest sector consists of several "chains," each composed of many interconnected and divergent activities performed by different business units. In most of these divergent processes, alternative production recipes can be used to generate different baskets of products from the same sets of inputs. For instance, in the sawmilling activity, different cutting patterns can be applied to a given log to generate different volumes of raw lumber and chips. Each of these decisions affects the amount of materials for many processes further down the chain. As such, there can be substantial benefits in taking into account how one's decision affects other components of the value chain.

The concept of value chain is also critical because of the very nature of the Canadian forest sector. In the business literature, the notion of "value" is sometimes reduced to its financial dimension as attributes that the end consumer is willing to pay for. However, a broader view is necessary to understand the sector. Forest value chains incorporate many actors with different roles, such as companies, governments, landowners, nonprofit organizations, and cooperatives. These actors often have very different views on the notion of value. While a company may emphasize profits, service levels, and product quality, governments might see value in employment levels, environmental protection, or impacts on specific groups such as forest-dependent communities and the First Nations.* The concepts of "value chain" are broad enough to encompass all these dimensions.

For example, one value chain might include the producers who manage a complex sequence of tasks in the forest industry, for example, the various entrepreneurs who convert the trees into logs or chips, the sawmills that cut the logs into boards or dimension parts, the pulp and paper mills that use the wood chips to create reels of paper that are then cut into smaller rolls or sheets, and the pellet mills that transform bark and sawdust into fuels. Furthermore, many value chains often mesh together, such as pulp and paper mills that produce energy, biofuels, or specialty chemical products in addition to the more traditional products. These varied many-to-many processes make the task of integrating the procurement, production, distribution, and sales activities very complex, particularly given that these activities are always bound by the tradeoffs between yield, logistical costs, and service levels, and must meet environmental and social objectives.

A full understanding of the value chain is needed to synchronize and optimize forest and business decisions so that market opportunities are exploited while sustaining and improving forest assets. Since the modern forest value chain extends to value-added wood products as well as to a broad variety of fiber products and biofuels/biochemicals, this added management complexity has the potential to increase the benefits derived from the forest value chain. As the supply of wood fiber is heavily constrained in volume and characteristics, new market

---

* The dependence of the forest industry on Crown Land in Canada requires an explicit recognition of governments' role as public stewards. In addition, governments define the various requirements of sustainability for the sector: the Canadian Council of Forest Ministers Criteria and Indicators (CCFM C&I) of Sustainable Forest Management. The CCFM C&I are organized under the six main headings of biological diversity, ecosystem condition and productivity, soil and water, role in global ecological cycles, economic and social benefit, and society's responsibilities.

opportunities must be seized through reengineering the value chains and aligning all its partners to extract the maximum possible value from the forest and to do so in a manner that preserves the forest as a sustainable resource for future generations. Forest value chains are industrial ecosystems that provide a variety of social and ecological values and continue through the many productive possibilities of building materials, fiber-based paper and other products, energy sources, and biofuels/biochemicals.

2.2.2 DECISION-MAKING IN FOREST VALUE CHAINS

Effective, coordinated decision-making in forest value chains faces a number of significant challenges. The sheer number of different actors in the chain is a first source of complexity. This phenomenon has been magnified by the current trend to see each business unit as a center-for-profit, effectively encouraging decoupled and myopic decision-making. The different nature of the various actors (government, landowners, and many types of companies) and the various missions they pursue is also a source of complexity. Effective, value chain–based decision-making is difficult because it requires coordination of decisions with different scopes, levels of aggregation, and planning horizons for each of the activities.

For instance, the process of growing and harvesting trees has its own complexities requiring an integrated view from short-term local harvesting decisions to long-term regional forest strategies. It implies a complex network of decisions over a variety of time scales by different actors. At the same time, the value chains involved in the production and distribution of timber and fiber products are at least as challenging to manage. As already mentioned, each step requires disassembly of products emerging from previous activities: from forest stands to trees, from trees to logs, from logs to boards, cants, chips, and sawdust, with the chips eventually broken down into their cellulose, hemicellulose, and lignin components, with alternative processing recipes and uncertain yields at every stage.

The industrial and market landscapes are also rapidly and constantly changing. Important new potential fiber uses have recently been identified as well as processes that transform wood fiber into their biochemical constituents, which allows for an even larger number of new potential products (Mansoornejad et al., 2013). These fiber and biochemical products present different market opportunities, as well as new sources of risks and uncertainty. A rapidly changing environment also raises the need to revise the design of the value chains at a more fundamental level.

Traditionally, the forest and business networks have been treated as decoupled networks driven by different values and with limited opportunity for one to influence the other (D'Amours et al., 2008). Even within the forest and business networks, decisions have been made at a localized (unit-based) level, which have disregarded the interdependencies and the opportunity to add value through potential integration and/or collaboration. Fortunately, of necessity, the mindset of decision-makers is fundamentally changing (Ouhimmou et al., 2009). One contributor is the recognition that there are tremendous opportunities to be achieved, even in environments where centralized control is undesirable or unworkable through collaboration, information sharing, and proper allocation of the resulting benefits. The benefits of better coordination in decision-making, either through centralized or collaborative planning, are being recognized, but are yet to be fully captured.

2.2.3 Value Chain Optimization

More integration in decision-making raises many opportunities, but capturing these opportunities is neither obvious nor straightforward. An interesting approach is to harness the power of optimization to accurately model and capture these trade-offs. It effectively allows sorting through large numbers of potential solutions and selecting the set that captures the most value.

In this context, value chain optimization refers to using optimization models and their solutions to support decision-making in value chains. Over the past 50 years, optimization has been used in many contexts for decision support purposes, including forest operations planning (Rönnqvist, 2003) and supply chain management (D'Amours et al., 2008). Many optimization models have been proposed for logistics planning, facility location, capacity expansion, and inventory management. However, just as marginal analysis of individual products is unlikely to yield system level optimization, it is unlikely that optimizing a single set of decisions for a given business unit or activity in the chain would result in maximization of overall benefit to the chain. Thus, the challenge in bringing value chain optimization to the organization involves the integration of many decisions into a set of broader models. This implies taking into account multiple activities and decision-makers, different planning levels and horizons, as well as different objectives and notions of value.

The value chain approach will find fundamental new applications in answering these new questions. Individually, some of these products/markets have limited environmental, social, and economic impacts, but taken together, along with existing products and processes, "integrated" value propositions provide synergies that can augment the impacts considerably. The overall design of the new technological possibilities raises the challenge of bundling those new product offers into coherent value propositions. Value chain modeling and optimization can contribute to the decision-making process by creating an integrated systems' viewpoint of the different processes from forest to market.

The aim is therefore increased "value" and a more integrated "chain" in the forest sector. Value chain optimization provides the appealing concept of a chain of value creation opportunities from the forest to the consumer, particularly looking backward from the final products and values to the tree growing in the forest. However, looking forward, what we are really dealing with is a complex network of activities ranging from management and harvest of the forest through many joint production and distribution activities in the modern forest products industry.

As is evident from the aforementioned outline, implementing this vision requires facing many challenges due to the unique aspects of the forest sector. Some aspects are worth reemphasizing:

- Society shares diverse values. People want not only economic growth and high-quality jobs but are also very concerned with environmental issues. The development of any value chain in the forest industry must meet the high environmental standards for future generations.

- The value chain is decoupled in ways that pose additional challenges in terms of integration mechanisms and decision support systems. The ownership structure varies among countries, and rarely does the whole value chain fall under one unique ownership structure. As an example, in Canada, the forest is mainly public (Crown Land) and the industry is privately owned.
- The forest structure is not uniform across the world. The boreal forest in Sweden and Finland is mostly privately owned and in comparative terms may be considered highly managed for wood production purposes. In Canada, the forest is diverse: many species are often found in mixed stands, each providing different wood attributes and being used in different production processes (such as lumber or pulp production). There is also a marked regional distribution of species at a variety of spatial scales. Forest managers need to deal with large forestland areas, and very diverse forests, terrains, and climate conditions. In the boreal forest, growth is slow and rotations are long. These facts pose challenges in terms of forest planning, forest allocation, road construction, and harvest planning. Forest conditions pose a constraint on the potential set of value propositions and how these can be deployed.
- The manufacturing and transformation processes of the industry are mainly divergent/disassembly processes. This means that policy design, business models, and production planning for various products cannot be carried out independently. This significantly increases the complexity of the planning process, particularly in the context of strong competition for various markets from international producers.
- Due to varying weather and seasonal conditions, the operational context changes significantly during the year. For instance, wood transportation and hauling may only be possible during summer and winter. In addition to increasing planning complexity, these factors also lead to increased inventories along the value chain, which may in turn lead to increased operating costs and reduced quality of raw materials and products.

The changes required to implement such a vision are substantial and challenging. The following section proposes a framework designed for value chain optimization in the forest bioeconomy.

The key objectives of the proposed framework are as follows:

- Support decision-makers in an effort to design new, optimized, and sustainable forest bioeconomy networks
- Increase value gain from forest and asset utilization by developing decision support tools for integrating the whole value chain
- Improve competitiveness through a structured and coherent implementation of new and optimized value propositions and business models
- Improve agile execution and value capture throughout the business networks
- Develop a culture of analytical decision-making in the forest bioeconomy

2.3 A HIERARCHICAL FRAMEWORK FOR VALUE CHAIN OPTIMIZATION IN THE FOREST BIOECONOMY

Optimization of the forest value chain implies optimization of design, planning, and execution. These optimization processes are typically tackled with different methodologies. Design is often supported by engineering approaches including advanced operational research models and tools. Design is a creative process that searches for creative design alternatives, often motivated by business knowledge and prescriptions, as well as a set of potential scenarios that define possible environment conditions and outcomes. Planning is mainly driven by the use of advanced planning systems built on an explicit representation of the resources and activities to be planned, the objectives to be fulfilled, and algorithms that propose an optimal plan for execution. Finally, optimizing execution relies on the ability to follow the plans, control and feedback to the planning function, and the implementation of best practices.

Following Silver et al. (1998) we can address three types of activities within a company: (1) strategic design, (2) tactical planning, and (3) operational control. These are distinguished by managerial level, time horizon, and uncertainty in the decision environment. Strategic design decisions establish and modify the resources available to the enterprise. These include not only physical sources of supply, processing facilities, and other physical infrastructure, but also the resources in terms of special skills, systems, contracts, and other long-term commitments. Strategic design decisions are the role of top management, occur over long time horizons often spanning from 5 to 20 years, are typically based on highly aggregated data, and involve considerable risks and uncertainties. Tactical planning involves plans for making the most effective use of the resources available. Typically, these imply developing time tables for planned production, transportation, workforce, sales, and the inventory levels that will optimally match production to sales. Tactical planning typically involves plant level management, monthly or annual time scales, moderate levels of data aggregation, and manageable levels of uncertainty. Enterprise level profit optimization and other measures of performance consistent with strategic goals are the focus of tactical planning. Tactical plans provide feedback to the strategic design process. Operational control involves the detailed execution of plans and schedules by all parts of the organization in such a way as to achieve the profit and performance goals established by the plans. This involves detailed short-term decisions of what to produce, when, where, how much, and by whom/what. The time frame is very short-range (e.g., weeks and days), the managerial level is at the supervisory level, and the degree of uncertainty is relatively low.

While the classical hierarchical planning framework described above is useful, several adaptations are required to better fit the needs of forest value chains decision-making. First and foremost, the proposed framework for the forest value chain optimization requires a distinction between two levels of strategic decision-making. The first of these addresses the need for coordination and strategy at the highest level. It involves both governments and companies, and aims to address policy-making and industry-wide strategies. Sustainability of the forest resource requires very long-term perspectives, as well as the coordinated efforts of both governments and a substantial number of companies and organizations at the national level; it is not something than even corporate-level executives can implement alone. In Canada, government-level decision-making representing society's interest in the forest is unavoidable because of the significant level of land ownership by provincial governments. Long-term decisions that define acceptable forest harvest levels are affected not simply by the timber growth capabilities of the forest, but also by the complex interaction of policies. At the national level, these can be seen in the Canadian Council of Forest Ministers Criteria and Indicators (CCFM C&I). While some policy goals (such as constant or increased forest growth) are common among various governments, every government has its own legislation and policy framework. Through the design of regulations, industry-wide norms, policies, and subsidies systems, decision-making at this level has the potential to drive whole value chains toward the configuration maximizing their competitiveness. It can also provide a basis to align the objectives of various actors, or at least settle compromises between conflicting value definitions. For instance, the investment in physical capacity both for conventional sawmills and pulp mills, and for the new types of capacity for advanced wood products, for new types of fiber products, and for biofuels and biochemicals, will require long-range time horizons and may involve both government policy-makers and top-level corporate executives. This will be referred to as the Integrated Forest and Industry Strategies level within the proposed framework.

The second, medium-term strategic design level involves decisions that define the business model and its strategies, as well as the value chain capacities. The business model and its strategies provide the coherent logic among the value proposition, the forest management, procurement, production and distribution strategies, as well as the innovation strategy. The value chain capacities are also critical in synchronizing and coordinating the business model and forest management. These include design of access to the forest in terms of roads and transportation infrastructure, design of wood handling and merchandising facilities in terms of their nature, location, and capacity, and the design of receiving and processing machinery at the conventional and new mills. Not all of these strategic decisions involve physical facilities. They can also involve decisions to adopt certain types of business strategies, the subsequent commitments and contracts for integrated operations, and the design of the information and control systems that facilitate collaboration. These medium-term decisions can be made within time frames ranging over a few years. The key distinction is that forest growth and regeneration plays only a limited role in this time frame, whereas market perspectives, industry investment capacities, and forest access drive the decision process. This will be referred as the Strategic Network Design level within the proposed framework.

Tactical planning involves making the resource allocation decisions that optimize performance of the enterprise value chain within the constraints defined by the long- and medium-term strategic designs. This involves annual planning of sales, workforce and contractor plans, harvest plans, and product flows from forests to customers. It also involves monthly planning of production and crew schedules both in the forest and in the transportation and mill sectors. Tactical planning has extended over recent years to include "cross chain coordination," providing new capabilities to the enterprise through collaboration with other value chains. This explains why this level of decision-making within the proposed framework is referred to as the Integrative Planning level, aiming to integrate both vertically and horizontally the value chains of forest companies.

Operational planning and control involves the processes that guarantee operational excellence in achieving or exceeding the value proposition goals established in the tactical plans while keeping costs and negative impacts on other components of the value chain as low as possible. These short-term decisions include order promising, order fulfillment, product allocation, product distribution, production scheduling, procurement planning, harvesting (including assortments to produce), transportation and inventory control, as well as the forest treatment schedule. In the framework, these planning and control activities are referred as the Operative Planning and Control level. In this framework proposal, decision processes are structured to show how decisions at all levels influence each other as shown in Figure 2.1. Decisions are taken in an asynchronous manner, that is, when they are needed by the people. The upper-level decisions always frame the lower-level decisions once they are made. However, the lower decision processes are always feeding back updates and critical information that can trigger review of upper-level decisions. The process is not linear, but a constant feedback process in which the upper-level decisions provide a context for the lower level, while the lower level also provides constant checking on the feasibility and optimality of the decisions

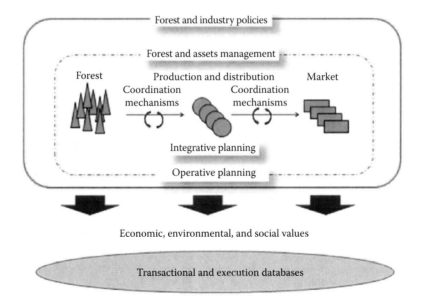

FIGURE 2.1   Coordination systems.

taken at the upper level (Schneeweiss, 2003; Gunn, 2007). A deep understanding of the lower level also provides potential options or solutions for the upper level, as well as grounds for a robust evaluation of the impact of a decision. The decision processes are also tightly linked to the execution and a formal control loop is required. This raises issues in terms of planning methods, engineering, and implementation approaches. Moreover, one needs to recognize that the value chain is almost always decoupled, through a distributed control structure, where direct control is not possible.

At the more long-term strategic level, scenarios are to be defined; they will include a complex mix of potential forest and industry strategies as well as some characterization of plausible economic and forest futures. These scenarios should support the strategic planning process. At the mid-term strategic and tactical level, agile logistics and manufacturing systems are to be proposed and engineered into integrated value chains (e.g., well-blended planning and production systems). These integrated value chains are characterized by different network models building on bounded business logic and productive systems. Finally, at the operational level, the components of the value chain define the elements used to conduct the planning and to execute the plans. They compose what we call the value chain knowledge base. Although the decision processes discussed in this framework have different scopes and horizons depending on the level at which they are performed, they share some common features. All the decision processes consist of analyzing a set of potential solutions or prescriptions. Optimizing and integrating the value chain at all levels requires a set of objects and models, as well as access to significant amounts of data (e.g., resources, capacities, recipes, and flows). These may be discrete or continuous and serve to set either the decision variables or constraints of the planning problem. These decisions can also be supported by decision support systems.

Typically, value chain analysis has emphasized operational excellence and it is appropriate to find ways of achieving good operational control even if the resources and the plans for the use of these resources are expected to change. Put simply, operative planning seeks to achieve the best performance with the existing facilities, policies, and environmental conditions. Similarly, tactical planning takes place within the limits imposed by resource availability and network structure (both physical and organizational). Regardless of the quality of decision-making at the upper levels or the amount of unexpected changes in the environment, it is still possible—and important—to make good tactical and operational plans and to make sure these plans are implemented into the correct actions. Another important aspect of operational control is to provide relevant feedback loops to the decision-makers at upper levels.

The framework as just described is shown in Figure 2.2. The aim is to integrate and optimize the value chain as a whole. To link high-level forest and industry strategies down to operational execution, the framework emphasizes the contribution of decision support methods and tools, scenarios, prescriptions and knowledge representation, as well as best practices in implementing value chain optimization.

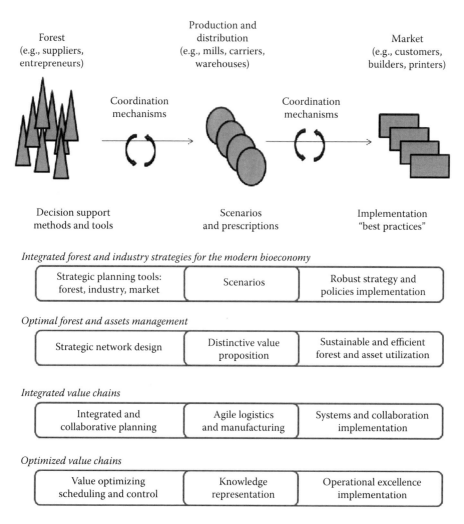

FIGURE 2.2 Hierarchical framework for value chain optimization in the forest bioeconomy.

2.4 DESCRIPTION OF FRAMEWORK FOR VALUE CHAIN OPTIMIZATION

The planning and implementation processes can be described through a sequence of integrated steps. Many stakeholders are involved in planning the forest and the industry and these planning processes fall under the responsibilities of different actors who may have different aims across the forest value chain. The policy level is driven by government, while the other three levels are driven by industry. In between, we find mostly a mixed structure of influences where government, companies, and communities are tightly involved in making the best decisions in a context of negotiation and compromise. In the coming sections, the different levels are explained in broader detail and supported by a literature review of key contributions to the field. While hierarchical planning has often been used in the forest industry, we propose to replace the traditional (decoupled, company-based) decision structure with a more integrative and value chain–based hierarchical planning.

2.4.1 INTEGRATED FOREST AND INDUSTRY STRATEGIES FOR THE MODERN BIOECONOMY

2.4.1.1 Robust Strategies and Policies Implementation

The first level we refer to in this section is integrated forest and industry strategies and policies. At this level, governments are typically the ones making the decisions on how the forest should be managed, protected, and used, understanding that the forest provides many different values, such as economic, environmental, and social values, to its owners and the forest-dependent communities. Whatever the ownership structure of the forests, the public officers will set a series of policies or tools to govern and control the protection and use of the resource. In the context of Crown forests, the policies can state, for example, that a triad approach is to be applied and a certain percentage of forest should be kept as a reserve. A triad approach sets one of the three management approaches to a forest area: ecosystemic forest areas, intensive silviculture areas, and reserved areas (set aside areas). The policies will normally define the accepted silvicultural prescriptions within an area. In the context of privately owned forests, the governments would normally use a mix of fiscal means and regulations to protect the forest and support a constant fiber flow toward the domestic industry. For example, they can set a fiscal advantage to support intensive silviculture. The forest policies are linked to other natural resources policies to ensure proper use and protection of biodiversity over a defined area.

At this level, industrial policies are also defined. These public actions aim to direct private and public initiatives to support restructuring, diversification, and technological development in a way markets could not. They relate to different aspects of the industry such as work force (training, reorientation), capital investments and subsidies, environment regulations (certification, carbon footprint), international trade, and innovation. The policies and public incentives can be in the form of subsidies, tax credits, or direct support of research and development (e.g., research funding, research infrastructure).

These types of decisions have been made by governments and industries for many years. However, a value chain approach changes both the scope and the perspective from which to address the challenges. While the typical discussion at this level has been around sustainable (nondeclining) yields in terms of aggregated volumes, value chain considerations also put the emphasis on the characteristics of raw materials needed for the forest value chains to be competitive, to develop new products, or to tap into completely new markets. The second involves strategies for appropriate investments in processing capacity in both traditional solid wood and fiber products, and advanced wood products, green fiber products, biofuels, and biochemicals that will characterize the new forest bioeconomy. In contrary to other industries, where companies have freedom over what markets they choose to serve, actors in a given forest value chain are highly dependent on one another. As the past paradigm has been of efficiency in raw materials use, developing new products or increasing processing capacities often means restricting another firm's access to fiber sources. As a result, it is very difficult for any company to adapt its structure to major changes in the environment. This change can often be facilitated through proper incentives from governments. That being said, there is little research done on the coordinated design of industrial policies and the participating firms' supply chains.

There is a long history of forest strategy formation in Canada with the first strategy document occurring in 1982 [A Forest Sector Strategy for Canada: Discussion Paper (1981–1987)] and the fifth in the series in 2003 (National Forest Strategy 2003–2008, A Sustainable Forest: The Canadian Commitment). Although the first concentrated on the industry structure, the primary focus of the remaining five reports was on the Canadian Commitment to Sustainable Forest Management as embodied in the Montreal Process documents (1999) and in the CCFM Criteria and Indicators (2003, 2006). These strategies primarily addressed the role of Canadian forests in the midst of worldwide concern for sustainable development and environmental stewardship. There was little focus on the design of the value chain and the economic sustainability of the Canadian industry and the communities supported by it. The most recent document, A Vision for Canada's Forests 2008 and Beyond, prepared by the CCFM using a different process than the four previous documents, has a different focus.

Among the issues tackled at this level, one is a particularly relevant example of the need to act at a national level: climate change. Forest ecosystems and forest value chains can play a significant role in adaptation to and mitigation of climate change. However, it is difficult for one actor in a value chain to act alone without risking hurting its competitiveness. Investments in R&D on new technologies may lead to groundbreaking improvements reducing fuel consumption and greenhouse gas emissions. Coordination at a policy level ensures that all actors are required to participate and share the burden.

The process of managing strategy involves "managing stability ... detecting discontinuity ... reconciling change and continuity" (Mintzberg, 1989). All strategic decisions are made in an environment of uncertainty, but in the current situation, policy-makers and corporate managers are faced with significant changes occurring in the world economy (and in the forest sector in particular), in the environment, in policies which affect the environment, and in the technologies for making increased use of renewable biomaterials to replace metals and fossil fuels. An essential requirement of a strategic approach is that it be economically effective in the near term and capable of coping with significant change in the longer term. It is impossible to foresee exactly what changes will occur in such a complex environment.

2.4.1.2 Scenarios

Scenario planning methodologies are based on the concept that it is possible to craft scenarios consisting of alternative plausible futures spanning a wide variety of possibilities. In attempting to design strategies that are robust in their performance against these diverse scenarios, it should be possible to identify the critical design issues and how to deal with these. The development of this set of realistic scenarios and the testing of various strategies of forest management and capacity investment against these scenarios remains a promising and highly relevant area for further research.

The scenario planning process should aim at striking a balance between taking into account enough factors to be realistic and keeping the number of scenarios at a manageable level to reduce complexity. Among the factors that may need to be reflected in the scenarios, we note the following:

1. Changing worldwide demand patterns for broad (national or regional) markets and product categories
2. Constraints on availability of the most critical resources (timber, biomass, capital, and workforce)
3. Development of major technological advances directly impacting the forest products sector, and in the energy and chemical products sectors
4. Changes in types of raw materials and products needed by each actor in the value chain
5. Changes in forest management policies and tenure situations on Crown Lands and society's attitude toward the role of forests (requirements of sustainable forest management in an environment where increased use of forest timber and biomass may lead to reductions in greenhouse gas emissions and in the use of fossil fuels and scarce materials)
6. The rate and extent of climate change
7. The possibility of large-scale forest disturbance (e.g., insects, disease, storm, and/or fire)

Given the complexity of these potential changes in the competitive environment for the forest bioeconomy, full-dimensional factorial approaches to scenario design are not possible.

2.4.1.3 Strategic Planning Tools

There is also a need for significant development of strategic planning and analysis tools, both as part of the scenario planning exercise and as decision support tools for policy-makers and corporate executives. In strategic forest management, there is a well-established tradition of using long-term strategic models to examine the availability of supply under alternative assumptions of land base availability and silviculture investments. However, these models have traditionally been oriented to wood volumes with limited analysis of product markets. Current approaches to meeting strategic management commitments typically include constraints such as the following: (1) old growth policies; (2) reserves and protected areas; (3) biodiversity/species at risk; (4) soil preservation and biomass removal policies; (5) watershed and riparian zone policies; and (6) location and extent of special management zones. Each of these can have spatially specific impacts on forest growth and yield. Typical forest strategic models have focused on growing stock management including the extent and location of intensive management zones, silvicultural policies including natural regeneration, species mix, exotic species, and various types of even and uneven age harvesting methods. The output of these models has typically been phrased in terms of wood supply but this wood supply is not specific either in location or in terms of its capabilities to meet market needs of specific products. From a viewpoint of both sustainable forest management and the design of the supply chain, it is evident that spatial issues matter and that we need different types of strategic forest management models that account for location, transportation, and market requirements. Additional elements such as carbon sequestration, land use change impacts, and multiple uses of the forest increase complexity beyond simple nondeclining volume yields that have been widely used for decades in strategic forest management.

2.4.2 FOREST AND ASSETS MANAGEMENT

Strategic design of the supply chain is about more than choices of facilities and locations. It is about deciding what the fundamental value proposition is in terms of products and markets, how existing and available resources can be directed to meet market demands, and how to do it in a sustainable way. It requires the capability to analyze issues of capacity, location, and resource availability, the ability to examine strategies of information gathering and sharing, and the structuring of collaborations all along the chain. In the last 10 years, significant efforts have been made to design efficient supply chains at the enterprise level, such as for pulp-and-paper (Martel et al., 2005a) or lumber production (Vila et al., 2006). However, value chain optimization requires bringing the issue to a new level of thinking and modeling beyond the traditional borders of corporate strategic planning. This means revisiting classical issues as well as extending traditional strategic planning in several ways (Gunn and Martell, Chapter 5, this volume).

Capacity planning issues are extremely significant in the strategic adaptation of value chains in regard to the new forest bioeconomy. Sawmills and pulp mills have significant economies of scale both in terms of capital and operating costs. New types of capacity in the forest bioeconomy will undoubtedly have similar economies of scale associated with them. This presents a situation where some types of capacity are likely to experience diminishing demand, while others are faced with a growing demand, where there are significant transportation cost issues for the input product and where there are location issues associated with the output production. There are also many questions that involve the competitive and complementary nature of new capacity. Some of this involves questions about developing new capacity as either add-on or replacement capacity at existing sites (e.g., replacing paper machines with biorefineries) versus closing existing mills and using new green field sites for the new capacity. The issues of technology choice, sizing, location, and timing, together with raw materials acquisition and product distribution issues, must be addressed at the value chain level rather than on a location-by-location basis or on the company level. In addition, increasing competition in major markets for wood products also requires value chains to develop and implement a distinctive value proposition to markets and customers. Planning at the value chain level may help in identifying promising value propositions that would be impossible for a single company to develop and that will be difficult for competing value chains to replicate.

Thus, network design of the value chain is aimed at identifying the mix of product types, production capacities, and production locations that can derive sustainable value from our forests. As forest resources are limited and take a long time to regrow, designing the value chain so it creates the most value in an efficient way makes sense from both a sustainability and a competitiveness perspective. For example, components of this value chain that are capable of replacing fossil fuel products in the form of plastics, petrochemicals, and petrofuels in the form of advanced fiber products, biochemicals, and biofuels will significantly impact climate change policies. A change of such magnitude is less likely to happen if the network design is addressed at the level of the company rather than at the level of the value chain.

There are certain competitive issues for access to the raw resource and there are also complementary issues in reducing overall net costs for some products by providing a market for previously underutilized wood species and quality classes.

2.4.3 INTEGRATIVE AND COLLABORATIVE PLANNING

In the forest products industry, the value chain brings together a large set of business units in need of synchronized plans. When the ownership of the business units of the value chain is divided, collaborative planning is needed. This relates to new information sharing and logistics, benefit sharing, and incentive approaches. In this distributed context, planning methods to support integration between forest and mills, and between mills and customers, are required. Concepts of collaborative planning and logistics solutions, including approaches to allocate costs or saving, have been proposed in recent years (Frisk et al., 2010), but there is still much to do in this area.

The tactical level involves an intermediate time horizon and focuses on tactical issues pertaining to aggregate workforce and material requirements. Depending on whether a forest management problem or a production/distribution planning problem is addressed, the tactical decision-making will be slightly different and may span over different planning horizons.

In forest management, hierarchical planning approaches are widely implemented as they permit the strategic planning problem to be initially addressed without taking spatial issues into account. Once this has been done, the problem is then tightly constrained spatially. While strategic forest management planning problems generally span a long period of time (e.g., two rotations possibly up to 100 years and over), tactical planning problems are often reviewed annually over a 5-year planning period (Martell et al., 1998; Epstein et al., 1999; Rönnqvist, 2003; Gunn and Martell, this volume).

Production/distribution tactical planning normally addresses the allocation rules that define which unit or group of units is responsible for executing the different network activities or what resources or group of resources will be used. It also sets the rules in terms of production/distribution lead times, lot sizing, and inventory policies. Tactical planning allows these two types of rules to be defined through a global analysis of the value creation network. Tactical planning also serves as a bridge between the long-term comprehensive strategic planning and the short-term detailed operational planning that has a direct influence on the actual operations in the chain (e.g., truck routing, production schedules). Tactical planning ensures that the subsequent operational planning conforms to the directives established during the strategic planning stage, even though the planning horizon is much shorter.

One challenge for tactical planning relates to the integration between business functions such as procurement, production, distribution, and sales. Sales and operations planning (S&OP) expresses the need for efficient tactical planning and has been around for 20 years, but it only recently began to be addressed through the paradigm of operations research-based models (Feng et al., 2008). One reason for this is that the implementation of efficient S&OP processes requires adaptation of the models to the production and distribution systems in the network. For instance, make-to-stock and make-to-order production strategies involve different methods for allocating supply/capacity to demand.

Mid-term sales-related decisions involve high-level demand planning, including forecasting, available-to-promise (ATP) planning, and the allocation of ATP to the targeted customer segments. The time horizon to be included in the advanced planning effort required for distribution depends on the transportation mode. For example, ship, rail, and intermodal transportation typically needs to be planned earlier than direct truck transportation.

Fast-changing market conditions may create some disturbances in integrated plans but they also convey opportunities to capture additional value or benefit from temporary conditions. For this to happen, the value chain must develop agile logistics and manufacturing capabilities. In this context, agility can be defined as the ability to adapt to and respond quickly to sudden or unexpected changes in the business environment. According to Audy et al. (2012b), the agility capabilities need to be closely tailored to supply and market characteristics for maximum impact to be achieved.

Another important reason for tactical planning is tied to the seasonality of the value creation network, which increases the need for advance planning. Seasonality has a great influence on the procurement stage (i.e., the outbound flow of wood fiber from the forests). One reason for this seasonality is shifting weather conditions throughout the year and between years, which can make it impossible to transport logs/chips during certain periods due to a lack of load-bearing capacity on forest roads caused by the spring thaw. In the northern countries, for example, a relatively small proportion of the annual harvesting is done during the summer period (July–August). During this period, operations are instead focused on silvicultural management, including regeneration, precommercial thinning, and cleaning activities. A large proportion of the wood is harvested during the winter when the ground is frozen, thus reducing the risk of damage while forwarding the logs out of the forest. However, in other areas, for example, winter weather can prevent harvesting in the mountainous areas. Seasonality can also affect the production stage (e.g., in areas experiencing severe winters, lumber drying times can vary over the year) or the demand process (e.g., again in areas experiencing severe winters, most construction projects are not conducted during the winter period).

An additional area in which tactical planning can be useful is budget projection. Most companies execute an important planning task when projecting the annual budget for the following year, deciding which products to offer to customers and in what quantities. Companies need to evaluate the implications of their decisions on the whole network (sales, distribution, production, and procurement) with the aim of maximizing net profits. Therefore, it becomes interesting to take into consideration these activities as well as to extend the planning horizon to a multiperiod (multiseasonal) horizon, to identify how the budget has to be defined for each of the business units within the value chain.

Integration may also create favorable conditions for a high level of collaboration, more specifically at the logistics level, what we would refer to as "Collaborative Logistics" (D'Amours and Rönnqvist, 2010). Collaboration is needed when business units involved in the production/distribution processes are independent or driven as profit units. In such cases, sharing resources or information becomes compulsory to integrate the value chain. The well-known bullwhip effect is an example of the impact of not sharing information within the value chain. The consequences of this, such as more numerous backlogs and greater inventories, can reduce the performance of the enterprise. Implementation of collaborative logistics approaches can also help achieve significant savings in logistics costs (Frisk et al., 2010).

2.4.4 Value Optimizing, Scheduling, and Control

The primary focus of this decision level is the immediate control of the flow of materials through a complex network to minimize the costs of meeting the market commitments given the availability of forests, equipment, and installations. The main challenge at every node of this network is to optimize production and allocation decisions so that overall value is maximized. There is a constant tension between immediate local optimization and the integration, both spatially and temporally, of the control decisions that realize maximum overall value.

Technology developments have made local optimization increasingly possible. Geographic information systems and enhanced stand/forest inventory systems have made it possible to identify the value of certain stands given a set of harvest prices and to identify those with highest value with respect to those prices. Alternative harvest technologies are available ranging from full-tree logging to a variety of cut-to-length systems. Bucking optimization systems are available that can maximize the value from each tree given the log prices. Given unit transportation costs and mill prices, it is possible to calculate product allocations that maximize the value of products produced on a given stand. At sawmills, given product prices and the ability to scan log dimensions make it possible to optimize the log breakdown into various product classes. At every level of the value chain, local optimization is possible. Some of these optimization problems are easily modeled and solved, while others may be more algorithmically challenging. However, the key point is that the sum of all these local short-term optimal decisions is unlikely to be even feasible for the overall problem of allocating the available immediate wood supply to meet market commitments. In fact, these infeasibilities are often masked by inventories (of both raw materials and finished products) in every part of the value chain. In all these local problems, it is the integral of the decisions made that produce value and enable the overall system to meet the requirements of equating supply with demand. Optimal control of these integral processes—that is, the planning and management of these processes so that the value chain creates as much value as possible—is a significant challenge.

The first challenge is related to data visibility and availability. It is impossible for any part of the value chain to align its processes for maximum value creation if it does not know what is required of it. In particular, information about end-market current and future requirements and expectations has to permeate the value chain without being overly distorted through the lens of specific actors. In the absence of appropriate information about the requirements of end consumers, businesses will typically and logically revert to locally optimal control policies while aiming to fulfill the demand of its immediate consumers. Joint planning-based approaches such as collaborative planning, forecasting, and replenishment (CPFR) are particularly relevant to overcome this challenge (Lehoux et al., 2010).

A second challenge may be tied to the complexity of optimal control. Forest value chains are composed of a large number of divergent processes, each involving a large number of decisions. In the lumber value chain, sawing, drying, and planning have to be planned jointly to reduce inventories and lead times. Integrated planning of these processes has shown to be challenging but feasible (Gaudreault et al., 2011). Similar challenges exist in the planning of harvesting and transportation operations and in the pulp and paper value chains (Weigel et al., 2009).

A third challenge is tied to the tools and methodologies used for optimal control. Complex processes yield difficult optimization problems, which needs state-of-the-art optimization techniques to be solved. However, implementing these techniques in practice requires significant effort. For these tools to be used across all links and elements of the value chain, proper integration of state-of-the-art models and algorithms into appropriate decision support systems is paramount. Proper mechanisms are also needed to make sure decision support tools are shared along with best practices and key performance indicators.

While these challenges are being tackled, each business unit in the value chain still has to address issues regarding short-term operational planning and control problems, aiming to meet product commitments while respecting the constraints on the supply of materials available. Supervisors must make decisions about individual production activities that, taken together, meet these supply and production constraints while creating maximum value and keeping costs under control.

2.4.4.1 Decentralized Control

Various approaches can be taken for coordination between the members of the value chain. First, coordination mechanisms can be derived by the shadow prices associated with decision variables, such as in the Dantzig–Wolfe decomposition scheme (Schneeweiss, 2003). However, even if appropriate shadow prices are available, decentralized control based on these shadow prices is not necessarily feasible. Mathematical programming can also be used along with game theory concepts so as to obtain decisions that represent the optimal compromise for the value chain, while also providing information on how to share cost or savings resulting from collaboration (Audy et al., 2012a). However, much research remains to be done to apply these concepts throughout complex value chains. Contracts have also been proposed as an effective supply chain coordination mechanism (Cachon, 2003). They can be used to set the background and rules for coordination, but handling the day-to-day control decisions through contracts would be too slow or too rigid. In addition to proper collaboration mechanisms, effective integrated control requires accurate real-time information about demand and inventories, as well as key performance indicators and metrics. Design and implementation of control mechanisms that strike the correct balance between centralized and decentralized control is thus a delicate task, as the proper set of tools and mechanisms depends on the context and business units that are part of the chain. Coordination and some degree of centralization are necessary to capture the maximum value but full central control is not possible. Effective operation requires that the individual operators be given the flexibility to use their own creativity and the incentives to achieve good performance. In many cases, the industry structure itself with independent contractors and different enterprises necessitates a decentralized approach. The requirement is for information and the optimization capabilities to make such a decentralized approach produce close to the maximum achievable value.

Achieving this level of integration without direct control requires not only the proper algorithms and coordination systems, but also effective change management and determination of the value chain actors to initiate information exchange and work toward process integration. This requires a significant paradigm shift given that actors in a value chain typically compete to access the best sources of raw materials.

2.4.4.2   Knowledge Representation and Management

The forest value chain is a complex network of units with different goals and beliefs. For these units to work together effectively, a common representation and understanding of the different components of the value chain is required to permit evaluation of forest and industry strategies as well as value chain configurations and planning approaches, and facilitate information sharing and collaboration between partners. Information exchange and proper language for communication are required at the horizontal level (between different business units of the value chain) and at the vertical level (between the different levels of decision-making in the proposed framework). Figure 2.3 displays the various types of information exchanged between different levels of the value chain. Information passed from top to bottom includes normative elements such as policies and resource allocation decisions for the lower levels to implement. In contrast, information transmitted to the upper levels mostly includes statistics and metrics on resource allocation for compliance purposes and to drive further decision-making.

Several frameworks exist for knowledge management in forest value chains. PapiNet (www.papinet.org) is a standard to exchange information at the operational level across all components of the forest value chain. At the academic level, some standards have been recently proposed to model transformation and market capacity (Azouzi and D'Amours, Chapter 8, this volume). However, standards at the aggregated level are still uncommon, even though aggregated information is used at the strategic and tactical planning level extensively. These are needed to support the hierarchical as well as the collaborative planning approaches. The challenges and frameworks for knowledge management have been the subject of recent academic research (Mosconi et al., 2011) and these elements are discussed in much more detail in Chapter 8 by Azouzi and D'Amours (this volume).

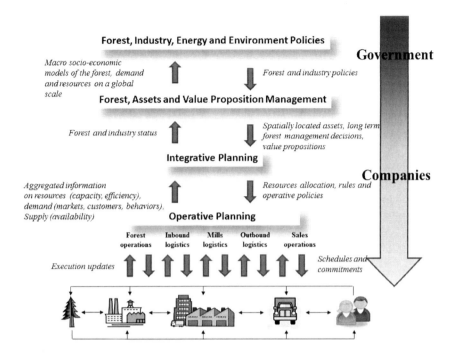

FIGURE 2.3   Knowledge and information exchange between value chain actors and between planning levels.

2.4.4.3   Managing under Uncertainties and Risks

Although information sharing and collaboration mechanisms can help in reducing uncertainty, many decisions in the value chain must be made before accurate and exact information is available at each level of the proposed framework. The types of risks, their sources, as well as the appropriate methodologies required to take these uncertainties into account vary considerably from one level to another.

At the policy-making level, where integrated forest and industry strategies are designed, much of the future is uncertain. In addition to the uncertainties present in the business environment of any industrial sector, the forest bioeconomy depends on a resource that is renewable but which regenerates slowly. At this level of decision-making, uncertainty can be tackled through the generation of proper scenarios representing alternative plausible futures.

The strategic network design level is characterized by forest management decisions as well as asset and capacity management decisions on the industrial side. Risk management principles are well known and are part of much of the discussion in the strategic network design and asset management literature. At this level, it is possible to incorporate a set of alternative quantitative scenarios within the optimization methodologies, effectively replacing deterministic models with stochastic or robust optimization models. However, effective approaches to manage complexity often have to be used, as the number of plausible scenarios is likely to be extremely large or even infinite (Santoso et al., 2005).

At the integrative planning and optimal control levels, the sources of uncertainties are more well-known and defined. Uncertainty can come from unexpected shifts in demand patterns or disruption in supply (in quantities or wood quality). In addition to robust optimization and stochastic programming, simulation can be used to assess the robustness of a plan or to design more robust plans.

Uncertainty can take many forms, from random events for which much information is available (such as demand for each product) to rare events that can have catastrophic impacts (major forest fire, wind storms, etc.) At each level, it is impossible to analyze, characterize, and take into account each and every potential source of risk. The identification of the risks that are of critical importance for a value chain at each decision level is thus a critical task for risk management and mitigation strategies to work effectively.

2.5 CONCLUSIONS

This chapter presented a framework for value chain optimization in the forest bioeconomy. This framework is based on a set of decision levels organized under the paradigm of hierarchical planning. This approach is widely used in the supply chain literature (Stadtler and Fleischmann, 2012). At each decision level, we ask the following two questions: (1) how can we create the maximum possible value from the forest value chain as defined at this level and (2) how can we redesign the forest value chain to enable the creation of more value? Throughout this, the major focus is meant to be on the value chain and the network of values and not just its links. The value chain is a conceptually attractive thought process whether it is seen as a series of flows (forest–trees–logs–mills–products–customers) or seen as a chain in time of management decisions (strategic–tactical–operational). However, this can also be deceptive. The key thing that distinguishes the forest industry is that of divergent flows, especially in the Canadian forest industry, where the forests host a very diverse mix of commercial forest species and are highly heterogeneous across the country. Thus, for any product, we can see a chain linking it back to its original source in the forest; but looking forward from the forest there is a very complex network of value creation challenges and opportunities. The other problem caused by the chain concept is that the linkages in the forest value network are not just physical and neither is it in one direction. The idea of a chain pulling on its links needs to be replaced as a coordinated set of actions involving multiple stages of the chain collaborating to achieve and share the opportunities to create value. This creates the challenge of developing and sharing information that permits the identification and exploitation of the opportunities for collaboration, while at the same time minimizing the risks of such tight integration, where a market disruption in one part of this divergent network causes disruptions in the ability to source materials in other parts of the network. The integration benefits and potential supply disruptions caused by sawmill chips as a source of supply for the pulp mill sector have long been known. As we begin to develop more complex value networks linking the forest with energy and chemicals, advanced fiber products, and advanced wood products, we need to ensure that these networks achieve benefits of collaboration and integration without creating undue risks of supply disruption to one product sector due to market disruption in another sector. Divergent processes have common roots, and disruption of a product pull in one branch of the value network can cause an interruption of supply to other value network branches and instability in the overall system.

REFERENCES

Audy, J.-F., Lehoux, N., D'Amours, S., Rönnqvist, M. (2012a). A framework for an efficient implementation of logistics collaborations. *International Transactions in Operations Research*, 19, 633–657.

Audy, J.-F., Pinotti, M., Westlund, K., D'Amours, S., LeBel, L., Rönnqvist, M. (2012b). Alternative logistic concepts fitting different wood supply situations and markets. CIRRELT, CIRRELT-2012-24.

Cachon, G. (2003). Supply chain coordination with contracts. In Graves, S.C. and de Kok, T. (eds), *Handbooks in Operations Research and Management Science: Supply Chain Management*, Horth-Holland: Springer.

D'Amours, S., Rönnqvist, M. (2010). Issues in collaborative logistics. In Björndal, E. et al. (eds), *Energy, Natural Resources and Environmental Economics*, Springer, pp. 395–409.

D'Amours, S., Rönnqvist, M., Weintraub, A. (2008). Using operational research for supply chain planning in the forest products industry. *INFOR*, 46(4), 265–282.

Epstein, R., Morales, R., Seron, J., Weintraub, A. (1999). Use of OR systems in the Chilean forest industries. *Interfaces*, 29(1), 7–29.

Feng, Y., D'Amours, S., Beauregard, R. (2008). The value of sales and operations planning in oriented strand board industry with make-to-order manufacturing system: Cross functional integration under deterministic demand and spot market recourse. *International Journal of Production Economics*, 115(1), 189–209.

Frisk, M., Göthe-Lundgren, M., Jörnsten, K., Rönnqvist, M. (2010). Cost allocation in collaborative forest transportation. *European Journal of Operational Research*, 205, 448–458.

Gaudreault, J., Frayret, J. M., Rousseau, A., D'Amours, S. (2011) Combined planning and scheduling in a divergent production system with co-production: A case study in the lumber industry. *Computers and Operations Research*, 38(11), 1238–1250.

Gunn, E. A. (2007). Models for strategic forest management. In Weintraub, A., Romero, C., Björndal, T., Epstein, R., Miranda, J. (eds), *Handbook of Operations Research in Natural Resources*. New York, NY: Springer, pp. 317–341.

Lehoux, N., D'Amours, S., Langevin, A. (2010). A win-win collaboration approach for a two-echelon supply chain: A case study in the pulp and paper industry. *European Journal of Industrial Engineering*, 4(4), 493–514.

Mansoornejad, B., Pistikopoulos, E. N., Stuart, P. R. (2013). Scenario-based strategic supply chain design and analysis for the forest biorefinery using an operational supply chain model. *International Journal of Production Economics*, 144(2), 618–634.

Martel A., M'Barek, W., D'Amours, S. (2005a). International Factors in the Design of Multinational Supply Chains: The Case of Canadian Pulp and Paper Companies, CENTOR Working Papers, Université Laval.

Martel, A., Rizk, N., D'Amours, S., Bouchriha, H. (2005b). Synchronized production-distribution planning in the pulp and paper industry, In Langevin, A., Ripel, D. (eds), *Logistics Systems: Design and Optimization*. New York, NY: Springer, pp. 323–350.

Martell, D. L., Gunn, E. A., Weintraub, A. (1998). Forest management challenges for operational researchers. *European Journal of Operational Research*, 104(1), 1–17.

Mintzberg, H. (1989). *Mintzberg on Management: Inside Our Strange World of Organizations*. New York, NY: The Free Press.

Mosconi, E., LeBel, L., Roy, M.-C. (2011). Knowledge management as a means to improve performance in the forest industry value chain. *Proceedings of the 34th Council on Forest Engineering (COFE)*, Quebec City, 1–12.

Ouhimmou, M., D'Amours, S., Ait-Kadi, D., Beauregard, R., Chauhan, S. S. (2009). Optimization helps Shermag gain competitive edge. *Interfaces*, 39(4), 329–345.

Porter, M. E. (1985). *Competitive Advantage: Creating and Sustaining Superior Performance*. Boston, MA: Free Press, p. 557.

Rönnqvist, M. (2003). Optimization in forestry. *Mathematical Programming, Ser. B*, 97, 267–284.

Santoso, T., Ahmed, S., Goetschalckx, M., Shapiro, A. (2005). A stochastic programming approach for supply chain network design under uncertainty. *European Journal of Operational Research*, 167, 96–115.

Schneeweiss, C. (2003). *Distributed Decision Making*, 2nd Edition. Berlin: Springer.

Silver, E. A., Pyke, D. F., Peterson, R. (1998). *Inventory Management and Production Planning and Scheduling*, 3rd Edition. New York, NY: Wiley, p. 754.

Stadtler, H., Fleischmann, B. (2012), Hierarchical planning and the supply chain planning matrix, In Stadtler, H., Fleischmann, B., Grunow, M., Meyr, H., Suric, C. (eds), *Advanced Planning in Supply Chains: Illustrating the Concepts Using an SAP APO Case Study*. Heidelberg: Springer, pp. 21–34.

Vila, D., Martel, A., Beauregard, R. (2006). Designing logistics networks in divergent process industries: A methodology and its application to the lumber industry. *International Journal Production Economics*, 102, 358–378.

Weigel, G., D'Amours, S., Martel, A. (2009), A modeling framework for maximizing value creation in pulp and paper mills. *INFOR*, 47(3), 247–260.

3 Strategic Transformation of the Forest Industry Value Chain

Cédric Diffo Téguia, Virginie Chambost, Shabnam Sanaei, Sophie D'Amours, and Paul Stuart

CONTENTS

3.1   PART I: KEY CONCEPTS AND TOOLS

3.1.1   Introduction

Over the past decade, the forest products industry in North America and Europe has been impacted by a decline in certain market segments, such as newsprint and printing and writing papers (Pöyry 2015; Schaefer 2015), while other segments, such as tissue and board, continue to experience stable growth. Although there has been some recovery in recent years, the North American housing market has been marked by a severe crisis between 2006 and 2010 that has caused a major reduction in demand for construction materials, leading to a difficult period for the entire forest industry (United Nations Economic Commission for Europe 2015). Coupled with low-cost competition from emerging countries in Asia and Latin America, these factors have been the driving forces that have led many forestry companies to consider revenue diversification and transforming of their business models.

Cost-cutting strategies have been employed to maintain the competitive position of forest-based products; however, over the longer term, a profound transformation of the business model is essential to guarantee a sustainable competitive position (Stuart 2006). Forest product companies should evolve from a commodity-centric culture that focuses on key performance indicators (KPI), to a margins-centric culture, as they diversify their traditional product portfolio to incorporate added-value products. This implies improving existing processes while new technologies are implemented, optimizing access to new biomass demands, and systematically developing new product supply chain strategies. In the manufacture of traditional pulp or paper products, for example, the fractionation of wood is accomplished by the kraft pulping process to recover cellulose. Residual streams including hemicellulose, lignin, and other chemicals remain underexploited and are typically burned in the recovery boiler for steam and power generation. The biorefinery concept aims to integrate conversion routes into and alongside existing processes for the production of fuels, bioenergy, and biochemicals from biomass (Axegård 2005). Today, opportunities to produce new bioproducts that compete economically with existing fossil-based products are being explored by forest product companies, examining the potential to create a competitive value proposal.

This chapter is divided into two parts that address the question of how forest products companies might consider strategic transformation and product diversification. Part I is a review of key concepts and tools for the assessment and evaluation of biorefinery business models. It highlights strategic planning concepts, specifics of market assessment, and decision-making from a multidisciplinary perspective. Part II proposes a systematic approach for the identification of promising transformative business models and illustrates it through two case studies in the pulp and paper sector.

3.1.2 Strategic Aspects of Biorefinery Implementation

Rapid changes in the business and technology environments are affecting the competitiveness of forest product companies. Historically, companies used operational effectiveness to create operating cost advantage. Nevertheless, Porter (2008) demonstrated that in turbulent economic, social, and political environments, successful operational effectiveness strategy should be supported by efforts concerning the company's strategic positioning. This approach includes synergies along the company value chain for creating *unique* competitive advantage.

Rather than targeting a first or second quartile position in market segments where they already compete, a better approach for forest product companies might be strategic (re) positioning, which implies enterprise transformation (ET). Two major ET approaches can be used: "inside-out" transformation and "outside-in" transformation (Chambost, McNutt, and Stuart 2009). The former relates to seeking bottom-line improvement by considering synergies in terms of work and process focusing on cost-competitiveness achievements across the value chain. In contrast, "outside-in" ET involves a core transformation of the vision and mission of a company—a makeover from the top down. For example, Georgia Pacific implemented an "inside-out" ET by focusing on improving its core business, while reworking how the company functions to deliver its product portfolio to the market. Another US-based forest products company, Potlatch, opted for an "outside-in" approach, transforming itself into a Real Estate Investment Trust or REIT to take advantage of favorable tax treatment and strengthen the company's position in timberland ownership and management. One of the most successful examples of ET has been that of DuPont De Nemours, which combined both inside-out and outside-in approaches. DuPont has punctually reinvented itself by adapting its core business to market needs to ensure profitability and has grown its market share in profitable new businesses over the long term. The company recently again diversified its product portfolio and divided its business into five major areas to improve net sales consolidation. DuPont acquired bioproduct manufacturing competency through strategic mergers and acquisition, such as Danisco, in recent years. Dupont recently merged with Dow Chemical Company and has planned a 2017 spin-off into three companies that include the following: genetically modified seeds and pesticides, plastics and other commodity materials, and high-tech specialty products. Each of the three would be stitched together from Dow and DuPont. This undoubtedly is considered a high-risk transformation of Dupont.

If forest product companies are to remain competitive by implementing the biorefinery, then clear diversification targets and new product portfolio definitions must be set ("outside-in" ET), as well as unique operations and supply chain strategies ("inside-out" ET). Diversification of the product portfolio implies targeting new customers from commodity-driven to specialty-driven markets, including proactivity, efficiency, responsiveness, and flexibility in the business model.

3.1.2.1 Phased Implementation of the Biorefinery

Chambost, McNutt, and Stuart (2009) recommended a phased approach for the retrofit implementation of the biorefinery into and alongside existing mills for risk mitigation and incremental transformation of the business model taking into account short-, mid-, and long-term goals (summarized in Figure 3.1).

Phase I of biorefinery implementation focuses on reducing risks related to the implementation of new processes and products in the shorter term and sets the path for a longer-term vision defined by the corporation. It typically involves products that have lower market risk such as bioenergy and biofuels that are often used internally to displace fossil fuels and lower operating costs or sold in the market to generate early cash flow. The production of building block chemicals might be envisaged to create opportunity for early access to market for value-added derivatives in subsequent implementation phases. The latter typically includes products that have higher technology risk, as well as specific market penetration strategies. The preliminary phase must be supported by long-term access to a large volume of low-cost biomass.

Phase II (and its subphases) aims to create improved margins through the manufacture of added-value bioproducts as part of the expanded product portfolio. Once operational effectiveness is achieved, the forest product company should build competitive advantage through the strategic positioning of the product portfolio on the

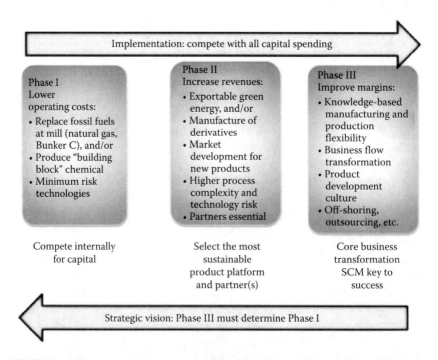

FIGURE 3.1 Phased implementation of the forest biorefinery. (Adapted from Chambost, V., J. McNutt and P.R. Stuart, Partnership for successful enterprise transformation of forest industry companies implementing the forest biorefinery. Pulp and Paper Canada, May/June 2009.)

market. Definition of process/product combinations coupled with product delivery and market penetration strategies are key drivers for the new business model definition. Gradual development of the product portfolio is essential to reduce risk and accommodate market price volatility in a robust business model, which should involve collaborations and partnerships that are critical to minimize technical, commercial, and financial risks.

Phase III targets margins maximization and improvement of bottom line results through the reengineering of supply chains, advanced information systems, manufacturing systems that exploit production flexibility, and new delivery mechanisms. Phase III is the consolidation of the "outside-in" transformation. Creating a unique product portfolio and associated supply chain is essential to the long-term competitive position of the forest products company.

The phased implementation of the biorefinery implies having clear short- and long-term objectives along with an incremental technology implementation strategy to build a flexible and diversified product portfolio. Figure 3.2 summarizes some key drivers for implementing a biorefinery strategy at an existing pulp and paper mill. The box entitled "value chain planning" presents an overview of the biorefinery business strategy definition using a step-by-step approach. Both technology and business disruptions are implied with the definition and the implementation of the biorefinery strategy. The *technology strategy* at the *facility level* should be supported by the definition of the *business strategy* at the *corporate level*. For an effective "outside-in" transformation, it is critical that the mission and vision of the company are adapted to reflect transformational objectives.

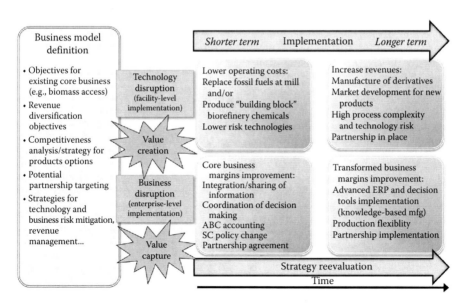

FIGURE 3.2 Transformation planning and implementation using a phased approach.

3.1.2.2 Strategic Interaction between Business and Technology Plans

3.1.2.2.1 *Process- and Product-Centric Approaches for Biorefinery Strategy Definition*

Defining successful biorefinery strategies requires interaction between the product-market-business strategy and technology-process strategy, coupling process and product design methods.

Process design is a well-known field that defines technically and economically feasible processes for the manufacture of products, including the assessment of technological, environmental, and economic risks at different production scales. Product design for the biorefinery is critical and perhaps less recognized. Methods have been developed for identifying promising biorefinery product platforms using process-oriented approaches. For example, the National Renewable Energy Laboratory (Werpy and Peterson 2004) and Pacific Northwest National Laboratory (Holladay et al. 2007) have presented methodologies to identify promising bio-based and lignin-based products. The methodologies are driven by the review of preliminary economic and technical potential, a screening based on chemical functionality, the determination of technical barriers based on best practices, and the potential for each building block chemical to produce a range of derivatives. In contrast, Penner (2007) used a product-centric approach to develop a road map of the most promising value chains in Canada. This approach was mainly based on market drivers and product feasibility. It included national and regional analyses characterizing opportunities from product and market perspectives, supporting the potential to develop company-specific supply chain strategies.

More recently Forest Products Association of Canada, in collaboration with Canada's national forestry research center FPInnovations, led a project (entitled *The Bio-Pathways Project*) to investigate opportunities for producing a range of bioproducts from wood fiber (Forest Products Association of Canada 2010). The methodology involved the following: (1) examining the potential economic, social, and environmental benefits of a set of technologies; (2) defining a set of promising biopathways considering their potential to create employment and enhance sector competitiveness; and (3) considering the global market potential of emerging bioenergy, biochemical, and bioproduct companies. Key drivers for establishing market value, fostering innovation, and deploying technology were examined.

Forest product companies that are considering biorefinery implementation should employ both market-driven and process-driven approaches within the strategy-building framework (Chambost and Stuart 2007). Specific characteristics such as the potential to penetrate markets, potential to create synergies in existing supply chains, the selection of partners for market risk minimization, and the definition of a value proposal are equally as important as technology identification, process design, and implementation when it comes to shaping business models at the corporate level (Chambost, McNutt, and Stuart 2009).

3.1.2.2.2   Product Portfolio Definition

Figure 3.3 illustrates the complexity of defining a product architecture based on the interaction among product functionalities (what the market expects), technology constraints (capacity limitations, yield, etc.), and product strategies (single product versus multiple product strategies) (Batsy et al. 2012). The challenges are associated with the definition of a multiproduct strategy that ensures profit maximization considering product and process design limitations, for example, by technology pathways (Sanderson and Uzumeri 1995).

When commodity products such as biofuels are targeted, limitations in production volume, price volatility, and uncertain energy policy may endanger the viability of the business model. A driver for the success of the biorefinery implementation is the development and management of a robust product platform (Chambost, McNutt, and Stuart 2009). Inspired from the experience of conventional refineries that yield a good deal of their margin through the sale of a relatively small amount of product, the biorefinery product family includes the production of building block chemicals as well as the production of added-value derivatives (Figure 3.4). The systematic development of the biorefinery product portfolio enables margin maximization, margin stabilization through production flexibility, and risk mitigation considering the phased implementation of the biorefinery.

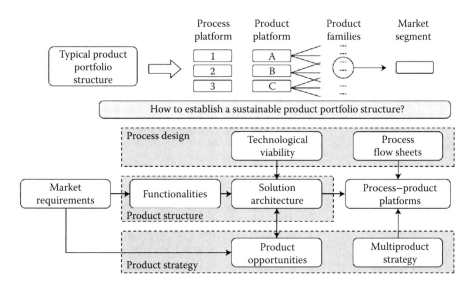

FIGURE 3.3   Drivers of product portfolio definition. (From Batsy, D.R. et al., Product portfolio selection and process design for the forest biorefinery. In M.M. El-Halwagi and Stuart, P.R. [eds.], *Integrated Biorefineries: Design, Analysis, and Optimization*, Boca Raton, FL, CRC Press, 2012.)

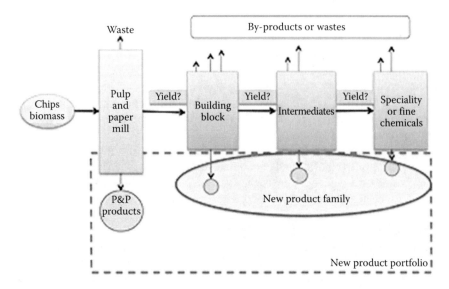

FIGURE 3.4 Generic representation of product portfolio resulting from biorefinery implementation. (Adapted from Chambost, V., J. McNutt and P.R. Stuart, Partnership for successful enterprise transformation of forest industry companies implementing the forest biorefinery. Pulp and Paper Canada, May/June 2009.)

3.1.2.2.3 Some Critical Elements of Transformational Biorefinery Business Models

Business models for successful biorefinery implementation should consider external factors for attracting potential investors:

- Transformational strategies that imply minimum technology risk, mitigated through the systematic appraisal of technology and process maturity.
- A technology plan that explicitly serves a strong business plan, which is implemented in a phased manner through the development of the product portfolio.
- Transformational strategies that rely on a secure and long-term plan for fiber supply agreements based on volume and price.
- Market strategies that support the gradual development of product portfolio profitability, including long-term off-take agreements for the products or partnership definition for value chain penetration.
- The financial situation of the proponent company needs to be robust, relative to the risk implied by the strategy.

Internal success factors should be defined to minimize the likelihood that the new business model becomes obsolete over the longer term. Even if technology selection may lead to *short-term competitive advantage* in terms of first-of-kind implementation and production costs, *over the longer term*, competitive advantages will be driven by the market and supply chain synergies that enable value chain penetration for each product of the biorefinery portfolio.

For building robust transformative business models that take into account external and internal success factors, an interaction between technology and business plan development is essential (Figure 3.5).

3.1.3 MARKET AND COMPETITIVE ASSESSMENT OF TRANSFORMATIVE BIOREFINERY STRATEGIES

3.1.3.1 Market Value Potential

The primary objective of a market assessment is to identify product/market combinations that will potentially lead to value creation over the longer term. The success of new product development relies on the fine-tuning of the market assessment accomplished through an iterative process. It involves reevaluation of the strategy as the product is developed over time to ensure (1) functionality and fit on the market, (2) optimum pricing and distribution strategy, and (3) the ongoing recognition and mitigation of market risks and uncertainties.

3.1.3.1.1 Strategic Corporate Planning as a Driver

Strategic planning is essentially driven by the objective of targeting a corporate vision and mission, setting specific benchmarks for success in specific time frames. Ranging from low-risk objectives to higher-risk diversification targets, product/market strategies associated with long-term business growth are presented in the Ansoff Matrix (Ansoff 1980) (Figure 3.6). This matrix is used as a marketing tool and characterizes four ways of business growth while considering the level of risk associated with each strategy—low-risk strategies are associated with business-as-usual while higher-risk strategies involve market expansion and/or new product development.

In the case of forest company transformation through biorefinery implementation, corporate drivers can be difficult to define. The culture in place in the forest products industry must emphasize short-term objectives that have a positive impact on the

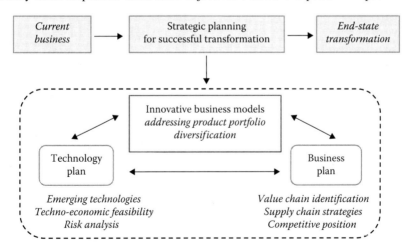

FIGURE 3.5 Concept of integrated business and technology plan development for innovative transformative business models.

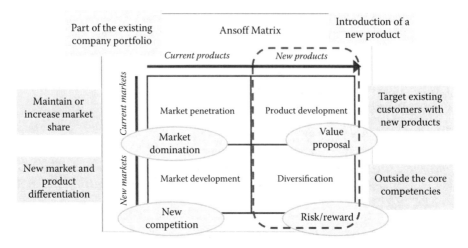

FIGURE 3.6 Ansoff Matrix for product/market strategies for targeted corporate growth objectives.

core business, and it is difficult to justify riskier transformational strategies seeking revenue increase and improved returns over the longer term. Since transformation to the biorefinery is not yet considered as an essential strategy for many forest companies, biorefinery projects are in competition with core business improvement projects involving less risk and having competitive returns.

As introduced in Figure 3.2, the business model for the biorefinery should be supported by a robust market strategy. Identifying and developing the market strategy relies on two major activities: (1) value creation through the assessment of market potential over the short and long term and (2) value retention through the definition and maximization of competitive advantages associated with the product and/or process strategy (Figure 3.7) (Forest Products Association of Canada 2011).

3.1.3.1.2 Types of Bioproducts and Associated Market Strategies

As part of the business model, the marketing plan includes the expression of the market strategy over short, mid, and long terms and identifies risks and mitigation strategies. The methodology for the development of a marketing plan includes (1) characterizing potentially promising market segments, (2) identifying the potential for creating value on those segments, (3) determining the competitive position and associated competitive advantages that may enhance market positioning, (4) defining and selecting the best approach to penetrate identified market segments considering long-term viability and market uncertainty, and (5) expressing the chosen market strategy in terms of value proposal, pricing, and distribution strategies.

In addition to pulp and paper products, nontraditional biorefinery products comprising the new product portfolio will have different market potential and market strategies. Products can be characterized as *commodity*, *specialty*, or *niche* products in regards to the type of markets that are targeted, mainly in terms of differentiation potential and market volume (Figure 3.8). Competitive advantages that a company may build are typically the result of the following factors: (1) existing value chain

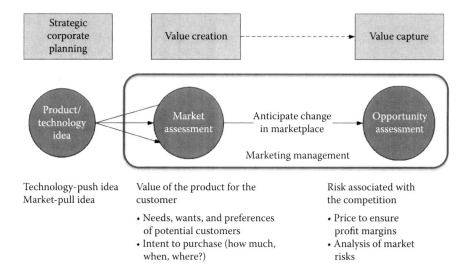

FIGURE 3.7   Value creation and capture through market management.

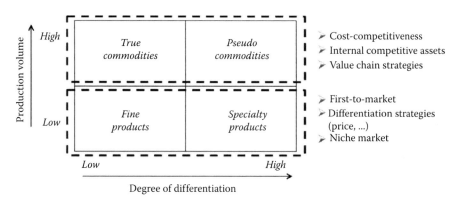

FIGURE 3.8   Market strategies for different product types.

drivers that affect the value proposition for each product and determine the potential of targeting commodity versus specialty markets, (2) the potential to create barriers-to-entry for prospective competitors in specialty markets considering that their presence would ultimately turn today's specialties into commodities in the future, and (3) "cost leadership" for commodities and "unique differentiation" for specialties and niches, enhancing the influence on the market (Chambost, McNutt, and Stuart 2009).

Biorefinery products are also be characterized as being *replacement, substitution,* or *breakthrough* products in regards to what they replace on targeted markets. Replacement products have identical chemical composition and target existing mature supply chains, such as biopolyethylene. Products are considered substitution products if they have similar or enhanced functionality/performance compared with products currently in the market, while having a different chemical composition, such as polylactic acid as a substitution for polyethylene terephthalate. Breakthrough products target enhanced performance or new functionality compared to existing products, such as the case for biocomposites compared to existing composite materials. Regardless of the type of new products, existing value chains must be assessed to evaluate new product positioning in the market, possible point-of-entry in the value chain, and the potential for competitive advantage.

3.1.3.1.3   *Product Portfolio Considerations during Market Assessments*

Using a market-centric approach, Chambost and Stuart (2009) defined a step-by-step methodology for the systematic assessment of potentially promising biorefinery product portfolios (Figure 3.9). It consisted of four main activities: (1) assessment and characterization of markets for individual products comprising the product portfolio, (2) assessment of potential value creation through the development of a "product family" of bioproducts, (3) definition of a company-specific supply chain for the sale of the product portfolio including the core business product portfolio, and (4) identification of possible partnerships that mitigate market risk and facilitate product positioning.

Classical indicators needed to assess the market potential associated with a product include the following: (1) market volume, value, and potential growth; (2) price volatility associated with the substituted product on the market; (3) market penetration options including potential for partnership; (4) "green" market potential and associated penetration potential in terms of market share and revenues; and (5) market risks.

The identification of potential competitive advantage is a prerequisite for the adequate definition of market strategies for new bioproducts. This can be achieved through the assessment of different points-of-entry in existing value chains, along with an understanding of competitive positioning.

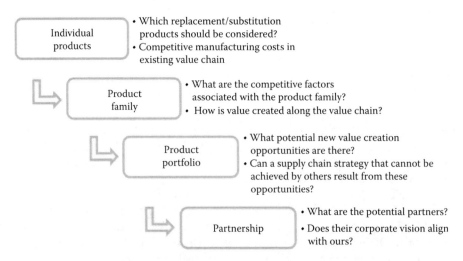

FIGURE 3.9   Systematic methodology for biorefinery product portfolio assessment.

Assessing the life cycle of a product from a market perspective helps to better understand market segments, potential market value, and the arena of competitors. This consists of identifying the stage of development achieved by the product on a given market segment—introduction, growth, maturity, or decline. Stages are characterized by different levels of competition and risk due to market volume, barriers-to-entry for new competitors, and price rivalry, implying different strategies depending on the maturity of the value chain.

Given that product families resulting from biorefinery processes are comprised of combinations of commodity and specialty products, approaches for mitigating market price volatility risks are different from one product to the other. Commodity markets are characterized by high volume and low margins, and prices tend to rise or fall with business cycles due to fluctuations of raw material prices in a highly competitive environment (Regnier 2007). Specialty products, however, are subject to less volatility and lead to a more stable and higher margins (Price Water house Coopers 2009).

Developing a biorefinery product portfolio that combines commodity and specialty products should be seen as an opportunity for developing certain transformation strategies that mitigate market price volatility risk and stabilize margins.

- *Manufacturing flexibility* allows for change from one manufacturing regime to another, adapting the product portfolio yields with market volatility. This strategy requires higher capital and operating costs; however, it leads to substantially higher margins under all market price volatility scenarios (Mansoornejad, Chambost, and Stuart 2010).
- *Combining contract and spot sales strategies* enables enhanced price control and margin generation. Supply chain operating strategy taking into account spot/contract sales strategy and manufacturing flexibility is part of the value proposal representing the potential for competitiveness improvement (Dansereau, El-Halwagi, and Stuart 2009).

Besides optimizing operational effectiveness and flexibility, the supply chain strategy, in many cases, involves the creation of quality partnerships that help to retain value, mitigate supply and distribution risks, facilitate value chain penetration, and create competitive advantages.

Important drivers to be considered for creating viable and sustainable partnerships include the following: (1) the strategic compatibility of business models and visions between the two partner companies, as it is not obvious to identify a balanced business model serving both interests, each company having a different level of risk acceptance, financial capacity, and objectives over the short, mid, and long term; (2) access to the long-term capital resources required for developing and sustaining the biorefinery strategy; and (3) manufacturing flexibility needed to respond to or anticipate market and/or technology disruption, allowing the product portfolio to evolve with time and achieve higher competitive advantages (Chambost and Stuart 2007).

3.1.3.2 Competitive Analysis

As it is essential to identify the value that a product can bring and how the combination of several products within a portfolio can generate more value, it is also crucial to define the potential of a company to retain this value through its unique competitive position. Performing a competitive analysis requires that the company characterizes the drivers for profitability within the targeted value chain and provides a framework for securing its profitability over time relative to the market place (Porter 2008). To characterize potential competitive advantages, it is necessary to understand such factors as (1) the competitive environment associated with the targeted value chains, (2) the drivers that impact the value chains and product price, and (3) the risks and mitigation strategies associated with penetrating the value chains.

An approach commonly used for analyzing the competition is the one defined by Porter (2008) based on the identification of Five Forces that shape the competitive landscape: (1) direct competitors, (2) new entrants, (3) power of suppliers, (4) power of buyers, and (5) substitute products.

An analysis of the direct competitors provides an understanding of the market share potential, which is also an indicator of the value achievable by targeting a specific market segment. The intensity of the competition is assessed considering the number, relative size of competitors, market growth, and the presence of high barriers to enter. The type of product also plays a major role in identifying the drivers of the competition—generally being cost driven for commodity products and performance/quality driven for specialty products.

The threat of new entrants comes from competition drivers such as market price relative to production costs and investment rate. The level of threat depends on barriers-to-entry created by existing players on the market including unique supply chain access, production economies of scale, capital requirements, or regulations. Besides barriers-to-entry, cost reduction strategies are a deterrent to potential new entrants to the market. The ability to moderate production levels, or even cease production, when market conditions are not favorable has to be considered as a competitive advantage (Oster 1999). This latter factor will be especially critical for substitution and breakthrough bioproducts.

The bargaining powers of suppliers and buyers have a similar impact on the potential competitive position of a company in a new market, especially if companies are vertically integrated. For example, forestry companies who have retained their cutting rights and who manufacture wood-based as well as pulp and paper products have a considerable competitive advantage when transforming to the bioeconomy. Further along the value chain, the concentration of suppliers to a targeted market segment, as well as the switching costs to change suppliers for market intermediates, enables the definition of value chain dynamics and associated competitive factors. From a buyer perspective, concentration as well as the potential to switch from one supplier to another due to standardization of the product may lead to a strong pricing negotiation position. Since any player on a value chain is successively a buyer or a supplier depending on its position on the value chain, the *supplier matrix* (Figure 3.10) can be developed to assess the company's competitive position relative

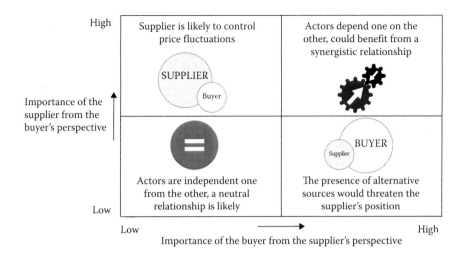

FIGURE 3.10 Supplier matrix to help assess competitive position in the supply chain. (Adapted from Oster, A.M., *Modern Competitive Analysis*, 3rd Edition, New York, Oxford University Press, 1999.)

to other actors upstream and downstream, where a monopsony is essentially where the buyer drives a monopoly.

The last competitive force is associated with substitute products, that is, the threat of a larger offering fulfilling partially or completely the same needs on the market. The threat is high if a substitute product offers an attractive price-performance trade-off, coupled with low switching costs for the buyers (Porter 2008). An example of this may be that of biodiesels made from different second generation biomass feed-stocks such as algae having different performance. Whether buyers actually switch to substitute products or not, the presence of many substitutes reduces profitability by implicitly imposing a ceiling on prices. A parameter that represents this limitation is the cross-elasticity of demand, represented by the ratio of the percentage change in demand for one product, in response to an increase of one percent of another product's price. Cross-elasticity generally has positive values for substitute products implying a competitive environment (Oster 1999), leading to lower profitability.

Based on the competitive assessment using the Five Forces framework, a set of key variables should be defined and used to evaluate the value proposal against the competition, and to position the biorefinery product portfolio over the long term. Using this analysis, forestry companies assess barriers-to-entry via production cost advantages, unique product attributes/quality, unique value chain potential, and so on. Different price scenarios considering the competitive landscape should be defined. Most importantly the competitive assessment leads to the identification of certain market and business risks associated with the biorefinery implementation.

Market potential analysis combined with competitiveness analysis leads to the determination of adapted strategies for the penetration of value chains. For each potential biorefinery option that a forestry company may be considering, a system-atic evaluation of the Strengths and Weaknesses along with potential Opportunities and Threats from the external environment—widely known as SWOT analysis—is an important complementary tool to understand the competitive position of the company and shape a strategy from there. Through the creation of additional advantages that balance identified weaknesses, the SWOT analysis leads to the definition of a more robust transformative strategy.

3.1.4 Practical Approach for the Evaluation of Biorefinery Strategies

Evaluating the gamut of potential biorefinery strategies available to forest product companies to determine which of these offer the greatest competitive advantage is a daunting task. Product design involves the screening of less viable ideas among the set of possible alternatives, with the aim of identifying the few that can potentially be converted into successful products (Cussler and Moggridge 2011). It is critical that forestry companies use product design in a structured manner to evaluate the set of promising transformation strategies in conjunction with opportunistically testing technologies and their business and economic advantages one-at-a-time.

Three main product evaluation approaches are identified from the literature (Killen, Hunt, and Kleinschmidt 2007; Cooper 2001b) namely, (1) benefit measurement techniques, (2) economic models, and (3) project portfolio optimization methods.

The first approach recognizes the lack of accurate data at the early stage of the design process and is often based on the usage of subjective criteria such as the alignment of the project with the corporate global strategy, the order-of-magnitude capital cost associated with the project, the attractiveness of potential markets, or the potential competitive advantages the project could provide to the corporation. This approach typically includes methods such as checklists and scoring models, which allow a preliminary and quick screening of possible options.

Economic models consider each project as an investment for which economic targets can be estimated. The decision is based on indicators such as net present value, discounted cash flow analysis, or internal rate of return (IRR). The availability of uncertain data at the early design stage is the main limitation of these methods, and statistical methods such as Monte Carlo analysis can be coupled with economic models to support economic estimations under uncertainty.

The third class considers that projects are not evaluated individually but rather as part of an overall project portfolio of the company. Constraints such as budget are applied, thus turning the decision-making problem into a mathematical problem, where a parameter (IRR for example) is optimized using a set of constraints. The complexity of these methods has limited their applicability in practice (Enea and Piazza 2004; Kornfeld and Kara 2011). The main advantages and disadvantages of some commonly used methods are summarized in Figure 3.11.

3.1.4.1 Multicriteria Decision-Making

Transformational projects are distinct to investments in the core business. They represent a higher-risk investment, sacrifice capital that could otherwise be available for strengthening the core business, and a dilution of focus. In contrast, transformation is essential for every company in the longer term, especially when disruption occurs in the marketplace, such as is the case in the forest products industry today.

Type of Method	Subtypes Presented Here	Basic Purpose	Major Advantage	Major Disadvantage
Numerical ranking methods	Scoring models	Rank candidate projects in order of desirability. Manage fund projects in order until resources are exhausted.	Completely transparent, easy to use, readily understandable	May give impression of false precision. Requires significant input from higher management.
	Analytic hierarchy process		Allows criteria to be disaggregated into several levels.	Requires extensive input from functional and higher management.
Numerical, economic methods	Payback period	Evaluate economic payoff.	Simple to use and understand; very robust against uncertainties. Direct comparison with capital budgeting.	Does not account for time value of money. Required data may not be available for some projects such as basic research.
	Net present value Internal rate of return	Evalute economic payoff, including time value of money.	Easy to calculate using speradsheet; direct comparison with capital budgeting.	Required data may not be available for some projects such as basic research.
Numerical, optimization methods	Portfolio selection	Choose portfolio of projects that maximizes some measure of payoff.	Allow use of multiple criteria for selecting an entire portfolio of projects.	Extensive computations required for large project portfolios.
Real options	Projects as options	Reduce risk by selecting best combination of alternatives.	Reduces both downside and upside risk associated with projects.	Requires extensive data and analysis.

FIGURE 3.11   Summary of project selection approaches. (From Martino, J.P., Project selection. In D.Z. Milosevic [ed.], *Project Management ToolBox: Tools and Techniques for the Practicing Project Manager*, 19–66, 2003.)

For transformation decision-making, a multidisciplinary approach is needed considering a wide range of market, technology, environment, and other factors. There exists risk in each of these decision-making elements due to scarcity and uncertainty of data. Multicriteria decision-making (MCDM) is an increasingly used, appropriate tool to assist in making well-informed and balanced decisions considering this environment. The MCDM approach requires that a series of multidisciplinary analyses be conducted on each transformational product/process option, which includes the following: (1) competitive assessment of each product portfolio, (2) techno-economic assessment, and (3) environmental analysis, that is, life cycle assessment. MCDM consists of two phases as illustrated in Figure 3.12: (1) the prepanel activity and (2) the panel activity. Prepanel activities include definition of decision-making objectives, presentation of candidate biorefinery strategies, and introduction of the set of decision criteria. The MCDM panel session is a one-day activity during which the meaning and significance of each criterion is discussed leading to the elicitation of panel member preferences and the determination of a weighting factor for each criterion. The process of gathering panel member preferences and the mathematical method to calculate weighting factors differ with the type of criteria that might be used. Common MCDM methods include the "multiattribute utility theory" and the "analytical hierarchy process" (Wang et al. 2009).

The MCDM panel session is conducted with a multidisciplinary group of ideally six to eight participants (decision-makers) who have a wide range of company perspectives (business, technology, financial, environmental). The outcome of the MCDM activity is to allocate a score to each biorefinery strategy. The score is calculated based on two parameters: (1) utility value (the normalized value of each criterion) and (2) weighting factor (a value between 0 and 1 that represents the preference of decision makers). Utility values are calculated using the results of techno-economic, market, and environmental analyses, while weighting factors are estimated through values expressed and justified by MCDM panel

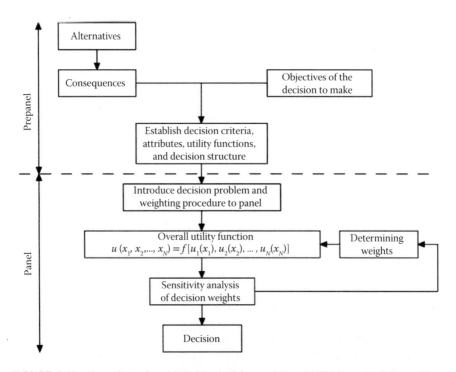

FIGURE 3.12 Overview of multicriteria decision-making (MCDM) methodology. (From Janssen, M. et al., MCDM methodology for the selection of forest biorefinery products and product families, International Biorefinery Conference, 2009.)

members. Figure 3.13 summarizes how criteria are organized for different biore-finery alternatives.

Sanaei (2014) presents a review of cases in which the MCDM tool has been employed for decision-making in the biorefinery context. In occurrence, Cohen et al. (2010) and Quintero-Bermudez et al. (2012) used it for analyzing emerging ethanol production technologies. As well, Sanaei, Chambost, and Stuart (2014) used it for assessing alternatives for the development of greenfield triticale biorefinery.

Financial and capital performance metrics are typically used for capital spend-ing decisions but have limitations when considering the unique strategic biorefinery investment. Competitive and environmental metrics need to be considered to ensure the sustainability of the decision-making (Batsy et al. 2012). A social aspect could also be incorporated in the definition of decision criteria in some cases; however, this perspective has not yet been widely covered in the literature and remains an area for future development.

3.1.4.2 Techno-Economic Criteria

Besides financial and capital performance metrics such as return on investment or internal rate of return (IRR), Hytönen et al. emphasized the importance of a bet-ter analysis of the process and economic risk associated with a specific technology along with the business transformation potential that the project may imply. More than one profitability metric is typically considered to accurately express the under-lying risks and uncertainties of different capital spending scenarios.

In this regard, Sanaei and Stuart (2014) suggested dividing techno-economic indi-cators into two categories, where profit-oriented and strategy-oriented criteria are distinguished (Figure 3.14). A subset of these six criteria is typically used to reflect downside return or whether the strategy can be substantive enough to reach revenue diversification objectives.

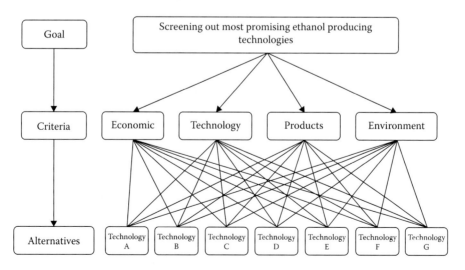

FIGURE 3.13   Summary of MCDM criteria organization.

	Criteria	Interpretation
Profitability-oriented criteria	Internal rate of return	Profit/risk ratio associated to the business model, estimated under normal market conditions.
	Downside internal rate of return	Viability of the business model under the worst predicted market conditions.
	Return on capital employed	Value the project generates from the capital invested.
Strategy-oriented criteria	Revenue diversification	Ability of the project to improve its margin, to mitigate volatility on a specific product, and to provide a backup for the core business product.
	Ability to respond to unknown changes	Capability of a project to maintain a positive free cash flow under unpredictable changes in the business environment.
	Resistance to supply market uncertainty	Robustness of the strategy in regards to highly volatile raw materials, especially petroleum-based raw materials and energy cost.

FIGURE 3.14 Profitability- and strategy-oriented techno-economic criteria. (Adapted from Sanaei, S. and Stuart, P., *Biofuel. Bioprod. Biorefin.*, DOI:10.1002/bbb.1499, 2014.)

3.1.4.3 Market and Competitiveness Criteria

Besides economic risks that are inherent to every transformation and investment project, issues related to transforming the company business model over the longer term in terms of impact on the core business, portfolio diversification, market penetration, and partnership options are considered as part of the decision-making framework. For example, from a financial point-of-view, access to capital can represent a major limitation for a project appropriation. Several indicators can be defined to illustrate this, such as (1) the amount of capital required for the investment, (2) the potential to attract investors or to involve stakeholders in the project, (3) the potential to secure financial support from the government in the form of a subsidy or grant to lower the investment risk, or (4) the possibility to divide the project into phases of lower investment and risk.

Important issues considered on the market side concern mainly the interaction with actors upstream and downstream on the value chain and the potential to capture or enhance competitive advantages. Both types of issues need to be addressed regarding the chances of success in the implementation of the project and the potential to retain value over the long term. Performing a competitive analysis helps to identify issues specific to the targeted market and to determine drivers for competitive advantage. An example of market-oriented decision metrics was presented by Diffo and Stuart (2012) for assessing triticale biorefinery alternatives (Table 3.1), including short- and long-term interpretations of each criterion.

3.1.4.4 Environmental Metrics

Reasons to consider environmental decision criteria are numerous; however, these are not always central in the decision-making process. The rise of environmental concerns in response to global warming issues and the predicted shortage of fossil

TABLE 3.1

Some Market-Oriented Decision Metrics

Criteria	Competitiveness Issue	Interpretation
Competitive access to biomass	Is there a long-term biomass strategy in place?	Potential (1) to provide a competitive value proposal to suppliers and (2) to secure economically viable and long-term access to the feedstock
Product portfolio positioning	Is there good potential to secure a competitive position on the market and keep it over the long term?	Potential (1) to capture market share on existing markets and (2) to implement first-to-market strategy on growing future markets
Competitiveness on production costs	Do production costs provide competitive advantages?	Potential (1) to compete on a market-price basis and (2) to create margins for mitigating price volatility
Margins under price volatility	Is there potential to remain competitive under price volatility?	Potential (1) to survive under unfavorable conditions and (2) to generate high margins under favorable conditions
Technology strategy related to market competitiveness	Is there potential for a clear and incremental technology strategy that serves the business strategy?	Potential (1) to face short-term capital investment constraints and (2) to easily adapt the strategy when required

Source: Diffo, C., and Stuart, P., Market competitiveness assessment for forest biorefinery product portfolios selection. Presentation at the BIOFOR Conference, Thunder Bay, ON, May 14, 2012.

resources resulted in (1) higher expectations in terms of green standards from customers and (2) development of governmental policies. It is unlikely that a project having relatively poor environmental performance would be selected against other alternatives.

Environmental impacts of a product or a project are commonly evaluated using life-cycle assessment (LCA) (Mu et al. 2010). LCA considers the inputs and outputs implied across the life cycle of the product from raw material extraction to end-of-life and evaluates their potential impact (International Standard Organization 2006). LCA provides an understandable vision of the environmental trade-offs occurring at different steps of the product life cycle and is suitable to generate decision metrics to differentiate between prospective transformation strategies. Gaudreault, Samson, and Stuart (2009) developed a set of criteria adapted to the biorefinery context using end-point impacts of the IMPACT 2002+ software (namely, human health, ecosystem quality, climate change, and resources) as described by Jolliet et al. (2003), and additional indicators including aquatic acidification, fresh-water input, and cropland occupation.

However, environmental indicators present the challenge of being complex and, in the case of MCDM panels, not easy to interpret for decision-makers who are not familiar with LCA. Liard et al. (2011) highlighted the importance of normalization and the definition of normalization factors specific to the context to have a better perspective of spatial scales of the impact categories in MCDM, as summarized in Table 3.2.

TABLE 3.2
Description and Scale of Environmental Criteria

Criteria	Description	Scale
Human health	Damage to human health that any substance can cause due to its toxic and respiratory effects, as well as ionizing radiation and ozone layer depletion it creates.	Local
Ecosystem quality	Degradation of ecosystems relative to ecotoxicity, acidification and nutrification of terrestrial systems, land occupation, ecotoxicity, acidification and eutrophication of aquatic systems, ozone layer depletion, photochemical oxidation, and land occupation.	Regional
Greenhouse gas emissions	Amount of greenhouse gases emitted at each step of the life cycle accounted in the analysis for a given product, in CO_2 equivalent.	Global
Nonrenewable resources	Energy used at each step of the life cycle accounted in the analysis for a given product and generated by primary nonrenewable resources and minerals.	Global
Aquatic acidification	Amount of acidifying substances emitted in aquatic systems at each step of the life cycle accounted in the analysis for a given product, in SO_2 equivalent.	Regional
Fresh water input	Amount of fresh water used at each step of the life cycle accounted in the analysis for a given product. It excludes water that is used and recycled within process loops.	Regional
Cropland occupation	Area used by the biorefinery activities that would normally be used for crops culture, or advantages associated to growing biorefinery feedstock on marginal lands.	Regional

Source: Liard, G. et al., Selection of LCA-based environmental criteria for multi-criteria decision support in a biorefinery context. Presentation at the SETAC Europe 21st Annual Meeting, Milano, Italy, May 2011.

3.1.5 CONCLUSIONS

Many forest product companies seek to define long-term strategies for improving their competitive position, while considering diversification and new business model definitions including outside-in transformation models. There have been many lessons learned, even as the forest sector begins to transform. Biorefinery strategies are identified, defined, and assessed considering a phased approach to identify and mitigate risks inherent to the transformation. The phases accomplish the short- and long-term corporate goals for the transformative strategy and imply both technology and business disruptions that have good potential to capture, create, and retain value.

The success of a biorefinery strategy is highly dependent on the targeted product portfolio strategy and its development over the short and long term. The definition and selection of the product portfolio involves both product and process design tools and requires systematic market and process evaluations. The market-driven analysis evaluates the strategic positioning of the new product portfolio considering market potential and uncertainties, while process design tools support product portfolio development and optimization through scale-up phases and incorporating manufacturing flexibility. Decision-making strategy to identify the most preferred biorefinery strategies based on preliminary technology and business plans uses a set of sustainable criteria considering economic, competitive, and environmental criteria.

Part II of this chapter presents biorefinery case studies, considering the fundamental concepts presented here.

3.2 PART II: TRANSFORMATION METHODOLOGY AND CASE STUDIES

3.2.1 INTRODUCTION

Since the early 2000s, first-generation biorefineries (based on sugar, starch, and oil feedstocks) have been implemented. Second generation biorefineries (based on lignocellulosic feedstocks) are more recent (Demirbas 2010). As a result, the interest in transformation via the manufacture of bioproducts from wood has increased significantly in recent years on the part of forest product companies. Success is not obvious. Recent examples of failed ethanol projects highlight the difficulties that will be encountered to overcome technology challenges and to penetrate commodity-driven markets while relying on capital subsidies.

However, the biorefinery offers a plethora of technology pathways and product–process combinations, implying different risk levels (Agbor et al. 2011; Maity 2015a, b). These are expressed in terms of a range of characteristics such as maturity of the selected process pathways, product portfolio options over the short and long term, market positioning, and competitiveness of the product portfolio. From the investor perspective, it is essential that the risks and uncertainties associated with the biorefinery be identified and mitigated as part of the development of transformational strategies. There is surprisingly little information in the literature offering a clear pathway for the definition of such strategies in the context of the biorefinery.

Part I of this chapter introduced a set of fundamental concepts critical for the evaluation of transformative strategies. For example, forest product companies need to implement transformative business models considering a phased approach to mitigate risks (Chambost, McNutt, and Stuart 2009). Product design tools are used, which consider a market-driven approach to build a successful business plan through market analysis, product life cycle assessment, value chain assessment, and competitive analysis (Ansoff 1980; Chambost and Stuart 2007; Porter 2008). The Ansoff Matrix, SWOT analysis, and Porter's Five Forces model are among some of the recommended tools critical to gathering and critically analyzing market and business-driven information. The outcome of this assessment is a clearer understanding of the value proposal of products within the biorefinery portfolio, with the objective of defining a robust product portfolio. Product design tools are supported by process design tools for the definition of a technology plan.

MCDM is increasingly employed to guide decision-makers in selecting promising strategies (Janssen, Chambost, and Stuart 2009). MCDM implicates a range of stakeholders relative to the decision and is supported by (1) a systematic decision-making procedure and (2) the use of a multidisciplinary set of criteria. Evaluation metrics are calculated for each transformation option in order (1) to obtain a contextualized weighting of the criteria and (2) to calculate overall scores for each of the options. These scores are used to rank the strategies and to identify the most promising ones for further investigation. Part I of this chapter also provided examples of environmental, techno-economic, and market and competitiveness criteria that could be used in an MCDM.

Part II of this chapter aims to introduce a systematic methodology for defining a forest product company's business model for the implementation of the biorefinery, considering both technology and market drivers. Many different approaches can be and indeed are being used by forest product companies to determine preferred transformation options. This chapter presents a systematic methodology based on extensive work in this domain that is proving successful with forest product companies.

The methodology considers the entire design process starting with the identification of possible product–process combinations through to the implementation of the business model. It integrates a phased-implementation approach, the interaction between business and technology plans, the product design and process design assessment tools, as well as the MCDM tool. First, a stage-and-gate model is introduced to capture the vision that brings these elements together into a single methodology. Two case studies treating (1) the early stage evaluation and decision-making of lignin-based biorefinery options and (2) the value proposal definition for torrefied pellets production are used to illustrate certain critical steps of the methodology in the context of biorefinery implementation in retrofit to forestry companies. We have not elaborated on critical elements of market assessment in this chapter including, for example, the Ansoff Matrix, SWOT analysis, and Porter's Five Forces.

3.2.2 Framework for New Product Development

In the context of an outside-in enterprise transformation, product diversification implies novelty and risk at several levels of the corporation. The introduction of biorefinery products into the existing company product portfolio requires (1) new technologies and processes to be implemented, (2) penetration/creation of value chains new to the company, and probably (3) new partnerships required to penetrate nontraditional market segments with the new products. Cussler and Moggridge (2011) developed a methodology involving the interaction between market-driven and process-driven factors for new product development. An important distinction was made between industrial and chemical product designs, considering different approaches to product development and structure of the value chain. Industrial product design focuses on *assembled* products, which consist of a combination of solid-state components such as razors, bicycles, and computers (Favre, Marchal-Heusler, and Kind 2002), whereas chemical product design focuses on *formulated* products, which can be represented by a single chemical or a mixture of chemicals resulting in a set formulation or a microstructure (Hill 2009). Considering the context of the biorefinery, a focus is made on chemical product design.

Several approaches have been presented in the literature to discuss the chemical product design process through a logical sequence of key steps (Costa, Moggridge, and Saraiva 2006; Cussler and Moggridge 2011) generally including (1) the identification of new customer needs, (2) the generation of ideas to fill these needs, (3) the selection of the most promising ideas, and (4) the development of the product and the associated manufacturing process. Most of these approaches do not systematically consider the simultaneous development of the product from a technology and a business perspective. In this regard, Seider et al. (2009) proposed a stage-and-gate approach incorporating the conventional chemical product design approach with the Stage-Gate® Product Development Process developed by Cooper (2001a) summarized in Figure 3.15.

The Stage-Gate® process consists of a combination of blocks of activities—called *stages*—separated by decision points where the evaluation and approval on whether to pursue a given project is made—called *gates*. It is traditionally constituted of five stages and five gates, as illustrated on the top-arrow in Figure 3.15. Projects are run by multidisciplinary teams within the organization including product and technology development, manufacturing, and marketing divisions. Activities within stages (see more detail below) lead to specific deliverables used to support decision-making regarding the continuation of the project. The evaluation of deliverables through a specific set of criteria is performed at each gate, and the decision is either made to (1) continue the project development, (2) kill/abandon/reject the project, or (3) put the project "on the shelf" until gate criteria are satisfied. This decision is usually based on showstopper criteria, while flexible or preference criteria are used to prioritize projects. Several evolutions to the traditional process have emerged with time to make it more flexible to the type and risk-level of the development projects, and to accelerate its time-to-market by overlaping stages and activities (Cooper 2014). Specific activities performed in the stages generally include the following:

- *Scoping:* The objective of this first step is to determine the value of the project from both a technology and business perspective. It consists mainly of rapid analysis and gathering of key data relative to market and technical feasibility such as market size and forecast, likely acceptance of the product by targeted customers, possible technology pathways, as well as market and technology risk.
- *Building the Business Case:* This point in the process is critical for the identification of a product concept and the definition of a business plan to

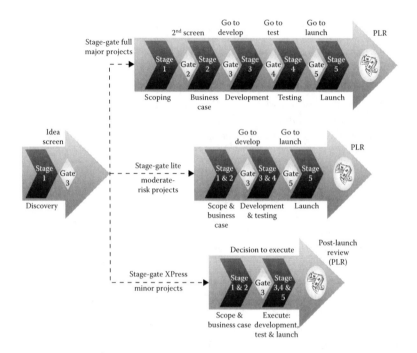

FIGURE 3.15 Traditional Stage-Gate® Product Development Process and variants. (From Cooper, R.G., *Research-Technology Management* 157, 2014.)

support it. Besides verifying the attractiveness of the project, it is critical at this stage to gather the "voice-of-customer"—market needs, desires, and preferences—and to convert this into product specifications in order (1) to freeze benefits that will be specific to the new product and (2) to define the value proposal and the product positioning strategy. This work, coupled with a detailed business and financial analysis, leads to the definition of the business case. At the end of this step, a clear definition of the product concept as well as a road map for subsequent development activities should be achieved.

- *Development:* Once product specifications are determined, further activities include the technical work necessary to convert the concept idea into a product that achieves expected performance and to perform preliminary testing for securing a competitive value proposal. The development stage is considered completed when there is assurance that the product can meet technical and market requirements under controlled conditions.

- *Testing:* Besides product testing, the production process must be designed and the associated economics refined to a greater degree of accuracy. A negative or marginal outcome at this stage is considered a showstopper, and implies a refinement of the project at the development stage. Typical activities might include (1) extended "in-house" tests to validate product quality, (2) testing of the product under functional use conditions to validate performance, (3) operation of the manufacturing process at the demonstration scale to identify and correct potential limitations, and estimate production costs more precisely, (4) validation of market penetration strategy, and (5) refinement of business and financial analysis to determine the project's economic viability.

- *Launch:* The marketing and production plans are implemented at the desired industrial scale during this stage. Efforts regarding market penetration should be supported by a supply chain strategy to enable manufacturing flexibility.

The Stage-Gate® process proposes a systematic approach for new product development; however, it is not obvious how it should be applied for the bioproducts portfolio, in the case of a forest products company implementing biorefinery processes. For instance, time-to-market and supply chain considerations are critical to capture early market share and achieve a competitive position for specialty products—but this is not the case for commodity products. Considering as well (1) the risks related to market penetration, (2) the relative lack of investment in new products research and development by forest products companies, and (3) the limited access to capital, a partnership strategy may be critical in the modified stage-and-gate process.

3.2.3 SYSTEMATIC METHODOLOGY FOR TRANSFORMATION THROUGH BIOREFINERY IMPLEMENTATION

The methodology proposed here follows a stage-and-gate approach and targets the development of innovative business models for the transformation of forest industry companies: (1) an iterative approach combining product and process designs is used for identifying, selecting, and designing robust product portfolios; (2) the technology plan supports the business plan to mitigate risk, to enhance margin creation, and to create long-term competitive advantages; and (3) the product portfolio market positioning should be defined through a systematic assessment of the existing value chains, market potential, and competitive position.

Three main steps are considered in the methodology as summarized in Figure 3.16, where P–P–P refers to product–process–partner combinations.

3.2.3.1 Step 1: Identification of Product–Process–Partner Combinations

A set of promising biorefinery product–process options needs to be identified, defined, and evaluated by the forest products company considering the following activities:

- Definition of the company prerequisites for implementing biorefinery options in terms of risk acceptance, existing assets and core business, and business diversification targets
- Identification of a set of biorefinery technologies that meet the company's prerequisites
- Identification of value chains in the vicinity of the mill(s) considered for the biorefinery strategy that can support the development of product portfolios

A set of "showstoppers" need to be defined based on process and market perspectives, to narrow the number of options and focus on company-adapted strategies that meet corporate needs over the short and long term. For the first screening of biorefinery options, typical showstoppers include the level of technology maturity, the order-of-magnitude of capital investment required, the volume of biomass feedstock

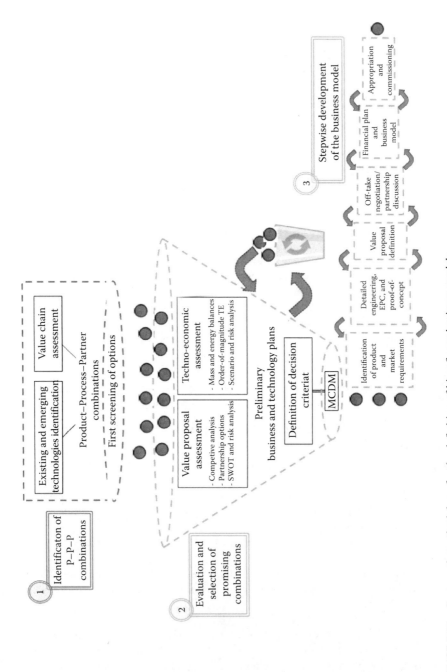

FIGURE 3.16   Systematic methodology for the definition of biorefinery business models.

required to have attractive economies-of-scale, the magnitude of expected growth for targeted market segments, the order-of-magnitude of the return on investment, and the existence of potential partners to create synergies.

3.2.3.2   Step 2: Evaluation and Selection of Promising Combinations

Preliminary business and technology plans are developed during this step to characterize the economic viability of each product–process–partner combination while identifying preliminary business strategies. The identification of market and technology uncertainties and risks, along with the characterization of the competitive environment, are critical activities.

At such an early stage of the design process, there remains significant uncertainty in cost estimates. A "large block analysis" cost estimating method can be employed, that is, a base case product–process combination is agreed on for which costs are developed, and alternatives are compared to the greatest extent possible to obtain relative cost estimate for the different options. By using this approach, the costs between processes are rendered comparable for decision-making.

A rapid market analysis is performed for the products defined within the portfolio, considering especially the local and regional value chains potential. Market parameters are estimated, including uncertainties in terms of future pricing, future supply, and demand for each product. Sensitivity analyses are performed for technology and market uncertain parameters, and the potential impact on profitability. Multiple criteria can be calculated to express the techno-economic, technology, market, and environmental risks, issues, and challenges identified along the phased implementation process, considering short- and long-term perspectives.

An MCDM panel with project stakeholders can then be organized in order (1) to fix a set of criteria that would be considered practical and complete enough in the context of the decision to be made, (2) to appreciate biorefinery options in regard to the set of criteria defined, (3) to weight criteria based on their relative importance in the context, and (4) to compare biorefinery option scoring and relative ranking, and elaborate on the most promising ones.

3.3.2.3   Step 3: Stepwise Development of the Business Model

A limited number of biorefinery options—typically not more than three—are retained at the outcome of the MCDM panel. A systematic process includes business and technical milestones such as (1) the negotiation of a letter of intent (LOI) for the off-take of bioproduct(s), (2) initiation of potential partnership discussions, if appropriate, (3) feasibility engineering for capital and operating costs, and (4) pilot and demonstration trials, if needed.

The business model is the formal statement of the business goals associated with the biorefinery strategy, along with the plan to implement the project considering the associated risks. The business model for the biorefinery transformation takes into account the following elements:

- *Executive summary*: key business plan elements and summary of risks and key project economics.
- *Business overview*: critical analysis of the current situation and unique value proposal of the project.

- *Market analysis*: market potential and future product positioning in the market as well as competitive analysis.
- *Sales and marketing plan*: market strategy for penetrating value chains while focusing on the off-take agreement/partnership, pricing strategy, and distribution strategy, that is, supply chain policy.
- *Operational plan*: operating strategy focusing on a review of the technology and the associated risk.
- *Project management plan*: project team and organizational structure.
- *Action plan*: project development schedule through to completion of the project highlighting key milestones.
- *Financial plan*: historic and projected financial performance through balance sheet and bottom line analysis, cash flow projection, and scenario and sensitivity analyses including identification of potential funding sources.

An emphasis should be put on the definition of a value proposal supporting the creation of competitive advantages over the long term. This value proposal should result in the negotiation of a memorandum of understanding and/or a letter of intent with potential off-takers and/or partners, establishing the basis for future terms and conditions for the sale of the biorefinery product(s). It is also critical to consider that the financial plan of the project be well-defined involving (or not) governmental subsidies, bank loans, or private investment.

The following case studies underline the (1) assessment and decision-making and (2) value proposal definition steps of the methodology. These two steps are of particular importance for the definition of successful biorefinery business models. Biorefinery project selection is sometimes done opportunistically or on an *ad hoc* basis by forest product companies, evaluating one transformation possibility at a time, without a consistent process to systematically identify the set of possible options and to aggregate results of their assessments from different perspectives. Another emphasis in the case studies concerns the value proposal definition step to highlight its importance on the creation of competitive advantages and the generation of possible partnerships.

3.2.4 CASE STUDIES

3.2.4.1 Case Study 1: Evaluation and Decision-Making of Lignin Biorefinery Options

The context of this case study is a pulp manufacturing facility, manufacturing softwood kraft pulp. The objectives of the case study were (1) to identify opportunities suitable for product diversification in the context of the company considering its current assets (process and unused feedstock), (2) to identify potential partners for implementing the technology and business strategies, and (3) to determine under which conditions each biorefinery alternative would be economically preferred. For this purpose, the company identified promising product–process–partner combinations internally, and then assessed the value proposal associated with each strategy.

3.2.4.1.1 Identification of Biorefinery Process Alternatives

The preliminary analysis consisted of (1) identifying process streams within the existing facility from which value from lignin products could potentially be created without negatively impacting the quality and quantity of pulp and (2) determining the quality and quantity of biomass potentially available for the transformed site. Process considerations are not the central point of this chapter and are presented only in a general context.

Possible feedstock–process transformation combinations were triaged based on two drivers:

- The biorefinery project must not impact the quality or the quantity of market pulp produced.
- The technology risk related to the project should be mitigated via a strategic alliance with the technology provider.

Potential biorefinery feedstocks included hardwood chips, and the black liquor stream produced and recovered by the kraft pulping process. Several pretreatment processes exist to extract hemicellulose and/or lignin from chips prior to pulping; however, the pretreatment processes generally degrade wood components including cellulose to a certain extent depending on operating conditions and were considered likely to impact pulp quality. Extracting lignin in black liquor on the other hand has less direct impact on pulp quality, since it affects primarily energy requirements at the mill and amounts of make-up chemicals needed for cooking in the kraft process digester. Lignin precipitation technologies can be used to extract lignin from black liquor to target potentially interesting bioproducts. It was estimated that up to 2000 tons per day of hardwood was potentially available. To exploit this, a biomass fractionation plant could be built in parallel to the existing pulping process based on this feedstock supply, from which lignin could be isolated and hemicellulose- and cellulose-based bioproducts sold.

Two commercial lignin precipitation processes were retained for further analysis: (1) black liquor acidification using carbon dioxide to precipitate lignin, which is separated from the remaining liquid by filtration and washing; and (2) black liquor oxidation prior to acidification, to facilitate the filtration and washing steps. Principles of operation of the oxidation and acidification process and its integration with the existing pulp producing process are presented in Figure 3.17. The delignified black liquor and the washing liquor used to purify lignin are recycled in the chemical recovery unit in the kraft process.

Among other potential technology pathways, a solvent fractionation process was considered that separates wood components into a cellulose-rich stream, and a mixed stream containing lignin and hemicelluloses solubilized in an organic solvent. From these streams, modules of the process include (1) an ethanol production line from the cellulose-rich stream, (2) a solvent recovery unit, and (3) a separation/purification unit that isolates a hemicellulose-rich syrup and extracts lignin of distinctive grades (Figure 3.18).

3.2.4.1.2 Definition of Product Portfolios

A gamut of bioproducts ranging from commodities to added-value products can be manufactured from wood-based feedstocks, such as cellulose, hemicellulose, or lignin. Product–process combinations are defined using a phased-implementation strategy: Phase II portfolios including added-value products are initially targeted, then

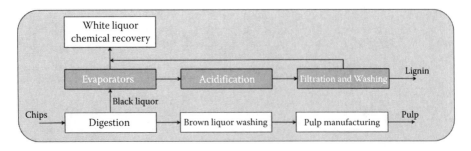

FIGURE 3.17 Simplified lignin precipitation process.

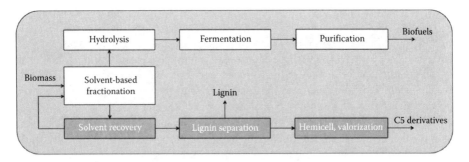

FIGURE 3.18 Simplified solvent-based biomass fractionation process.

Phase I intermediates are identified to mitigate short-term technology and economic risks.

In the case of the solvent pulping process, ethanol is considered the cellulose derivative for Phases I and II, while for the hemicellulose stream, a pentose-rich product sold as animal feed was considered during Phase I, prior to being converted into furfurals during Phase II. Formic and acetic acids are obtained as coproducts of sugar conversion into furfurals (Xing, Qi, and Huber 2011). Lignin was to be sold as a polyacrylonitrile replacement in carbon fiber or as a phenol replacement in resin production during Phase I. In Phase II, lignin was to be converted into carbon fibers or phenolic resins on site. Figure 3.19 summarizes the biorefinery options considered.

Certain technical differences characterize the conversion of lignin into added-value products from (1) kraft softwood lignin and from (2) solvent pulping of hardwood. In the case of phenolic resins production, the reactivity of depolymerized lignin is critical and determines the degree of substitution of phenol that can be attained by lignin (Wang, Leitch, and Xu 2009). Although solvent pulping lignin is relatively more reactive than precipitated lignin, the lignin extracted from both the precipitation and solvent pulping processes would be chemically modified to increase its reactivity before being used in resin formulations. Among possible pretreatment techniques, phenoylation was selected to meet required resin specifications related to processing (e.g., curing time and strength of resulting particleboards) while maintaining a low amount of free phenol and free formaldehyde (Çetin and Özmen 2002). The resin precursor to be commercialized in Phase I was thus phenoylated lignin.

Lignin precipitation options

Option #	Process	Products
1	Lignin precipitation	PF resin
2		Carbon fiber
3	Black liquor oxidation + lignin precipitation	PF resin
4		Carbon fiber

Solvent pulping options

Option #	Process	Products
5	Solvent pulping	PF resin — Ethanol, Furfurals, Acetic acid, Formic acid
6		Carbon fiber — Ethanol, Furfurals, Acetic acid, Formic acid

FIGURE 3.19 Summary of product–process options of the lignin biorefinery case study.

Regarding carbon fiber production, the process considered includes (1) spinning for turning lignin into lignin fibers, (2) stabilization to give thermoset properties to the fibers and prevent fusing at high temperatures, (3) carbonization to form tightly bonded carbon crystals that are aligned more or less parallel to the long axis of the fiber, and optionally (4) graphitization to improve fiber properties depending on the targeted application (Figure 3.20). Due to its highly branched structure, softwood lignin is difficult to spin, and usually chars instead of spinning (Kadla et al. 2002) unless it is acetylated (Eckert and Abdullah 2008). In contrast, hardwood lignin is less reactive and therefore has a much longer stabilization step than softwood lignin (Nordström et al. 2013). Whether lignin is extracted from black liquor precipitation or biomass solvent pulping, it is essential to add a plasticizer to obtain final properties closer to existing carbon fibers. Pellets of lignin mixed with plasticizer are commercialized in Phase I as a carbon fiber precursor.

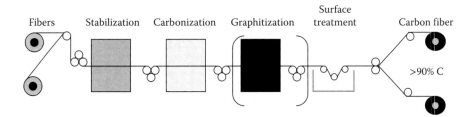

FIGURE 3.20 Carbon fiber manufacturing process. (From Axegård, P. et al., *Road Map 2014 till 2025: Swedish Lignin-Based Carbon Fibre in Composite Materials of the Future.* Sweden: Swerea SICOMP, 2014.)

3.2.4.1.3   Sustainability Assessment and Decision-Making

Using the perspective of conducting a sustainable assessment of biorefinery options, a multidisciplinary approach has been applied to this case study (Sanaei, Chambost, and Stuart 2014).

From an economic/competitiveness perspective, the following criteria were considered:

- *IRR* representing project profitability under expected market conditions.
- *Competitiveness on production costs (CPC)* measuring the potential to penetrate the existing market while facing price competition and to create margins when facing market volatility. This criterion is calculated based on the difference between the product portfolio market price, adding 10% discount on the minimum available market price, and the production cost of a given product family.
- *Phase I implementation capability (PIC)* representing the implementation capability of the technology at full-scale while considering (1) level of technology maturity, (2) process scalability, and (3) ability of implementing Phase I over a 2-year construction period.
- *Downside economic performance (DEP)* measuring the ability to sustain cash flow for a finite period of time under poor market conditions. This criterion is calculated as the operating earnings before interest and taxes (EBIT) considering minimum market price (Sanaei and Stuart 2014).
- *Return on capital employed (ROCE)* measuring the value of the business gains from its investment. This criterion is critical for evaluating economic risk associated with capital-intensive investments.

From an environmental perspective, the following three top-ranked criteria were considered:

- *Greenhouse gas emissions (GHG)* representing the carbon footprint of each biorefinery option measured as CO_2 equivalent compared to the competing processes to manufacture the functionally equivalent product portfolio.
- *Nonrenewable energy consumption* representing the level of dependency of the biorefinery option on fossil-based fuel/energy.
- *Respiratory organics* measuring the potential impact of contaminant emissions into the air affecting human health considering the relative amount of equivalent ethylene emitted into the air compared to the competing processes to manufacture the functionally equivalent product portfolio.

MCDM considering this set of criteria was conducted to evaluate the sustainability associated with each lignin-based biorefinery option. Based on the weighting factors obtained during the MCDM, IRR, GHG, PIC, and CPC were considered the most important criteria for decision-making. Applying the criteria utility values, a score was determined for each biorefinery option leading to the identification of the most promising options (Figure 3.21).

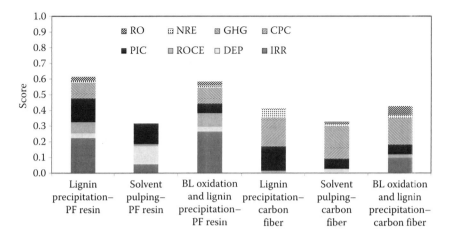

FIGURE 3.21 Lignin-based biorefinery options scoring, resulting from MCDM panel.

Lignin precipitation processes for the production of phenolic resins were the most preferred biorefinery options by the stakeholder group participating in the MCDM process. The MCDM result is similar for all solvent pulping options, regardless of the lignin derivatives. Based on the results, recommendations were made to consider several alternatives at the next level of the business model definition: (1) differentiation of lignin product specifications between lignin precipitation processes and solvent pulping, especially for carbon fiber and (2) risk assessment to examine ways to reduce capital and operating costs for the solvent pulping options.

This case study underlines (1) the importance of considering the set of possible conversion routes and product combinations that can be considered for the biorefinery objectives, (2) the importance of using restrictive showstoppers at the early stages in process design considering company/mill business requirements/constraints, (3) the role of decision-making criteria through the MCDM process to guide decision-makers regarding the selection of the most preferred options, and (4) the importance of the context in the MCDM results, since criteria weights and trade-off between options depend on the decision-making objective function.

3.2.4.2 Case Study 2: Value Proposal Definition for Torrefied Pellets

The case study considers the implementation of a torrefaction process in retrofit to an existing forest products company. As part of the definition of the most promising strategies considering short-, mid-, and long-term objectives, the forest products company identified several potential biorefinery pathways involving a range of emerging biorefinery technologies targeting commodity biofuels and bioenergy production. Based on economic, market, and technology-driven criteria, decision-making was conducted at the corporate level to identify the most preferred option, which was the production of torrefied pellets to be sold in Europe for electricity production. The criteria considered included (1) DEP reflecting how unfavorable market conditions may impact the financial situation of the company over the long term, (2) technology and project risks measuring the maturity of the technology, its scalability, and the ability to execute the project over a short-term period, (3) IRR determining whether the project should go ahead considering the relative potential financial return, (4) total capital investment costs representing the capital spending required for the project, and (5) competitive access to biomass representing the margin gained for each ton of biomass.

Business and technology plans were developed for the production and sale of torrefied pellets in the European market. Key elements covered within the business plan are presented below, focusing on the definition of the value proposal and the determination of key drivers for the off-take agreement to build a viable business model.

3.2.4.2.1 Overview of Market Potential Assessment

Several market segments can be targeted for the sale of torrefied pellets on local and international value chains:

- Electricity and heat generation in Europe, considering cofiring torrefied pellets with coal in existing power facilities or dedicated biomass plants. Regulations in Europe are (currently) favorable for such projects and create the opportunity for competitive advantages against other sources of energy such as conventional wood pellets.
- Coal replacement for industrial thermal needs and as reducing agents in the metallurgical industry. However, challenges associated with quality issues and cost-competitiveness against coal made this segment less attractive.

The former market opportunity was targeted. The use of pellets as coal replacement in existing power facilities is increasing and demand is growing more rapidly than production. In 2010, Europe consumed 11.5 million tons of wood pellets (International Energy Agency 2011), representing 45% growth in 2 years. Growth in wood pellets demand is expected to remain strong (Figure 3.22) (RISI 2015). European markets are expected to benefit from the renewable energy directive (European Commission 2009) to reach 20% renewable energy by 2020, supporting renewable sources in the global European energy mix. Each member state is required to reach legally binding targets depending on their energy profile. The United Kingdom has set aggressive targets and has developed a system of renewable obligation certificates (Office of the Gas and Electricity Markets 2013) to reach 15% renewable energy by 2020 and up to 45% by 2030.

Torrefied pellets are new entrants to the wood pellet market. They could well be the next generation of wood pellets, offering better fuel characteristics and cost savings on transportation, storage, and handling. The production of torrefied pellets at a large scale has not yet been reached, with small volumes being tested at demonstration plants.

3.2.4.2.2 Overview of Competitiveness Assessment for Product Positioning

It was essential that the value proposal of torrefied pellets manufacture be built on competitive advantages against existing and substitute energy sources, considering (1) quality- and functionality–driven advantages and (2) cost-competitiveness. Table 3.3 illustrates the competitive positioning of torrefied pellets in terms of quality and functionality considering major drivers for energy production.

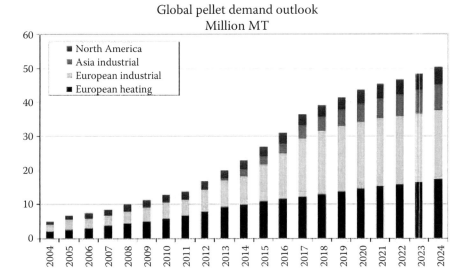

FIGURE 3.22 Historic and anticipated wood pellet demand, in millions of tons. (From RISI Inc., 2015 Global Pellet Demand Outlook Study, http://www.risiinfo.com/risi-store/do/product/detail/global-pellet-demand-outlook.html, 2015.)

TABLE 3.3
Competitive Positioning of Torrefied Pellets Against Substitutes

Main Fuel Properties	Wood	Wood Pellets	Torrefied Pellets	Coal
Water content (% wt)	30–45	7–10	1–5	10–15
Calorific value (MJ/kg)	9–12	15–16	20–24	23–28
Bulk density (kg/L)	0.2–0.25	0.55–0.75	0.75–0.85	0.8–0.85
Energy density (GJ/m³)	2–3	8–11	18–20	18–24
Hydroscopic properties	Hydrophilic	Hydrophilic	Hydrophobic	Hydrophobic
Biological degradation	Yes	Yes	No	No
Transport cost	High	Medium	Low	Low

Source: Kleinschmidt, C.P., Overview of international developments in torrefaction. Presentation at the Bioenergy Trade Torrefaction Workshop. Graz, Austria, 2011.

Fuel characteristics associated with torrefied pellets led to a price premium relative to conventional wood pellets. Major savings can be achieved due to lower cost transportation and storage and handling on site since torrefied pellet handling does not require the same level of investment when used in existing coal-based power plants (Roberts 2012).

When considering the replacement of part of the existing coal supply to an existing power facility, limitations on cofiring rate should be considered. Tests made in Europe by leaders in the domain demonstrated that cofiring has reached 80% without requiring capital investment in new boilers, compared to a cofiring rate of 20% with conventional wood pellets.

3.2.4.2.3 Value Proposal Definition

The value proposal associated with torrefied pellets was built based on (1) existing market dynamics and the consideration of regulations in place, (2) intrinsic advantages of the torrefied pellets against other sources, and (3) cost–competitiveness.

Existing coal prices coupled with government incentives dictate the price of pellets. To build the proposal value and define a room for negotiation, the following elements were considered:

- Analysis of wood pellet and coal market prices and trends
- Identification of possible subsidies and financial incentives, including their sunset dates, resulting from policies for cost-competitiveness compared to coal
- Determination of potential cost savings on the part of the off-taker
- Evaluation of the production costs for torrefied pellets
- Definition of economic threshold for the company in terms of targeted IRR

At another level, the potential cost-competitiveness of torrefied pellets compared to traditional wood pellets was evaluated and is presented in Figure 3.23. The case of two plants located in the southeastern United States was taken into account, presenting

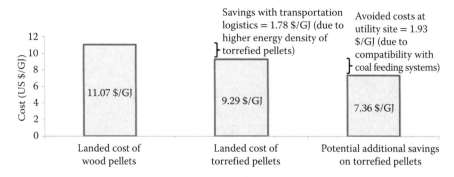

FIGURE 3.23 Cost comparison between wood pellets and torrefied wood pellets. (Adapted from Roberts, D., The forest sector's status and opportunities from the perspective of an investor (and what it takes to do more). Presentation at The Transformation of the Canadian Forest Sector and Swedish Experiences, Seminar, Stockholm, May 28, 2012.)

the same constraints in terms of production volumes for delivery to Rotterdam. The feedstock considerations included (1) the delivered chipped cost of whole logs for wood pellets and (2) whole logs and residues for torrefied pellets assuming 50% moisture content. Based on this analysis interesting savings can be achieved considering avoided cost at the utility site since torrefied pellets can be handled with the existing coal utilities. Additional savings can be achieved on landed costs associated with transportation due to the higher energy density of the torrefied pellets as well as, in this scenario, cheaper biomass cost due to the unprepared feedstock and access to residues. Thus, a room for negotiation of 3.7 $/GJ, that is, potential margin to be created, could be targeted during preliminary discussions with potential off-takers.

Some lessons learned from this case study included (1) a sound business plan to be served by the technology plan, (2) the value proposal to be identified early in the design of the biorefinery strategy, and (3) competitive advantages to be determined and maximized by robust market penetration strategies and partnership strategies.

3.2.5 Conclusions

This chapter has addressed the broad range of market- and technology-based assessments that need to be considered in a systematic manner by forest product companies to identify preferred transformation strategies.

To the extent possible, the many sources of risk must be identified and quantified. The magnitude of the impact of these risks on the economic, environmental, and competitive viability of the project requires a systematic methodology to guide decision-makers. The methodology presented in this chapter is founded on the combination of product design and process design tools and an effective interaction between business plan and technology plan development, considering a phased-implementation approach for transformation strategies. An approach inspired by the stage-and-gate model is employed to assemble these concepts. Stages are suitable to perform the analyses and assessments relevant to gather information on prospective biorefinery candidates prior to evaluating them. This is done during gate reviews incorporating proper milestones, such as MCDM activities.

The focus of this chapter was on value identification and capture in defining transformational business strategies. A practical approach based on value chain considerations has been developed. Whether it is via biomass procurement strategies, manufacturing flexibility, or optimization of the warehousing and product delivery network, supply chain management is crucial to take advantage of the uniqueness of biorefinery product portfolios and maintain a competitive position over the long term. These subjects are covered in greater detail in subsequent chapters of this book.

REFERENCES

Agbor, V.B., N. Cicek, R. Sparling, A. Berlin, and D.B. Levin. Biomass pretreatment: Fundamentals toward application. *Biotechnology Advances* 29, no. 6, 2011: 675–685.

Ansoff, H.I. Strategic issue management. *Strategic Management Journal* 1, no. 2, 1980: 131–148.

Axegård, P. The future pulp mill—A biorefinery? Presentation at the First International Biorefinery Workshop, July 20–21, Washington, DC, 2005.

Axegård, P., P. Tomani, Innventia, and H. Hansson. *Road Map 2014 till 2025: Swedish Lignin-Based Carbon Fibre in Composite Materials of the Future.* Sweden: Swerea SICOMP, 2014.

Batsy, D.R., C.C. Solvason, N.E. Sammons, V. Chambost, D. Bilhartz, M.R. Eden, M.M. El-Halwagi, and P.R. Stuart. Product portfolio selection and process design for the forest biorefinery. In *Integrated Biorefineries: Design, Analysis, and Optimization*, by M. M. El-Halwagi and P. R. Stuart. Boca Raton, FL: CRC Press, 2012.

Çetin, N.S., and N. Özmen. Use of organosolv lignin in phenol–formaldehyde resins for particleboard production I. Organosolv lignin modified resins. *International Journal of Adhesion & Adhesives* 22, 2002: 477–486.

Chambost, V., J. McNutt, and P.R. Stuart. Partnership for successful enterprise transformation of forest industry companies implementing the forest biorefinery. *Pulp and Paper Canada*, May/June 2009: 19–24.

Chambost, V., and P.R. Stuart. Selecting the most appropriate products for the forest biorefinery. *Industrial Biotechnology* 3, no. 2, 2007: 112–119.

Chambost, V., and P. Stuart. Product portfolio design for forest biorefinery implementation at an existing pulp and paper mill. Presentation at the Seventh Foundations of Computer-Aided Process Design (FOCAPD) Conference, Breckenridge, CO, June 7–12, 2009.

Cocchi, M. Global Wood Pellet Industry Market and Trade Study. 2011. http://www.bioenergytrade.org/downloads/t40-global-wood-pellet-market-study_final_R.pdf, Accessed April 22, 2016.

Cohen, J., M. Janssen, V. Chambost, and P. Stuart. Critical analysis of emerging forest biorefinery (FBR) technologies for ethanol production. *Pulp & Paper Canada* 111, 2010: 24–30.

Cooper, R.G. The new product process: The stage-gate game plan. In *Winning at New Products: Accelerating the Process from Idea to Launch*, by R. G. Cooper, 113–153. Cambridge, MA: Perseus Publishing, 2001a.

Cooper, R.G. Picking the winners: Effective gates and portfolio management. In *Winning at New Products: Accelerating the Process from Idea to Launch*, by R. G. Cooper, 213–251. Cambridge, MA: Perseus Publishing, 2001b.

Cooper, R.G. What's next? After stage-gate. *Research-Technology Management* 157, 2014: 20–31.

Costa, R., G.D. Moggridge, and P.M. Saraiva. Chemical product engineering: An emerging paradigm within chemical engineering. *AIChE Journal* 52, no. 6, 2006: 1976–1986.

Cussler, E.L., and G.D. Moggridge. *Chemical Product Design.* 2nd Edition. New York, NY: Cambridge University Press, 2011.

Dansereau, L.-P., M.M. El-Halwagi, and P.R. Stuart. Sustainable supply chain planning for the forest biorefinery. Presentation at the Seventh Foundations of Computer-Aided Process Design (FOCAPD) Conference, Breckenridge, CO, June 7–12, 2009.

Demirbas, A. *Biorefineries: For Biomass Upgrading Facilities.* London, United Kingdom: Springer, 2010.

Diffo, C., and P. Stuart. Market competitiveness assessment for forest biorefinery product portfolios selection. Presentation at the BIOFOR Conference, Thunder Bay, ON, May 14, 2012.

Eckert, R.C., and Z. Abdullah. Carbon fibers from kraft softwood lignin. US Patent 0317661, 2008.

Enea, M., and T. Piazza. Project selection by constrained fuzzy AHP. *Fuzzy Optimization and Decision Making* 3, 2004: 39–62.

European Commission. Renewable energy directive. 2009. https://ec.europa.eu/energy/en/topics/renewable-energy/renewable-energy-directive Link to official legislation document: http://eur-lex.europa.eu/legal-content/EN/TXT/PDF/?uri=CELEX:32009L0028 &from=EN, Accessed April 22, 2016.

Favre, E., L. Marchal-Heusler, and M. Kind. Chemical product engineering: Research and educational challenges. *Transactions of IChemE* 80A, 2002: 65–74.

Forest Products Association of Canada. Transforming Canada's Forest Products Industry: Summary of findings from the Future Bio-pathways Project. www.fpac.ca. 2010. http://www.fpac.ca/wp-content/uploads/Biopathways-ENG.pdf, Accessed April 22, 2016.

Forest Products Association of Canada. The New Face of the Canadian Forest Industry—The Emerging Bio-Revolution. www.fpac.ca. 2011. http://www.fpac.ca/wp-content/uploads/BIOPATHWAYS-II-web.pdf, Accessed April 22, 2016.

Gaudreault, C., R. Samson, and P. Stuart. Implications of choices and interpretation in LCA for multi-criteria process design: De-inked pulp capacity and cogeneration at a paper mill case study. *Journal of Cleaner Production* 17, 2009: 1535–1546.

Hill, M. Chemical product engineering—The third paradigm. *Computers and Chemical Engineering* 33, 2009: 947–953.

Holladay, J.E., J.J. Bozell, J.F. White, and D. Johnson. *Top Value-Added Chemicals from Biomass—Volume II: Results of Screening for Potential Candidates from Biorefinery Lignin.* Oak Ridge, TN: PNNL for US Department of Energy, 2007.

Hytönen E., and P.R. Stuart. Integrating bioethanol production into an integrated kraft pulp and paper mill: Techno-economic assessment. *Pulp and Paper Canada*, May/June 2009, 25–32.

International Standard Organization. ISO 14040: Management environnemental—Analyse du cycle de vie—Principes et cadre. Geneva, Switzerland: ISO, 2006.

Janssen, M., V. Chambost, and P. Stuart. MCDM methodology for the selection of forest biorefinery products and product families. International Biorefinery Conference, Syracuse, New York. October 6, 2009.

Jolliet, O., M. Marfni, R. Charles, S. Humbert, J. Payet, G. Rebitzer, and R. Rosenbaum. IMPACT 2002+: A new life cycle impact assessment methodology. *The International Journal of Life Cycle Assessment* 8, no. 6, 2003: 324–330.

Kadla, J.F., S. Kubo, R.A. Venditti, R.D. Gilbert, A.L. Compere, and W. Griffith. Lignin-based carbon fibers for composite fiber applications. *Carbon* (Pergamon) 40, 2002: 2913–2920.

Killen, C.P., R A. Hunt, and E.J. Kleinschmidt. Managing the new product development project portfolio: A review of the literature and empirical evidence. Presentation at the Portland International Center for Management of Engineering and Technology (PICMET), Portland, OR, August 5, 2007. pp. 1864–1874.

Kleinschmidt, C.P. Overview of international developments in torrefaction. Presentation at the Central European Biomass Conference, Graz, Austria, January 28, 2011.

Kornfeld, B.J., and S. Kara. Project portfolio selection in continuous improvement. *International Journal of Operations & Production Management* 31, no. 10, 2011: 1071–1088.

Liard, G., B. Amor, P. Lesage, P. Stuart, and R. Samson. Selection of LCA-based environmental criteria for multi-criteria decision support in a biorefinery context. Presentation at the SETAC Europe 21st Annual Meeting, Milano, Italy, May 2011.

Maity, S.K. Opportunities, recent trends and challenges of integrated biorefinery: Part I. *Renewable and Sustainable Energy Reviews* 43, 2015a: 1427–1445.

Maity, S.K. Opportunities, recent trends and challenges of integrated biorefinery: Part II. *Renewable and Sustainable Energy Reviews* 43, 2015b: 1446–1466.

Mansoornejad, B., V. Chambost, and P. Stuart. Integrating product portfolio design and supply chain design for the forest biorefinery. *Computers and Chemical Engineering* 34, no. 9, 2010: 1497–1506.

Martino, J.P. Project Selection. In *Project Management ToolBox: Tools and Techniques for the Practicing Project Manager*, by D. Z. Milosevic (Ed.), Hoboken, NJ: John Wiley & Sons, 19–66, 2003.

Mu, D., T. Seager, P. Rao, and F. Zhao. Comparative life cycle assessment of lignocellulosic ethanol production: Biochemical versus thermochemical conversion. *Environmental Management*, no. 4, 2010: 565–578.

Nordström, Y., I. Norberg, E. Sjöholm, and R. Drougge. A new softening agent for melt spinning of softwood kraft lignin. *Journal of Applied Polymer Science*, 2013: 1274–1279.

Office of the Gas and Electricity Markets. Renewable Obligation Annual Report 2011–2012. Office of the Gas and Electricity Markets, 2013.

Oster, A.M. *Modern Competitive Analysis.* 3rd Edition. New York, NY: Oxford University Press, 1999.

Penner, G. The future for bioproducts. Presentation at the BlueWater Sustainability Initiative Conference, Sarnia, ON, 2007.

Porter, M.E. *On Competition.* Boston, MA: Harvard Business School Publishing Corporation, 2008.

Pöyry P.L.C. Paper and paperboard market: Demand is forecast to grow by nearly a fifth by 2030. 2015. http://www.poyry.com/news/paper-and-paperboard-market-demand-forecast-grow-nearly-fifth-2030, Accessed April 22, 2016.

PricewaterhouseCoopers. *Managing Commodity Risk Through Market Uncertainty.* London, UK: PricewaterhouseCoopers, 2009.

Quintero-Bermudez, M.A., N. Janssen, J. Cohen, and P.R. Stuart. Early design-stage biorefinery process selection. *Tappi Journal* 11, no. 11, 2012: 9–16.

Regnier, E. Oil and energy price volatility. *Energy Economics* 29, 2007: 405–427.

RISI Inc. 2015 Global Pellet Demand Outlook Study. 2015. http://www.risiinfo.com/risi-store/do/product/detail/global-pellet-demand-outlook.html, Accessed April 22, 2016.

Roberts, D. The forest sector's status and opportunities from the perspective of an investor (and what it takes to do more). Presentation at The Transformation of the Canadian Forest Sector and Swedish Experiences Seminar, Stockholm, May 28, 2012.

Sanaei, S. Sustainability Assessment of Biorefinery Strategies Under Uncertainty and Risk using Multi-Criteria Decision-Making (MCDM) Approach. PhD diss., École Polytechnique de Montréal, 2014.

Sanaei, S., V. Chambost, and P. Stuart. Systematic assessment of triticale-based biorefinery strategies: Sustainability assessment using multi-criteria decision-making (MCDM). *Biofuels, Bioproducts and Biorefining* (accepted for publication). DOI:10.1002/bbb.1482, 2014.

Sanaei, S., and P. Stuart. Systematic assessment of triticale-based biorefinery strategies: Techno-economic analysis to identify investment opportunities. *Biofuels, Bioproducts and Biorefining* (accepted for publication). DOI:10.1002/bbb.1499, 2014.

Sanderson, S., and M. Uzumeri. Managing product families: The case of the Sony Walkman. *Research Policy* 24, no. 5, 1995: 761–782.

Schaefer, K. Outlook for the World Paper Grade Pulp Market. 2015. http://www.cepi.org/system/files/public/documents/events/EuropeanPaperWeek2015/Schaeffer%20-%20RISI.pdf, Accessed April 22, 2016.

Seider, W.D., J.D. Seader, D.R. Lewin, and S. Widagdo. *Product and Process Design Principles.* 3rd Edition. John Wiley & Sons, Inc., 2009.

Stuart, P.R. The forest biorefinery: Survival strategy of Canada's P&P sector? *Pulp & Paper Canada* 107, 2006: 13–16.

United Nations Economic Commission for Europe. Forest Products Annual Market Review, 2014-2015. 2015. http://www.unece.org/fileadmin/DAM/timber/publications/2015-FPAMR-E.pdf, Accessed April 22, 2016.

Wang, J.J., Y.Y. Jing, Z.F. Zhang, and J.H. Zhao. Review on multi-criteria decision analysis aid in sustainable energy decision-making. *Renewable and Sustainable Energy Reviews* 13, 2009: 2263–2278.

Wang, M., M. Leitch, and C. Xu. Synthesis of phenol–formaldehyde resol resins using organosolv pine lignins. *European Polymer Journal* 45, 2009: 3380–3388.

Werpy, T., and G. Peterson. *Top Value-Added Chemicals from Biomass Feedstock—Volume I: Results of Screening for Potential Candidates from Sugars and Synthesis Gas.* Oak Ridge, TN: NREL for US Department of Energy, 2004.

Xing, R., W. Qi, and G.W. Huber. Production of furfural and carboxylic acids from waste aqueous hemicellulose solutions from the pulp and paper and cellulosic ethanol industries. *Energy & Environmental Science*, no. 4, 2011: 2193–2205.

4 The Roles of FPInnovations, Governments, and Industry in Transforming Canada's Forest Industry

Jean Favreau and Catalin Ristea

CONTENTS

4.1 INTRODUCTION

The forest industry is an important component of the Canadian economy with Canada being the second largest exporter of primary forest products in the world. The nature and type of forest sector activities vary from region to region and from community to community and reflect the comparative advantages of these various areas. Communities in western Canada tend to specialize in the manufacturing of wood products (e.g., lumber), while those in Ontario, Quebec, and the Atlantic provinces are involved in everything from the manufacturing of softwood lumber to value-added products, pulp, paper, and newsprint. Canada's forest industry also plays a central role in many rural and remote communities, including First Nations communities. Natural Resources Canada reports that there are over 300 such communities across the country, which are economically dependent on the forest industry (Natural Resources Canada 2007).

The Canadian forest sector is facing many challenges related to renewing itself and increasing its profitability. Globalization of the world's economy in recent decades has affected the structure and business models of forest industries worldwide. Competition from Russia, India, China, and South American countries that provide cheap wood fiber to international markets has sent shock waves through many traditional industries, including the Canadian forest products industry. Canadian companies have started to realize the importance of having a better understanding of customers' needs. They are also realizing the importance of developing diversified value propositions and linking the entire value chain—from forest to market—by establishing efficient and sustainable partnerships that allow them to expand economic activities beyond the traditional business models. Many Canadian forest products companies are faced with developing new value propositions, that is, ones that are different from the historical low-cost commodity business model.

The value chain approach has tremendous potential, in light of the Canadian forest sector's high cost structure relative to most other Northern Hemisphere countries, the diversity of Canada's forests and fiber, and the fact that most of the Canadian forest industry's revenue still comes from commodity products. Other industrial sectors and some forest product companies in Europe, New Zealand, and South America have successfully implemented some value chain concepts in their business models. However, the Canadian forest sector struggles to increase the agility of its supply chains and to create innovative and efficient business models capable of overcoming competition from other countries. Table 4.1 summarizes the unique characteristics of Canadian forests and the Canadian forest industry. These characteristics differentiate the industry in essential ways from other countries and illustrate why Canada needs to create its own value chain approach. The challenges faced by the Canadian forest sector are shared by only a few other countries with similar environments, such as Russia, which is one of the largest forestry-reliant countries in the world.

In Canada, the forest industry is strongly dependent on Crown land. The current structure of the Canadian industry allows the optimization of individual processes at the expense of—or without consideration of—the potential value gains of upstream and downstream operations. The Canadian forest industry has historically been driven by a cost-reductive, economies-of-scale-oriented, and "push-to-market" business model. Fortunately, despite their different competencies regarding the forest

TABLE 4.1
Characteristics of Forests and Forest Industries, Canada vs. Other Countries That Have Cost Advantages: Summary

Canada	Other Countries
55 commercial species	Few species
Diverse forest types, essentially natural forest	Plantations
Wide range of fiber attributes	Narrow range of fiber attributes
Resource extraction affected by seasons	Mainly just-in-time delivery with few seasonal impacts
Low processing efficiency	High processing efficiency
High production costs	Low production costs
Old and dispersed industrial network	New and integrated processing systems

sector, the industry and government share common values and goals for developing a sustainable forest economy.

The opportunities for the Canadian forest sector to overcome its challenges include optimizing the traditional forest products value chains, developing new value streams for next-generation products and markets, and addressing the industry's environmental impacts while maximizing resource values in a sustainable manner.

There are many technological, systems, and cultural challenges that need to be addressed before the optimized forest value chain becomes a reality. Wood supply is perhaps one of the most difficult areas in the forest products value chain to optimize due to the heterogeneous raw material supplies and complex supply networks. Another challenge is the willingness of each independent business agent in the supply chain to work with others to pursue the benefits of maximizing the entire value chain. As an old proverb says, a chain is only as strong as its weakest link. A key objective of economic activity is to create value for the end customer, and this could be achieved by integrating the flows of information, raw materials, technology, and resources (Handfield and Nichols 2002). If the industry works simultaneously on reducing costs and maximizing value, the average wood net value could be significantly increased. However, the tools for supply chain optimization are not sufficient: there is a need for a different mind-set, one of collaboration and information sharing between business units and of deeper knowledge of the value chain (beyond each individual unit or chain link).

This chapter first introduces the national research organization collaborations that support Canadian forest industry transformations. Second, the roles of Canadian federal and provincial governments are discussed. Third, the challenges and opportunities for the Canadian forest industry are illustrated in the context of the complex issues of value chain optimization. The discussion includes success stories of federal and provincial programs related to the aspects of value chain optimization, as well as examples of applied studies and decision support tools that have been implemented in practical industrial applications. The chapter concludes with some ideas about how the government and industry may invest in, and benefit from, efficient and sustainable solutions for optimizing forest industry value chains so that the industry can deliver maximum value to the end customers while improving the well-being of society.

4.2 NATIONAL RESEARCH

The vision of enhancing the Canadian forest sector's global competitiveness and transforming the Canadian forest industry is shared by FPInnovations,* one of the largest private, nonprofit scientific research centers in the world.

FPInnovations is a large leading research organization focusing on research and development in the forest sector. Powered by creative people and world-class research, FPInnovations fuels the growth and prosperity of the forest sector by nurturing its people and scientific excellence within a diverse workplace; developing solutions to enhance competitiveness and sustainability; creating and seizing opportunities beyond traditional markets; and accelerating innovation and enabling partnerships among industry, governments, and academia. The organization supports the Canadian forest industry in developing innovative solutions based on the unique characteristics of Canadian forest resources, and in applying a sustainable development approach in close collaboration with universities, industry, and government through industrial membership and government-funded research.

FPInnovations has created a collaborative research model that responds to the needs of its industrial members and its university, research, and governmental partners. This model combines the strengths of its strategic research alliances from universities and collaborators, and develops products and services, along with licensing and joint venture agreements, while taking into account the needs of its members and customers. Projects defined under this model are transformed into deliverables and value propositions through the works of 18 research departments linked with more than eight university research networks.

By creating the connections between research and development needs and industry needs, FPInnovations supports the Canadian forest industry in envisioning new opportunities. An example in the area of value chain optimization research is FPInnovations' Modeling and Decision Support research program, which has the objective of creating decision support tools that enable forest products companies to provide the right volume of wood and products of the right quality to the right market at the right time.

4.3 CANADA'S FEDERAL AND PROVINCIAL
GOVERNMENTS: ROLES AND CHALLENGES

Given the rapid development of the bioeconomy, which has affected the business models and shaped the business structures of forest products industries worldwide, governments in Canada have the opportunity to support the Canadian forest industry in increasing the agility of its supply chains, and in creating innovative and efficient business models to improve global competitiveness. The Canadian forest products industry needs to expand its economic activities beyond the traditional business models through efficient and sustainable partnerships across the value chain. Key

---

* FPInnovations' staff number more than 525. Its research and development laboratories are located in Quebec City, Ottawa, Montreal, Thunder Bay, Hinton, and Vancouver, and it has technology transfer offices across Canada.

competencies regarding the value chain optimization of the forest sector are required. These include the following:

1. Wood supply and forest resources management that focuses on maximizing markets, sustainability, and value.
2. Systems that facilitate a faster response by the value chain to market signals, especially to signals from nontraditional and emerging markets in the new global bioeconomy.
3. A business-enabling economic system that assists the forest industry in demonstrating and deploying new value chain optimization systems.
4. An educated workforce that has practical skills suitable for implementing and adopting value chain optimization technologies.

The roles of Canada's federal and provincial governments in addressing these competencies are discussed next in detail.

4.3.1 MANAGING WOOD SUPPLY AND FOREST RESOURCES WITH A FOCUS ON MAXIMIZING MARKETS, SUSTAINABILITY, AND VALUE

In Canada, because the wood supplies of forest products companies are obtained mainly from publicly owned forests, setting up the appropriate federal and provincial forest policies related to tenure and access to the public resource are of prime importance for the renewal of the forest sector. Governments can act on forest tenure, renewable energy, and support for research and innovation, as well as on the development of markets overseas beyond the traditional North American markets.

In Canada, other industrial sectors are expected to increase their demand for accessing the land base. Oil and gas development activities in the western provinces will remain high in the coming decades, which will create significant competition for forestland uses. For example, the development of the oil sands in Alberta has the potential to expand oil sands' land use by an additional 350%, thus affecting an estimated 13.8 million hectares of boreal forest (Lisgo et al. 2009). In addition, hydroelectric power development in Canada is predicted to place increasing pressure on forestlands that have significant potential for additional production of this type of energy in Canada.*

Mining activity is also likely to increase as many areas of Canada have yet to be explored for minerals, and Canada is expected to have a promising future as a major producer of minerals (Lisgo et al. 2009).

In recent decades, the economic value of nontimber forest products has also gained increasing recognition. Maple sap products are the biggest earners (this industry accounted for $354 million in Canada in 2009[†]), followed by Christmas trees, mushrooms, plant extracts, resins, florist supplies, and craft-making materials, among others.

---

* http://www.nrcan.gc.ca/energy/electricity-infrastructure/about-electricity/7359 (accessed July 18, 2014).
† http://www.nrcan.gc.ca/forests/industry/products-applications/13203 (accessed November 22, 2015).

Fundamental responsibilities of the federal government include limiting the market power of "natural monopolies," encouraging market competition, and regulating prices or monopoly profits. For example, the forest industry is currently highly affected by the lack of competition associated with rail shipments of forest products. The government can improve the performance of transportation in rural regions, thereby contributing toward maximizing the value of timber harvested in Canada. One of the most important roles of the federal government is to support research and development—not just in transportation, but in all areas of the forest sector—to increase the total value created from Canadian forests. This can be achieved by finding innovative solutions for creating a competitive and sustainable forest value chain.

Major power shifts in international markets have impacted the Canadian forest sector as well as the federal government's fiscal policies. The Canadian Forest Service, a division of Natural Resources Canada, faces the ongoing challenge of responding, in a timely manner, to these emerging trends. By supporting the successful transformation of the forest sector, the Canadian Forest Service aims to foster a collective partnership that looks into how currently allocated energy and resources can play a more effective and substantive role in serving Canadians. It is the Canadian Forest Service's vision that efforts must be multifaceted and involve innovation, market expansion, optimization of fiber value, and a continued commitment to environmental leadership. The Canadian Forest Service's strategic priorities are to (1) support forest sector competitiveness, (2) optimize forest value, and (3) advance environmental leadership.* Its long-term objective is to develop new policies and programs that support the advancement of the next generation of industrial innovation. It also strives to facilitate the implementation of game-changing products and processes.

4.3.2 GOVERNMENT-FUNDED INITIATIVES IN CANADA

In recent years, Canada's federal and provincial governments have introduced several initiatives that support the forest industry at various levels of the forest value chain. One initiative that was successfully implemented is the Pulp and Paper Green Transformation Program (Natural Resources Canada 2012). The federal government invested Cdn$1 billion to improve the environmental performance of 24 Canadian pulp and paper mills in 38 communities, which created more than 14,000 jobs. It also helped in the process of building a more sustainable, prosperous, future forest industry sector.

Another government-funded initiative is the Investments in Forest Industry Transformation (IFIT) program, aimed at supporting Canada's forest sector in becoming more economically competitive and environmentally sustainable through targeted investments in advanced technologies.† The initial 4-year, $100 million initiative by the federal government supported forest industry transformation by accelerating the deployment of highly innovative, first-in-kind technologies at Canadian forest industry facilities. These projects included bioenergy, biomaterials, biochemicals,

---

* http://www.nrcan.gc.ca/forests (accessed July 9, 2014).
† http://www.nrcan.gc.ca/forests/federal-programs/13139 (accessed July 9, 2014).

and next-generation building products. The Forest Industry Transformation program was renewed in February 2014, with an additional $90.4 million provided for the program over 4 years under Canada's Economic Action Plan 2014. This continued commitment will help bring the next wave of innovation to market and will solidify Canada's position as a leader in forest industry transformation.

The third government-funded initiative is the Expanding Market Opportunities (EMO) program. This program aims to increase market opportunities for the Canadian forest industry in offshore markets and in nonresidential construction and mid-rise segments in North American markets.* Under Budget 2012, the federal government invested $105 million over 2 years to support Canada's forest sector. The funding is targeted at fostering innovation and expanding market opportunities for the sector. Additional funding of $92 million, over 2 years, was announced under Budget 2013 to further support market diversification and forest sector innovation. Most projects funded under the Expanding Market Opportunities program are cost shared with industry and other partners and with Natural Resources Canada providing up to 50% of eligible project costs.

The Forest Communities Program (FCP) is the fourth government initiative, which was developed to assist community-based partnerships in developing and sharing knowledge, strategies, and tools to support the forest sector transition and to take advantage of emerging forest-based opportunities.† The Forest Communities Program is a $25 million, 5-year program set up by the federal government that funds 11 sites across Canada as well as national projects. The FCP community partnerships are located in defined geographic areas on a regional scale and include a mix of urban, rural, and aboriginal communities.

The fifth initiative is the Forest Innovation Program (FIP), which focuses on supporting research development and technology-transfer activities in the Canadian forest sector. Under Budget 2012, the federal government invested $105 million, over 2 years. The funding is targeted on fostering innovation and expanding market opportunities for the sector. Additional funding of $92 million, over 2 years, was announced under Budget 2013 to further support market diversification and forest sector innovation. The Forest Innovation Program directly supports the Transformative Technologies Program (TTP) and the work of the Canadian Wood Fibre Centre (CWFC). The Transformative Technologies Program, in turn, supports precommercial research and development as well as technology transfer for innovative technologies and processes in the forest sector. The research, delivered by FPInnovations, is focused on four main areas: next-generation building systems, bioproduct development, integrated value maximization, and innovation deployment.

The CWFC works closely with FPInnovations to increase the economic return of Canadian forest resources by ensuring a coordinated, transformation-oriented approach to research and innovation along the entire forest value chain. It carries out research in the following three key areas: (1) improving the knowledge of forest inventory, including types and volumes of trees; (2) improving understanding about the production of trees with desirable characteristics; and (3) value chain

---

* http://www.nrcan.gc.ca/forests/federal-programs/13133 (accessed July 9, 2014).
† http://www.nrcan.gc.ca/forests/federal-programs/13135 (accessed July 9, 2014).

optimization from forest to the final product market (Canadian Wood Fibre Centre 2013).

The federal government and the British Columbia provincial government have supported the B.C. Coastal Hem–Fir Initiative with a $3 million annual research grant, supplemented by in-kind contributions from British Columbia's Coastal forest products industry.* In partnership with British Columbia's Coastal industry, the federal government, and the provincial government, FPInnovations delivers the program with the aim of increasing the competitiveness of the coastal sector and the value extracted from the coastal Hem–Fir resource.

The B.C. Coastal Hem–Fir Initiative has shown results in terms of cost savings and productivity improvements. Two industry project participants have reported combined benefits of $30 million to date (this is assumed to be over 3 years) as a result of improvements in recovery within sawmills—with little investment in capital equipment. Based on these data, the ratio of annual benefits realized to date by two companies through the federal and provincial investment is approximately 3:1.

One of the challenges affecting the effectiveness of forest value chain management of public forestlands is that of forest inventory management; Canada's forest resource inventory is not accurate or very detailed. Federal and provincial governments have the opportunity to provide better forest inventory data that can be used for long-term management of sustainable forests as well as short-term planning of forest operations. As mentioned earlier, the Canadian Wood Fibre Centre is working with FPInnovations and with provincial, industrial, and academic partners to develop an enhanced forest inventory aimed at providing precise information for the value chain optimization.† The system will have the potential to reduce costs and increase profitability, and it will position the forest sector for long-term economic sustainability. It will also benefit the forest industry communities across Canada. Enhanced forest inventory can be the key to effective forest monitoring systems, predictions, and forest planning. Benefits of the value chain approach include improved planning decisions and better wood allocations, both of which will lead to the generation of greater value per hectare of harvested forest. More accurate information about wood fiber attributes will increase the reliability of predictions and improve the values of harvested wood. It will also improve access to wood resources on public lands, foster effective and efficient forest management, support the development of innovative decision support systems, and promote better use of forest resources.

4.3.3 SYSTEMS FOR FACILITATING FASTER RESPONSES BY THE VALUE CHAIN TO MARKET SIGNALS, ESPECIALLY TO THOSE OF NONTRADITIONAL AND EMERGING MARKETS IN THE NEW GLOBAL BIOECONOMY

Provincial forestlands policies must enhance the Canadian forest industry's ability to adapt to rapidly changing market conditions and not impede restructuring and rationalization. This means allowing licensing, enabling more efficient production decisions to be made by streamlining how market signals affect the decision to extract

---

* http://www.bccoastalinitiative.ca/index.html (accessed July 9, 2014).
† http://www.nrcan.gc.ca/forests/inventory/13421 (accessed July 2, 2014).

raw materials, and encouraging the development of new business models and expertise as needed by the value chain approach. The allocation of resources must also be achieved in the most optimal way possible—from the perspectives of creating value for the end customer and reducing inefficiencies in the supply chain.

An important, but nevertheless challenging, area of the governments' responsibility is to address the issue of equity on stumpage fees and administrative fairness, especially in the context of exports to the United States. In market economies, some of the market failures occur when the market does not allocate resources efficiently or when there is imperfect information on timber pricing (Klemperer 2003). Governments have the ability to step in and intervene to promote efficiency and equity.

The arrival of new technologies for carrying out forest inventories bodes well for increased accuracy and more comprehensive information. However, the costs associated with these new technologies have prompted foresters to question the real value of enhanced forest inventory. FPInnovations, in partnership with Tembec and the Ontario Ministry of Natural Resources, have estimated the monetary impacts of forest operations and wood processing planning with the aid of an enhanced forest inventory system. The analysis done with FPInnovations' FPInterface simulation tool* indicated a net savings of $1.57/m^3 on wood net value when the operational harvesting plans were made with light detection and ranging (LiDAR)-enhanced forest inventory data (Lacroix and Charette, 2013).

Better planning tools are also needed to support and orient further decisions that are in line with the value chain approach to maximize the value extracted from the wood supplies based on market signals. Such tools would be able to provide the net values of forest stands, which will facilitate better wood allocation and a better timber appraisal system, as well as support the industry in bidding on timber sales based on fiber value instead of volume.

4.3.4 A BUSINESS-ENABLING ECONOMIC SYSTEM THAT ASSISTS THE FOREST INDUSTRY IN DEMONSTRATING AND DEPLOYING NEW VALUE CHAIN OPTIMIZATION SYSTEMS

An important competence area for governments is creating a business-enabling economic system. International investors believe that the business climate in Canada represents a competitive advantage over many other countries. For example, Canada's total stock of inbound, direct foreign investment as a proportion of gross domestic product is considerably high among industrialized countries, being more than twice the level in the United States and over 12 times the level in Japan (United Nations Conference on Trade and Development 2007). Recognizing that high tax rates affect competitiveness, especially in a context in which the Canadian dollar is almost at par with its U.S. counterpart, governments can encourage investments in the forest sector with effective tax rates that are lower than that in competing countries.

---

* http://fpsuite.ca/l_en/fpinterface.html (accessed July 18, 2014).

In Canada, provincial governments also play an important role in encouraging the industry to adopt a value chain approach. Provincial governments have a major contribution in expanding the meaning of the value chain approach by including values associated with public social and economic needs as well as with natural habitats and the environment. Policies and rules on tenures, in addition to wood access, allocations, and investments, can serve as the basis for a genuine value chain approach.

For many Canadian forest products, the cost of producing products from the raw material is high, representing more than 50% of the total cost, due to high raw material costs that are related to high forest operations and transportation costs. The cost of fiber is also affected by harvesting and processing methods. Through their forestland policies, provincial governments can impact forest resource management and, in turn, the costs and the values produced from forest resources. Rules and regulations issued by provincial governments may also have significant impacts on investments and industrial capacity to make optimal use of the fiber resources.

Another challenge for governments is the ability to provide a supportive policy framework for quantifying the carbon footprint across the value chain and tracing various product flows. This framework will enable the industry to realize its potential in contributing to greenhouse gas emission mitigation, providing information required for future certification, and supporting the mechanisms required for agile manufacturing.

Finally, another key responsibility of the governments is to enable the decision-makers in forest management planning to develop the plans that match wood supplies—by species and with spatial patterns—with the market requirements and manufacturing capacities. This includes complementary facilities of appropriate scale and location to generate sustainable cash flows while meeting the Criteria and Indicator Framework of the Canadian Council of Forest Ministers (2006).

4.3.5 AN EDUCATED WORKFORCE, WHICH HAS PRACTICAL SKILLS SUITABLE FOR IMPLEMENTING AND ADOPTING VALUE CHAIN OPTIMIZATION TECHNOLOGIES

One of the difficulties in understanding the benefits and challenges of the value chain approach lies in the technical skill of the human resources in the forest industry's traditional business model. The typical approach, based on the personal experience of industry personnel, is to optimize individual processes of business units, without considering the other potential value gains to be had from including the upstream and downstream operations.

In many forestry companies, forest operations planners do not consider the full impacts of their operations on manufacturing, transportation, or marketing strategies. In some cases, the information is simply not shared between units. The manager of each unit is often evaluated by how well that operation is performing, rather than on how the operation collaborates with other units to produce higher returns for the company as a whole. As a result, the experiences and skills of the current workforce are no longer able to address the complex issues of large supply chains. Its current knowledge has been acquired through many years of on-the-job training and experiences. While these experiences are very important and often sufficient for optimization of a particular unit, they are not sophisticated enough to account for the complex trade-offs that can arise between different activities along the value chain.

Given these needs, governments can invest and support educational programs that prepare the next generation of the workforce with the skills necessary for the development and management of these value-chain-wide information systems and for the demonstration and deployment of the decision support systems that optimize the industry's value chains.

4.4   ROLE OF THE CANADIAN FOREST INDUSTRY

As mentioned earlier, most producers in the Canadian forest industry are stewards of publicly owned forest resources. Forests need to be managed to maintain public confidence, nontimber values, ecosystem integrity, and the carbon sink role. A Canadian task force (Forest Products Industry Competitiveness Task Force 2007) raised four major issues regarding the renewal of the industry:

1. Delivered wood costs vary regionally from low to high by international standards.
2. The forest products industry has a unique capacity to self-generate energy from renewable sources. Canada's pulp and paper sector already self-generates 60% of its energy needs from renewable sources, and the forest sector as a whole has the potential to become a net source of green power.
3. Despite pockets of excellence, the capital stock of the industry as a whole is older and less productive on average than that of leading global competitors.
4. Across each segment of the industry analyzed by the task force, Canada's labor costs are the highest, or very near the highest in the world, and labor productivity levels in most sectors of the industry are well below world-class levels, thus creating a productivity gap.

Consequently, the Canadian forest industry faces many challenges in implementing new tools and decision support systems based on value chain concepts. Some fundamental issues, either reported by FPInnovations' industrial partners or observed by FPInnovations researchers during their activities with industry, include the following:

- Organizational structures that segregate different components of the value chain
- Incentives that are not always aligned with company-wide profit maximization goals
- A lack of communication and information exchange between supply chain participants
- Among managers, a culture of risk aversion
- Insufficient and inaccurate data
- A lack of understanding of the tools and a lack of qualified personnel in the field of value chain optimization, which hinders effective utilization of the decision support tools

- A general mistrust of technology
- Competing demands for people's time and resources
- Difficulty in clearly defining various decision levels along the value chain

The following issues are fundamental to the forest products industry's successful implementation of the value chain approach:

1. There is a need for an efficient fiber supply structure to meet market requirements.
2. Product flows along the value chain need to be improved.
3. The use of agile manufacturing is required to respond rapidly to forest resource disturbances and market dynamics.

Achieving these fundamental aspects is possible if companies are able to build optimal networks that connect their interdependent networks of activities with those of their suppliers and distribution channels (Porter 1990). Recognizing the importance of regional collaboration and supporting members of a company's supply chain (Lambert et al. 1998)—including companies that provide knowledge, as well as the supply chain of primary research and development, such as FPInnovations, FORAC Research Consortium, and CIRRELT–Interuniversity Research Centre on Enterprise Networks, Logistics and Transportation (Lambert et al. 1998) —are key issues.

4.4.1 A More Efficient Fiber-Supply Structure to Meet Market Requirements

The new international economic realities are challenging the Canadian forest products industry to reach beyond its traditional markets. New market opportunities for bioenergy, biofuels, and biochemical products are emerging, and the industry needs to engage their business activities with new and compelling value propositions. These value propositions should be based on value optimization from the forest value chains despite the challenges associated with the wood supplies being increasingly remote.

The evolving values that Canadian society associates with forests are changing the way forest "products" are defined. In recent years, forests have increasingly been seen as contributing to the benefit of society through the development of bioproducts and bioenergy. These new forest products represent opportunities for the forest products industry to lead into the new green bioeconomy, as these bioproducts are seen as sustainable substitutes to nonrenewable materials and fossil-fuel-based energy. For these new bioproducts to achieve their full potential and deliver increased values to business and customers, new business models are needed that can integrate the production of these new products in an efficient and value-maximizing manner. The various value chains need to be optimized as a whole to integrate the sustainable production of these new products in addition to, and in harmony with, existing production processes and transportation networks.

An efficient fiber supply structure needs to take into account growing market segments. In turn, it must be optimized to sort fiber by species and attributes. Current supply chains do not have the ability to systematically extract maximum values from the forest resource. In addition, it is important to note that fiber supply is a major cost component of sawmills, and kraft and thermomechanical pulp production, and it is a significant determinant of lumber and pulp qualities. There is still much more improvement needed. FPInnovations has compiled many case studies to optimize the upstream portion of the value chain in Quebec (Charette and Haddad 2013). FPInnovations assessed the gains of a new approach to select harvesting sectors, optimize silvicultural treatments, and improve sawmill agility and transportation logistics. This study showed a potential net gain of around $14 million if the provincial government and the industry applied technologies to optimize their decisions.

Another area of forest-harvesting research is bucking—that is, the breakdown of delimbed stems into shorter segments for transportation and sawmill processing— which has substantial impacts on the downstream value chain operations at sawmills. FPInnovations studied the importance of choosing the appropriate technologies according to the wood supply at a regional level (Québec Ministère des Ressources Naturelles 2002). The results indicated that bucking optimization can increase revenue by 3% and improve lumber yield by 4.6%. In the follow-up study, the bucking decisions made by the harvester operator with and without the aid of the bucking optimization software were compared (Corneau et al. 2005). The results showed that the value recovery achieved was $1.40/m^3 more when bucking optimization software was used. This represents a gain of $560,000/year for a sawmill with an annual production of 400,000 m^3.

It has long been recognized that successful supply chain management requires a change from managing individual functions to integrating activities into key supply chain processes (Lambert and Cooper 2000). As mentioned earlier, optimizing a company's value chain activities relies on two important abilities: (a) being able to connect a company's interdependent system or network of activities within itself and (b) being able to improve interdependencies between the company and its suppliers in supply channels. One of the major barriers to achieving the benefits of the value chain approach is the industry's unwillingness to change from the traditional business model. Optimizing the local individual business units offers limited gains compared with the greater value gains that can be achieved from value chain coordination, including the upstream and downstream operations in the decision-making process.

The planning of forest operations does not often consider manufacturing, transportation, or marketing strategies. Each business unit is often evaluated by how well the individual operation is performing, without considering the potential impacts of their activities on the upstream and downstream operations. The absence of a business decision and planning structure does not allow for optimization of all the business units. Information is not shared between business units and business decisions are made independently. Transportation costs and hourly rates are assessed without fully capturing their impacts on the value chain. This traditional business model misses out on potentially significant gains that could result from a more holistic approach.

Optimizing each link separately does not necessarily optimize the overall value chain. A major component of the value chain approach is the potential for synergies between the linked activities to be realized. Linkages occur when the way in which one activity is performed affects the cost or effectiveness of other activities (Porter 1990). These linkages often create trade-offs in the performance of different activities. A value chain in which each link, such as a sawmill, is individually optimized does not necessarily mean that it extracts the maximum value from the raw material or make the maximum profits. The trade-offs can be optimized only by planning all related activities and their respective links and by considering the linkages between them within an overall optimization framework. Linkages require activities to be coordinated. The coordination of linked activities enables the generation of better information for control purposes and it allows less costly operations in one activity to be substituted for more costly ones elsewhere. For example, an integrated forest company can reduce overall value chain cost by improving overall mill efficiencies and allowing the mill to process the right quality logs, extracting higher values and reducing costs associated with unnecessary log handling. The benefits for the overall supply chain can outweigh the increased costs of some of the links in the chain, for example, a more costly tree-bucking and log-management system at the harvesting blocks and a more costly log sorting, bundling, and transportation system, which occurs because the right log is supplied to the right mill with a more efficient log inventory management system. Similarly, at the sawmill, the use of a more expensive log inventory and management system that dynamically monitors deliveries and consumption of logs as well as their locations in the yard can reduce the costs of inventory by preventing logs from deteriorating. This could ultimately improve mill profits by producing higher quality lumber products to satisfy the needs of the customers.

Given the complexity of many forest products production chains, many linkages and the associated trade-offs are not obvious; reaping the benefits requires complex organizational supply chain coordination, which in turn requires continuous information flows. Each business unit in the supply chain has its own customers. For a forest company that controls activities from harvesting all the way to shipping dry lumber to final customers, planer facilities and kilns are among the sawmills' customers and sawmills can be regarded as the customers of the harvesting cut blocks. Optimizing such a value chain means finding the lowest cost sequence of operations, which results in logs with characteristics that satisfy each of the possible sawmills, that is, logs that meet manufacturing capacities and product demands specific to each mill. In this case, optimizing only the forest-harvesting activities to achieve the lowest cost may result in log products with an availability and timing that do not necessarily meet the overall goal of reducing the costs across the entire value chain.

4.4.2 IMPROVED PRODUCT FLOWS ALONG THE VALUE CHAIN

Market pressures, as well as constant manufacturing capacities and fixed log-supply contracts, can prompt cost-reduction strategies instead of value extraction maximization through increased quality and added value. To counter the potentially negative effects on the value chain, the industry needs to develop strategies for ensuring that the right raw materials are sent to the right mill at the right time and at the right cost.

A study on the importance of choosing the right log profile to supply to sawmills indicated that log profile has a significant impact on sawmill performance (Cool and Bédard 2012). The case study, based on an actual industrial operation, showed that choosing the right logs per sawmill line increased productivity by approximately 11% and increased lumber value by 7%.

In a market-driven economy, customers' needs and new market requirements are the core drivers of business decisions and planning activities. Thus, the market should "pull" the required forest products from the supply chains in such a way that it gears the supply chain's production activities, maximizes the value extraction, maximizes the profits of each business unit, and minimizes the time required to deliver the products from the forest to the markets. In the forest products industry, these concepts are complicated because of the divergent nature of material flows. For example, given the market demand for a particular lumber product of a particular grade, it would be inefficient and costly to harvest particular trees to produce that particular product. Sawmill processes will produce a variety of coproducts (e.g., other lumber products) and by-products (e.g., chips, sawdust, hog fuel) from the trees. It is difficult to understand and realize which combinations of products and coproducts are most efficient and profitable to manufacture. To achieve this, a value-driven approach is necessary. This does not have to be in complete antagonism with the traditional "push" approach and cost-reduction strategy. In fact, new business decision models are needed to combine these two different business approaches.

This creates the need for developing implementable planning tools that accurately calculate the net values of forest stands, which in turn would allow better wood allocation and support the industry in bidding on timber sales. FPInnovations documented the advantages of using ForestPlan,* a harvest-planning and log-allocation optimization technology, to assess the potential value-recovery gains that would result from applying optimal harvesting and log-allocation decisions (Mortyn 2014). The study compared the optimized plan generated by ForestPlan to the plan that used traditional methods for harvesting a cutting area in coastal British Columbia in 2014. Several opportunities were identified: (1) to identify which cut blocks are uneconomical to harvest (do not harvest these cut blocks); (2) to ensure that all log assortments have been recovered; and (3) to see if logs are allotted differently than was originally planned. The combined benefit of these opportunities was found to be greater than \$3.5 million, or \$17.4/m^3.

Product flow improvement also includes the supply of more uniform raw material to the processing mills to allow the production of better quality and higher value products. This could be achieved through merchandizing yards operating in such a way that the right log is supplied virtually on demand to the conversion facility. In merchandizing yards, roughly sorted raw material brought in from the forest is evaluated, separated into uniform bundles, and shipped on demand to various processing facilities. These concepts can provide a natural decoupling point between forest and mills and subsequently improve the control of raw material flow and quality throughout the supply chain. A feasibility study of a value-added hardwood merchandizing yard indicated that the use of a merchandizing yard for logs from tree-length tolerant hardwoods, based

---

* http://www.forestnet.com/LSJissues/2014_dec_jan/edge.php (accessed June 27, 2014).

on the use of advanced technologies and revamped marketing of the yard's products, has significant potential to increase revenues without interfering with the current wood supply to local wood-processing facilities (Hamilton and Favreau 2010). Depending on the investment and risk assessment, the reported net gain can range from \$5 to \$40/m^3 for high-quality tolerant hardwoods. Strongly dependent on a better supply from the forest, this market-strategy shift is especially critical for the future viability of saw-mills due to the worldwide decline of newsprint and pulp markets, which results in less demand for the wood chips that are still produced in high quantities using current saw-milling technologies. This strategy would also improve the transportation and storage of raw logs. In this regard, FPInnovations developed Opti-Stock, an economic model for estimating the effects of wood storage on wood supply and product-manufacturing costs; the model is based on data from recent Canadian studies in this area (Favreau 2001). The main objective of the model is to help mill managers evaluate the relative importance of the various elements that can influence the log-storage duration. The model can simulate a variety of storage scenarios and evaluate their consequences for softwood sawmills as well as various types of mechanical and chemical pulp mills. There are four cost categories in the Opti-Stock model: (1) logistics, (2) financial and operations, (3) product yield, and (4) transportation. The cost of choosing suboptimal storage duration can be substantial; it is often as much as \$0.40 to \$0.80 m^3/extra week.

The Canadian forest industry must increase the overall efficiency of fiber flows and operations, from forest to the markets. Expected benefits include the following: (1) improved fiber freshness with shorter storage duration, (2) collaboration among the various actors in transportation, (3) optimized inventories along the chain, (4) efficient ability to deliver the right fiber (sort, quality, species), (5) maximized trans-portation capacity by mode, (6) increased truck-full rates, (7) transportation costs reduced by 5 to 10% (i.e., a major component of delivered product costs), (8) reduced total number of truck trips, (9) decreased energy intensity, and therefore a smaller environmental footprint, and (10) improved customer service.

In the wood flow and transportation sector, creating efficient distribution networks is critical to adequately manage fiber and end-products inventories (low costs, prod-uct quality, and good service delivery). Integrated solutions would need to address concerns of midterm timber supply shortages and the need to bring fiber from remote areas. For example, backhauling opportunities in wood and bulk transportation have been reported to be economically attractive in the forest sector (Favreau and Gingras 2007). The study showed potential savings of \$5.8 million/year in Quebec's Abitibi region when seven companies worked collaboratively to capture the backhauling opportunities in bulk and roundwood transportation.

Efficient wood flow is also critical to adequately manage the raw material inven-tories, to offer good service delivery, to meet the expectations of mill consumption, and to meet raw material quality requirements. FPInnovations studies have indicated the benefits of using truck-scheduling tools to manage truck fleets on a weekly basis and under various operating conditions. For example, one study demonstrated the benefits for a forest company on Canada's west coast. The study found some poten-tial to decrease the number of trucks on the road by 2%, increase truck utilization by 5%, meet the weekly schedule 99% of the time, and reduce hauling costs and energy intensity by more than 5% (LePage and Ristea 2012).

4.4.3 AGILE MANUFACTURING FOR MAKING RAPID RESPONSES TO FOREST RESOURCE DISTURBANCES AND MARKET DYNAMICS

Agile manufacturing is another important element for the successful implementation of the value chain approach. Forest companies are often reluctant to move away from commodity production to more agile manufacturing due to perceived productivity losses, increased complexity, and higher handling costs and uncertainty about potential benefits. This indicates a need for a planning tool that would support managers in evaluating whether the net gains from higher product values outweigh the costs associated with smaller production orders and more frequent changeovers of the production line. It was shown, for example, that applying optimized production scheduling to a furniture maker reduced total operating costs by 22% (Ouhimmou et al. 2009).

Agility is key to improving profit margin in situations where markets and/or forest resource supplies experience disturbances. When change happens, the supply chain must be able to adapt to continue enabling value maximization across the supply chain. Real-time monitoring systems and adjustable planning systems, as well as flexible manufacturing, transportation, and distribution systems, are examples of how supply chains can react to market and natural disturbances in ways that do not diminish their capacity to maintain competitiveness and profit margins.

Achieving a good customer-focused system requires that information be processed both accurately and in a timely manner so that quick responses can be made to frequent changes in customer demand, manufacturing and transportation capacity availabilities, and fiber supply availability. Controlling uncertainty in customer demand, manufacturing, and transportation processes, as well as in fiber supply availability, is critical to effectively optimize the industry's value chains. It is clear that for all this to work properly, the existence of a company-wide (i.e., value-chain-wide), seamless, and timely information system is key. The kind of information that is shared among value chain business units and information updating frequency have a strong impact on the efficiency of the value chain (Lambert and Cooper 2000).

The existence of supply chain–level optimization tools does not negate the need for business unit–level planning and optimization tools. The decision support tools that address planning issues at the value chain level can be strategic or tactical in scope. The decision support tools for business centers (individual links in the value chain) can be operational in nature.

The Canadian forest industry has traditionally been reticent, in some aspects for obvious and valid reasons, about applying value-stream mapping and other lean manufacturing principles to their value chains. By efficiently linking wood yards and mills to the forest supply areas, the potential benefits could include reduced lead time and elimination of waste and nonvalue-added activities. Higher profits would be achieved as managers are able to assess whether or not to take on new orders. However, many questions are still to be answered before these benefits can be quantified and realized. Is there sufficient supply and can it fit into the manufacturing schedule? Is it profitable, and what is the break-even price? How can processing be more efficient in periods with limited capacity and high inventory sitting in yards? What disturbances in the log supply and/or in the markets are companies and their supply chains able to absorb by being more flexible and agile?

The bottom line is that more work is needed to develop implementable decision support tools to assist forest companies in understanding the need to become more specialized and to evaluate the range of benefits and costs of organizational changes and capital upgrades.

4.5   CONCLUSIONS

The globalization of the world's economy in recent decades has affected the structure and business models of forest industries worldwide. Canadian forest products companies have started to realize the importance of having a better understanding of customers' needs; of developing diversified value propositions; and of linking the entire value chain, from forest to market, by expanding economic activities beyond the traditional business models through efficient and sustainable partnerships. Many Canadian forest products companies are forced to develop new value propositions that are different from those of the historical low-cost-commodity business model.

The competencies of both federal and provincial governments regarding the forest sector are diverse and crucial in enabling a value chain approach. Key government roles that could have major positive impacts on the forest industry include the following:

- Crown land tenures with a market focus and value maximization objectives at the supply chain level.
- Support for the development of enhanced forest inventories that link wood fiber qualities of final products with fiber qualities in standing trees to optimize the value extraction from the forest resource.
- Forest management for the long-term sustainability of the wood supply, while satisfying the short-term market needs of the evolving bioeconomy.
- Exploring and facilitating access to nontraditional and emerging international markets in bioproducts and renewable energy sources.
- Making bridge funding available to industry to overcome the "valley of death" in the development and scale-up phase of technology transfer and implementation. This entails giving continued support for basic research at universities and research centers for the development of new technologies and decision support systems for value chain optimization, while at the same time, investing more in and focusing largely on the demonstration and deployment phase to close the gap between basic research and practical deployment, thereby ensuring that a larger portion of new knowledge actually gets implemented and adopted by industry.
- Investing in educational programs that prepare highly qualified personnel specifically for the demonstration and deployment phase of technological breakthroughs, which is a necessary complement of basic research and skill development, which are critical for the industry's successful adoption of value chain optimization systems.

At the same time, the forest sector has the ability to overcome its challenges by obtaining the greatest value from Canadian fiber attributes through the development of new value streams by optimizing forest products value chains for traditional and next-generation products and markets and by achieving environmental sustainability. Industry and government, at both the federal and provincial levels, have essential roles and responsibilities that can enable the adaptive restructuring and optimization of the Canadian forest industry's value chains. Important focuses for the Canadian forest industry are as follows:

- To consider realigning organizational structures in such a way as to link together the various components of the value chain
- To adapt the incentives offered to unit managers based on performance indicators that are aligned with company-wide value maximization
- To implement company-wide communication and information exchange systems that enable easy information flow between supply chain participants
- To offer incentives that encourage employment of qualified personnel in the field of value chain optimization to assist with the demonstration, deployment, and adoption of the developed decision support systems
- To introduce new supply chain planning positions that streamline and simplify the decision levels along the value chain

These challenges have the potential to create immense opportunities to strengthen collaborations between governments, industry, and research centers to enable the demonstration, deployment, and adoption by industry of value chain optimization technologies and planning systems.

ACKNOWLEDGMENTS

The authors thank Denis Ouellet, Darrell Wong, and anonymous reviewers for their helpful comments on draft versions of this chapter.

REFERENCES

Canadian Council of Forest Ministers. 2006. *Criteria and Indicators of Sustainable Forest Management in Canada: National Status 2005*. Ottawa, ON: Canadian Council of Forest Ministers.

Canadian Wood Fibre Centre. 2013. *Canadian Wood Fibre Centre: Who We Are and What We Do 2013–2014*. Ottawa, ON: Natural Resources Canada.

Charette, F., and S. Haddad. 2013. Optimisation de la chaîne de valeur forestière du Québec. Rapport de contrat—RC 8595. August 2013. Pointe-Claire, Quebec: FPInnovations.

Cool, J., and P Bédard. 2012. Détermination de la valeur nette de la ressource. Internal Report. August 2012. Quebec City, Quebec: FPInnovations.

Corneau, Y.C., F. Fournier, J. Favreau, and I. Makkonen. 2005. *Sawmill and Harvester Bucking Efficiency*. Lumber Manufacturing 05-02E. Sainte-Foy, Quebec: FERIC and Forintek Canada Corp.

Favreau, J. 2001. *Identifying the Cost Impacts of Wood Storage Using the Opti-Stock Model*. Advantage Report 2(60), November 2001. Pointe-Claire, Quebec: FERIC.

Favreau, J., and A. Gingras. 2007. *Étude sur le flux de fibre en Abitibi, Québec*. Internal Report RI-2007-05-16. Pointe-Claire, Quebec: FPInnovations.

Forest Products Industry Competitiveness Task Force. 2007. *Industry at the Crossroads: Choosing the Path to Renewal.* Ottawa, ON: The Forest Products Association of Canada.

Hamilton, P., and J. Favreau. 2010. Feasibility study of a value-added hardwood merchandizing yard in Huntsville, Ontario. Internal Report IR-2010-08-06. Pointe-Claire, Quebec: FPInnovations.

Handfield, R.B., and E.L. Nichols. 2002. *Supply Chain Redesign: Transforming Supply Chains into Integrated Value Systems.* Upper Saddle River, NJ: Prentice Hall.

Klemperer, W.D. 2003. *Forest Resource Economics and Finance.* ISBN 0-9740211-0-5, Blacksburg, VA: W.D. Klemperer.

Lacroix, S., and F. Charette. 2013. Better planning with LiDAR-enhanced forest inventory. Advantage Report 14(1), June 2013. Pointe-Claire, Quebec: FPInnovations.

Lambert, D.M., and M.C. Cooper. 2000. Issues in supply chain management. *Industrial Marketing Management* 29:65–83.

Lambert, D.M., M.C. Cooper, and J.D. Pagh. 1998. Supply chain management: Implementation issues and research opportunities. *International Journal of Logistics Management* 9(2):1–20.

LePage, D., and C. Ristea. 2012. Truck scheduler case study. Internal Report IR-2012-00. Vancouver, British Columbia: FPInnovations.

Lisgo, K.A., F.K.A. Schmiegelow, R.G. D'Eon, and P.N. Duinker. 2009. Competing natural-resource demands associated with Canada's forests. Report No. 10 in Drivers of change in Canada's forests and forest sector. Edmonton, Alberta: Forest Futures Project of the Sustainable Forest Management Network, University of Alberta, Canada.

Mortyn, J. 2014. Using ForestPlan to determine the value impact of different harvest and log allocation decisions. Report for the BC Coastal Forest Sector Hem-Fir Initiative. March 2014. Vancouver British Columbia: FPInnovations.

Natural Resources Canada. 2007. The state of Canada's forests: Annual Report 2007. Ottawa, Ontario: Natural Resources Canada, Canadian Forest Service.

Natural Resources Canada. 2012. Pulp and paper green transformation program. Report on results. Ottawa, Ontario: Natural Resources Canada, Canadian Forest Service.

Ouhimmou, M., S. D'Amours, R. Beauregard, D. Ait-Kadi, and S.S. Chauhan. 2009. Optimization helps Shermag gain competitive edge. *Interfaces* 39(4):329–345.

Porter, M.E. 1990. *The Competitive Advantage of Nations.* New York, NY: The Free Press.

Québec Ministère des Ressources Naturelles. 2002. *Comment accroître les revenus d'une scierie de bois d'oeuvre résineux?* Sainte-Foy, Quebec: Forintek Canada Corp.

United Nations Conference on Trade and Development. 2007. Annex table B.3. FDI flows as a percentage of gross fixed capital formation, 2004-2006, and FDI stocks as a percentage of gross domestic product, 1990, 2000, 2006, by region and economy. In *World Investment Report 2007, Transitional Corporations, Extractive Industries and Development* (pp. 259–270). New York, NY and Geneva, Switzerland: United Nations Conference on Trade and Development.

5 Decision Support Needs for Strategic Planning of Canadian Forest Value Chains

Eldon A. Gunn and David L. Martell

CONTENTS

5.1 INTRODUCTION

The Canadian forest industry recently emerged from a 30-year period that witnessed growing and widespread recognition that forests that have traditionally been viewed primarily as a source of timber for industrial purposes can serve as a source of many other values as well. The forest value chain is in fact now widely recognized as being much more than a supply chain; it is a value web through which a complex set of environmental, social, and economic values flow from Canada's forests. The Montreal process (Montreal Process Working Group, 2009) and the Canadian Council of Forest Ministers process (Canadian Council of Forest Ministers, 2003) for sustainable forest management (SFM) are accepted as the basis of forest management policy at both the federal and provincial government levels and within the forest industry itself.

Sound management of Canadian forest value chains calls for integrated strategies that address forest management and industrial capacity issues, both of which need to be examined in the context of environmental, energy, and industrial development strategies. Recent decades proved to be one of the most wrenching periods in the history of the Canadian forest industry, with more than 100,000 job losses, the closure of numerous mills, and recognition of the potential for new thrusts in the production of biofuels, biochemical, and other bioproducts. These changes will have a substantial impact on the nature of forest products capacity and the amount and quality of raw materials required and brought about changes in the National Forest Strategy (Canadian Council of Forest Ministers, 2008a). In this chapter, consistent with such change, we focus on longer-term strategic issues for the Canadian forest industry with an emphasis on economic issues and decision support requirements.

This subject is of interest worldwide. The European Union COST Action Forest Management Decision Support Systems (FORSYS)* project, aimed at documenting the use of decision support systems (DSS) in forestry, presented the results of a survey of 26 countries in which DSS were used in forestry. Martell and Gunn (2014) reviewed some of the forest management decision support systems that have been developed and used in Canada, where the complex settings and substantial regional differences must be addressed by forest management planners and their clients.

The main themes addressed in this chapter are strategic forest management planning, industrial capacity strategy, and the need for decision support tools that link them in a multistakeholder environment that is complicated by uncertainty. We begin by looking at strategy and the role of decision support in strategy development. We point out that in a forest industry environment with many players, it is quite possible for a globally rational strategy to be inconsistent and even in conflict with the different business interests of different players and stakeholders. We then look at simple modeling frameworks for industrial capacity strategy and for forest management strategy and use these frameworks to identify some of the decision support issues and needs.

5.2   STRATEGIC PLANNING IN THE FOREST SECTOR

5.2.1   Strategic Forest Planning in Canada

The Canadian forest sector is unique with extremely diversified species and forest types owned and managed predominantly by provincial and territorial government agencies, which distinguish Canada from many other countries. In addition, it is characterized by different forest management regimes with ownership/tenure structures and by environmental and ecological regulations that can vary dramatically not only from region to region but also within some regions (Martell and Gunn, 2014;

---

* FORSYS Cost Action website: http://pub.epsilon.slu.se/11417/

McGill University, 2011). This uniqueness affects fundamentally, both strategy and decision support needs.

Because strategic planning of the forest value chain involves many different players in many different business contexts, simple decision support systems cannot be developed and applied universally. Decision support for strategic planning is not about computing strategy, it is about helping decision makers assess the potential consequences of strategic business choices. (Anthony, 1965; Drucker, 1995; Mintzberg, 1987). Because these strategic choices involve decisions that will change the future by redefining the resources and opportunities available to the organization, strategic planning must be carried out from the perspective of the decision maker's values, objectives, and their future business anticipations (Anthony and Govindarajan, 2001 [Chapter 1]; Gunn, 2005). The planning of the land base, the forest that is growing on that land base, decisions concerning the installation or closing of processing facilities, and decisions concerning the construction of infrastructure for the transportation and distribution of forest products are examples of strategic planning activities. Other strategic planning focuses on the policies and contractual structures that permit, limit, and encourage the use of forest resources and the protection of the forest itself. Policies that address technology development, information infrastructure, or skills and trades training are part of forest strategy but beyond the scope of this review.

Strategic planning is usually subject to considerable uncertainty. Uncertainty is often associated with the production of raw materials and product markets, as well as competition and new technologies (Anthony, 1965; Anthony and Govindarajan, 2001 [Chapter 1]). One source of uncertainty that significantly complicates forest management in Canada is the impact of natural disturbance processes. Both human- and lightning-caused fires are common in many Canadian forest ecosystems. In the boreal forest region of the province Ontario, for example, in some years, the forest volume consumed by fire can exceed the total commercial harvest (Ontario Ministry of Natural Resources, 2013). Moreover, fire is not the only major disturbance on the landscape. Large insect outbreaks, such as the spruce budworm in eastern Canada (Maclean et al., 2002) and the mountain pine beetle in the west (Safranyik and Wilson, 2007), can significantly alter forest ecosystems, and insect outbreaks are expected to increase with climate change (Volney and Fleming, 2000). An important secondary impact of insect mortality is the heightened flammability and increased fire intensity and burned area associated with insect mortality, which together can have a very significant impact on both the availability and quality of the timber resource. In Canada, hurricanes play a significant role in shaping the forests (Dyer, 1979) on the Atlantic and Pacific coasts; windthrow effects can limit the types of silvicultural treatments that can be employed (Strathers et al., 1994).

Much of the investment in current forest industry capacity in Canada occurred many years ago. Increasingly, this means there is a need for upgrading or replacement, even without the consideration of new technologies. Technologies have improved in both sawmills and pulp mills over the years. In some cases, entirely new types of capacity and technologies have been installed, such as those used to process forest resources to produce biomass energy and feed biorefinery processes (Forest Products Association of Canada, 2010). Capital and labor economies of scale dominate the economics of new or upgraded capacity. Matching the input requirements of the manufacturing capacity and markets to the outputs of the forest management processes is an ongoing challenge. The species mix required for conventional and new bioeconomy forest products can be quite different from that produced previously. Decisions concerning forest strategy, industrial capacity strategy, and the logistics, systems, and processes that link these cannot easily be separated. Market uncertainty (e.g., prices and quality requirements) and uncertainty concerning the availability and cost of forest products call for decision support systems that can be used to help assess the robustness of strategies.

5.2.2 Whose Strategy?

In Canada, most forest resources are "owned" by governments, in particular, the provincial and territorial governments that bear the primary responsibility for natural resources both in terms of land ownership and environmental regulation, as well as regional economic development. Industrial strategy is typically seen as falling within the domain of forest companies that may or may not be led by management teams that share an integrated viewpoint internally within their company. In the new forest industry, with multiple interacting sectors such as solid wood, fiber, chemicals, and energy, interindustry strategic collaboration would appear to be a necessity. This suggests that collaborative strategic approaches between forest management and industry are necessary to ensure that the management of the forest resources is consistent with industrial requirements in species and quality, both spatially and temporally.

Government and industry are not the only players. Society, often represented through various nongovernmental organizations (NGOs), has a strong interest in the ecological and social values as well as the economic benefits provided by the forest (Adamowicz et al., 2003). Furthermore, not all forest land is owned and administered by governments. Canadian privately owned forest land, for example, although small in percentage, still exceeds the total forest areas of many significant forest nations such as Sweden and Finland (see Rotherham [2003]). New land tenure developments in many provinces are creating effectively private ownership situations on significant blocks of Crown land (Nelson, 2008). First Nations communities are prominent among these (Wyatt, 2008), but there are other community-based tenures as well. The need for decision-making in these complex, multistakeholder, and collaborative environments creates new decision support challenges.

When one refers to forest management strategy in such multistakeholder environments, a question that often arises is, "whose strategy?" This leads to a long-time concern of professional foresters, namely "high grading," usually described as taking the best and leaving the rest, and thereby rendering the remaining forest uneconomic for other, present and future, participants. High grading can be thought of, in a more general sense, as processes that damage the ability of other participants to sustainably reap value from the system, either in the present or in the future. Processes that require the participation of multiple participants but with benefits and costs distributed among the participants in ways that are perceived as being unfair and uneconomic are unsustainable. Logistics, such as wood allocation to mills, may minimize overall wood procurement costs while providing timber to some mills at very low costs and others at unsustainably high costs. Optimizing the harvest of low cost, accessible stands and the failure to harvest certain less accessible, poorly stocked, low quality, and/or over mature stands early in the planning horizon can mean that regeneration of future stands is inadequate. Conversely, justifying high present levels of harvest by assuming future high cost allowable cut or expensive silviculture, the well-known allowable cut effect (ACE) (Davis et al., 2001), can also be unsustainable. All of these can be thought of as high grading, a term that is more evocative to foresters than is the dry economic term of externalities.

Economic externalities (see Baumol [1977] and Hobbs [1996]) can arise when a participant's costs and benefits depend on the actions of others. The high grading examples discussed above involve externalities. Capacity decisions made by some participants can also produce both costs and benefits for other participants over time. Mills interact with each other in many ways ranging from harvest, transportation, and the production of by-products such as chips, sawdust, shavings, and bark used as input to other mills. Installing capacity, possibly of the wrong type, possibly in the wrong location, can impose future costs across the system. Installing capacity for certain types of low value products may mitigate against higher value products in the future. All material movements through the forest industry system typically involve transactions between buyers and sellers. Transaction costs (Hobbs, 1996; Kant, 2003) are another form of externality that can prevent the system from realizing its full economic potential.

Decision support models often have system level objectives. It is important not to lose sight of the fact that their solutions may involve "high grading" by certain participants both within a time period and over time. In some cases, "high grading" may be acceptable so present profitability may provide financial and infrastructure benefits that can be transferred to the future. However, if this is to be the result implied by our decision support models, we need to be careful that it be explicit and well understood and not an unintended consequence.

5.2.3 Environmental Strategy

Forest management strategy and industrial capacity strategy cannot be separated from environmental or other strategic considerations. An overview of Canada's environmental strategy can be found in the Canadian Council of Forest Ministers Criteria and Indicators of Sustainable Forest Management (CCFM C&I) (Canadian Council of Forest Ministers, 2008a). The six main criteria are (1) biodiversity, (2) soil and water, (3) ecosystem condition and productivity, (4) global cycles, (5) multiple economic benefits, and (6) accepting societies' responsibility for sustainable forest management. Within these criteria and the many indicators laid out in the CCFM C&I, there remains enormous strategic scope both in the indicator levels to be attained over time and how they can be achieved. In Canada, it is a provincial responsibility to decide how to apply the CCFM C&I within each province.

As Mintzberg (1994) points out, most strategy is not in the form of written documents, but is emergent as behavior by those who can implement strategy. The governments, both in their role as owners of the public lands and in their defining and implementing regulations and policy for the citizens of their province, have been the lead implementers of forest sustainability strategy. Corporations develop their own strategies concerning how they manage their land holdings, and how they produce and market their products, but a key aspect in Canada has been the adoption of certification to support the marketing of green products. There are three main certification schemes used in Canada (see Certification Canada [2011], Forest Stewardship Council [FSC], Canadian Standards Association CSA-Z809, and the Sustainable Forestry Initiative [SFI]). Although they differ in many ways, at their heart these have the same criteria as those found in the CCFM C&I. All three certification systems recognize that implementation must be through appropriate local level indicators and that one-size-fits-all certification is not appropriate given the diversity of forest types and conditions across Canada.

When considering environmental strategies, spatial issues matter because ecosystems are inherently spatial. Biodiversity considerations often lead to the development of protected areas or special management areas. Associated with biodiversity are old growth and old forest policies. Riparian zone policies respond to both biodiversity and water quality concerns. Policy choices, both in terms of the location and extent of protected areas, old growth areas, and riparian zones and in terms of allowable silvicultural activities and forest cover, remove or restrict certain locations from harvest. Biodiversity, soil and water, and ecosystem condition considerations impose spatial requirements not just on the harvest itself but also on the condition of the overall forest landscape postharvest. What the landscape scale should be is the subject of an ongoing debate.

Ecozones, and even ecoregions, tend to be extra-provincial in their extent. Some provinces have adopted the notion of an ecodistrict as the landscape basis for their strategic analysis (Ontario Ministry of Natural Resources, 2007b; Stewart and Neily, 2008). Biodiversity within an ecodistrict is usually difficult to measure or manage, particularly in terms of populations of fauna or nontree flora. Most people agree that at the strategic level, it is impossible to model population responses to forest management practices. About the best that can be expected is that one can model habitat availability. However, given the differing habitat requirements of different animals, finding a few proxy species can be challenging and modeling the many issues of habitat for each species can also be difficult (Davis and Barrett, 1992). As a result, specific issues of forest cover, such as the percentage of certain species and/or percentage by age class are often used as proxies for biodiversity.

Similarly, the hydrologic response to changes in forest cover is difficult to model (Moore et al., 2005; St-Hilaire et al., 2000). Moreover, watershed scale ranges from the complete watershed of major river systems to the 20-m riparian zones in use in many jurisdictions. Proxies such as the percentage of the area with a certain category of forest cover, defined in terms of stocking, height, or age, may be the best that can be used for strategy analysis over extensive watersheds.

Issues of tree biodiversity (examples include the Ontario Ministry of Natural Resources, 2010 or Stewart and Neilly, 2008) and riparian management (e.g., Blinn and Kilgore, 2001; Lee et al., 2004) can impose restrictions on harvest practices. Despite this modeling simplicity, managing biodiversity and watershed indicators will pose spatial restrictions on the location, amount, and type of forest harvesting activity. Restrictions on allowable harvest practices will influence harvest costs and forest product availability in a region.

5.2.4 Forest Management Strategy

One of the significant questions is the goal of forest management strategy, which should be centered on the ability to supply sustainably, in other words the wood required to meet the needs of the forest products industry in the area. The limitations of growth and yield models and models for environmental policies have significantly complicated forest management. There are also economic and social issues of what to supply, from where to supply, to which market, and for what use to create value and jobs for local communities. Today, many ecological services are of high social value. Although difficult to evaluate in monetary terms, this tendency of high social awareness will likely remain (Adamowicz et al., 2003; Liu et al., 2010; Raunikar and Buongiorno, 2007).

Even when the focus was on timber production, there was often a disconnection* that resulted from strategic forest management planning models that were structured to guarantee a supply of timber without adequately considering the delivered wood costs or the planned use of the timber supplied. Long-term supply is often described in terms of level flow or nondeclining yield. However, the resulting harvests may produce timber species and size/quality flows that are inconsistent with either the current or envisaged capacity of the industry to use the wood produced and its cost of production and transportation.

Another disconnect is the one between present and future responsibility. A key feature of forests, as a supplier of raw materials, is that trees grow, with the implication that forests are therefore renewable. This implies that current actions can affect the future supply and that assumptions about the future can influence the current capabilities. A plan to produce future supplies of timber may require the development and maintenance of a costly road network, expensive silviculture such as planting, harvesting stands of poor quality, or harvesting mature, but expensive to manage stands that these can be regenerated and become available at a future date. Conversely, in order for a plan to produce high present harvest levels, one may be willing to commit future forest managers to invest in silviculture and/or to harvest wood that may be of low quality, expensive to harvest, or far from mills. Most strategic forest management analysis has been carried out using wood volume as the goal with some variant of level harvest flow or nondeclining yield as a constraint. Even if economic goals are used, problems can arise from discount rates (Church and Daugherty, 1999; Daugherty, 1991). Using a 10% discount rate, costs that occur 50 years from now have a present value that is less than 0.8% of costs incurred now. Models that maximize net present value can produce solutions that will create

---

* This disconnect is well known among provincial resource analysts. It was identified repeatedly at the Value Chain Optimization workshop Strategic Management of the Forest Value Chain: Sustainable Forests that Support Sustainable Businesses, May 12-13, 2011 (see http://www.reseauvco.ca/en/activities-networking/workshops/strategic-forest-management/) particularly in the presentations by representatives from the provinces of British Columbia and Ontario.

very expensive future wood supplies. This changes the strategic perspective from one of growing to mining (Gunn, 2011). In some cases, this may be appropriate (Adamowicz et al., 2003) such as with low site class boreal forests or west coast old growth. If so, it should be intentional, not an accident of discounting.

5.2.5 INDUSTRIAL CAPACITY STRATEGY

Industrial capacity strategy is difficult to separate from either environmental or forest management strategies, particularly when energy and greenhouse gas emission mitigation are considered, which means that the life cycle criteria of SFM are increasingly important. Forest products are often seen as being CO_2 neutral, although this viewpoint depends on the planning horizon.

Unlike forest management strategy, which often entails decisions made within the public sector, industrial capacity strategy refers to decisions usually made by decision-makers in the private sector. These decisions can include whether or not to invest in a new technology and a corresponding facility. Despite the private nature of industrial business strategies, public policies can significantly impact industrial strategy, particularly when government financial assistance and incentives as well as regional development strategies that address access to raw materials become involved. Thus, even in a private sector decision-making framework, companies investing in facilities need to be aware of the potential supply and cost implications of forest management strategies. Meanwhile, for sustainable forest management and long-term wood supplies, forest managers must be aware of the potential industrial demand and cost associated with supporting industrial strategies.

The basic industrial strategy decision for the forest value chain involves deciding what businesses one wants to be in and why. In any large-scale process industry, there are at least three key aspects of strategy development that can benefit from decision support. One is economies of scale (Lieberman, 1987), a second is capacity location and allocation decisions (Melo et al., 2009), and a third is interoperability (Chen et al., 2008). Large plants typically require less capital investment per unit output and reduced labor costs. However, larger facilities may result in higher unit transportation costs as material has to be procured over longer distances. This can also mean higher levels of investment in silviculture to secure a satisfactory wood supply. Interoperability deals with a series of plants and processes that make use of different parts of the forest products supply stream. Pulp and paper production, for example, uses wood chips, a by-product supplied by sawmills, instead of processing chips directly from forest round wood. The opportunities for capacity interoperability were discussed in a recent biopathway study (Forest Products Association of Canada, 2010). Instead of building a "greenfield" facility, capacity can be added to an existing site for new bioproduct productions. Successful interoperability requires not only the coordination of mill operations but also the coordination of the material flows through the forest value chain. This includes the coordination of the harvesting of saw logs and pulpwood for conventional forest products production and traditionally underutilized materials for bioenergy production. Making all these processes work together calls for balancing product outputs with input needs, in terms of amount, cost, and quality. The overall economics of the system depends on the capital and operating costs of the facilities, the costs of transportation from facilities to markets, between facilities, and from the harvest areas to facilities, and the cost of harvest operations. All these costs are likely to vary over time. Nevertheless, capital investment strategies and forest management strategies involve fairly long-term commitments with respect to demand and supply levels, without easy recourse to changing market conditions.

5.3 INTEGRATED FOREST VALUE CHAIN STRATEGY

We now focus on forest management and industrial capacity in more depth. To make this more concrete, we use two mathematical models to illustrate and highlight some of the decision support issues associated with the development and management of integrated forest value chains. The first is a simplified capacity planning model that illustrates some of the aspects of the product flow supply chain and the second is a forest management planning model.

5.3.1 A CAPACITY PLANNING MODEL

Capacity planning models have a long history that can be traced back to the early work of Manne (1967). The more recent paper by Melo et al. (2009) provides both a review and places the capacity planning problem in a supply chain context. Santoso et al. (2005) indicate how such problems can be extended to deal with uncertainty. Vila et al. (2009) look at specialized models for the forest industry. Klibi et al. (2010) review strategic issues in terms of robustness and resilience.

The key aspects of strategic capacity planning in the forest industry involve the economies of scale associated with processing capacity, the configuration of the transportation system, and the interdependence of inputs and outputs. As discussed in Gunn (2011), economies of scale dominate process industries such as sawmills and pulp mills. Doubling a mill's size can reduce its capital cost per annual ton by more than 20%. Labor cost reductions can even be greater, particularly if the new capacity incorporates state-of-the-art automation.

Transportation from the regions where wood is harvested to the mills where it is processed is expensive because it typically involves transporting wet round wood with a moisture content as high as 50%. If a mill receives all of its material directly from the forest, the larger the mill capacity, the more raw material required, and therefore the higher the per unit transportation cost. In such cases, transportation costs are affected by the number of mills, their capacities, and their locations relative to the resources. Gunn (2011) illustrates how these costs can increase with mill size. However, integrated logistics systems can reduce such costs (Forsberg et al., 2005).

The cost of transporting products and by-products to markets and other processing facilities can also be significant. Material from the forest typically undergoes some initial processing in a facility such as a sawmill. The by-products, such as chips, sawdust, shavings, and bark, are then transported to a pulp mill, pellet mill, energy plant, or biorefinery for further processing. The transportation costs depend on the relative locations of the facilities and the nature of the material being transported.

Chips are easier to handle than round wood; moisture levels determine the mass being transported. Fuel costs are also obviously important. The facilities and the transportation modes connecting them constitute a network through which material flows. The scale and relative location of the facilities on the network determine the amount and costs of material flow. Similarly, the location of the facilities relative to markets and the nature of the product being transported determine the cost of transportation to markets.

Here we give a simplified formulation, reproduced from Gunn (2011), that illustrates several aspects of the capacity-planning problem in the context of the forest industry. We start with an assumption that we know the available wood supply that can be harvested over time from several forest regions d. We assume that the goal is to develop a capacity plan that makes the best possible use of this wood supply. The objective is to maximize the net present value of all cash flows using a specified discount rate i over a planning horizon with time periods indexed by t.

The structure of the model is determined by the wood flows where f indexes the wood flow type. The total wood flow of type f leaving region d is limited by the resource availability $A^{Res}_{f,d,t}$. Wood flows are differentiated by species and form (e.g., pulpwood, logs, chips, lumber, and other products). When a wood flow is processed as an input, other types of wood flows can arise as outputs. These outputs can be intermediate wood flow types, final products, or both. There are three types of flows for which one must determine transportation costs. These include (1) flows $WF_{f,d,l,c,t}$ of wood type f from district d to capacity type c at location l in period t, (2) flows $CF_{f,l_1,l,c,t}$ of intermediate product f from location l_1 to capacity type c, location l in period t, and (3) flows $MF_{m,f,l,t}$ of market product f to market m from location l in period t.

There are various types of capacities, indexed by c, which exist or can be established at various locations l. We let $X^{Cap}_{l,c,t}$ be the addition of capacity of type c in location l in period t. The capital cost is assumed to be a nonlinear function of capacity addition, $C\left(X^{Cap}_{l,c,t}\right)$, reflecting the capacity economies of scale (Manne, 1967). Although determining such costs is nontrivial, we do not deal with the process of estimating them here. Operating costs are assumed to be proportional to operating levels $OL_{l,c,t}$.

At each facility, the operating levels $OL_{l,c,t}$ are constrained by installed capacity $Cap_{l,c,t}$. Each facility is assumed to operate according to some recipes, which transform wood flow inputs to wood flow outputs. The recipe limits the proportion $(\gamma^-_{f,c}, \gamma^+_{f,c})$ of input of either forest wood inputs $WF_{f,d,l,c,t}$ or intermediate product inputs $CF_{f,l_1,l,c,t}$ and provides for fixed conversion factors $\beta_{f,c}$ of the incoming flows. The output of products is assumed to be in fixed proportion $\alpha_{f,c}$ to the operating levels $OL_{l,c,t}$. The market for products is represented by a price endogenous demand curve.

This suggests a mathematical model like the following:

Sets:

T set of time periods (1 ... T)
D set of districts
L set of facility locations

C	set of capacity types
F	set of wood flow types
M	set of markets
$P(m)$	set of price ranges for market m

Data:

A_{fdt}^{Res}	resource availability of wood type $f \in F$ in district $d \in D$ in time period $t \in T$
ρ_{pmt}	unit revenue at price level p in market m in period t
o_{lct}	unit operating cost at production location l, capacity type c in period t
τ_{dflct}^{W}	unit transportation cost from district d of type f wood to location l, which has capacity type c in period t
$\tau_{fl_1lct}^{F}$	transportation cost from location l_1 of type f wood to location l, which has capacity type c in period t
τ_{mflct}^{M}	transportation cost to market m of type f wood from location l, which has capacity type c in period t
$\gamma_{fc}^{+}, \gamma_{fc}^{-}$	maximum, minimum percentage of wood flow type f in operating levels of capacity type c, limited by the recipe
β_{fc}	conversion coefficients for input wood flow type f to operating of capacity type c
α_{fc}	amount of output wood flow type f for unit operating level of capacity type c
MA_{mpt}^{+}	maximum demand at price level p for market m in period t
MA_{mpt}^{-}	minimum demand at price level p for market m in period t

Variables:

$MREV_t$	total market revenues in period t
$OpCost_t$	total operating cost in period t
$TranCost_t$	total transportation cost in period t
$CapCost_t$	total capital cost in period t
X_{lct}^{Cap}	capacity of type c to be installed in location l at beginning of period t
Cap_{lct}	capacity of type c at location l available throughout period t
WF_{fdlct}	wood flow type f (typically logs) from district d to location l, which has capacity type c in period t
CF_{fl_1lct}	wood flow type f (typically intermediate products) from location l_1 to location l, which has capacity type c in period t
MF_{flmt}	wood flow type f (typically market products) from location l to market m in period t
OL_{lct}	operating level of capacity type c at location l in period t
M_{mpt}	market sales at price level p in period t

Model:

$$\text{Maximize} \sum_{t}(1+i)^{-t}\left[MREV_t - CapCost_t - OpCost_t - TranCost_t\right] \quad (5.1)$$

$$S.T. \quad \text{CapCost}_t = \sum_{l\,f} C\left(X_{lct}^{\text{Cap}}\right) \qquad t \in T \tag{5.2}$$

$$\text{MREV}_t = \sum_{m\,p} \rho_{pmt} M_{mpt} \qquad t \in T \tag{5.3}$$

$$\text{OpCost}_t = \sum_{l\,f} o_{lct} \text{OL}_{lct} \qquad t \in T \tag{5.4}$$

$$\text{TranCost}_t = \sum_{f}\sum_{dl}\sum_{c} \tau_{dflct}^W \text{WF}_{fdlct} + \sum_{f}\sum_{l_1l}\sum_{c} \tau_{fl_1lct}^F CF_{fl_1lct} \tag{5.5}$$

$$+ \sum_{m}\sum_{fl}\sum_{c} \tau_{mflct}^M \text{MF}_{mflt} \qquad t \in T$$

$$\text{Cap}_{lct} = X_{lct}^{\text{Cap}} + \text{Cap}_{lc(t-1)} \qquad l \in L,\, c \in C,\, t \in T \tag{5.6}$$

$$\sum_{l\,c} \text{WF}_{fdlct} \le A_{fdt}^{\text{Res}} \qquad f \in F,\, d \in D,\, t \in T \tag{5.7}$$

$$\text{OL}_{lct} \le \text{Cap}_{lct} \qquad l \in L,\, c \in C,\, t \in T \tag{5.8}$$

$$\left\{
\begin{aligned}
\gamma_{fc}^- \text{OL}_{lct} &\le \sum_{d} \text{WF}_{fdlct} \le \gamma_{fc}^+ \text{OL}_{lct} \\
\gamma_{fc}^- \text{OL}_{lct} &\le \sum_{l_1} CF_{fl_1lct} \le \gamma_{fc}^+ \text{OL}_{lct}
\end{aligned}
\right\} \quad f \in F,\, l \in L,\, c \in C,\, t \in T \tag{5.9}$$

$$\left.
\begin{aligned}
\sum_{f\,d} \beta_{fc} \text{WF}_{fdlct} + \sum_{f\,l_1} \beta_{fc} CF_{fl_1lct} &= \text{OL}_{lct} \quad l \in L,\, c \in C,\, t \in T \\
\sum_{c} \alpha_{fc} \text{OL}_{lct} &= \sum_{l_2c_2} CF_{fll_2c_2\,t} + \sum_{m} \text{MF}_{flmt} \quad f \in F,\, l \in L,\, t \in T
\end{aligned}
\right\} \tag{5.10}$$

$$\left\{
\begin{aligned}
\sum_{f\,l} \text{MF}_{flmt} &= \sum_{p} M_{mpt} \quad m \in M,\, t \in T \\
MA_{mpt}^- &\le M_{mpt} \le MA_{mpt}^+ \quad m \in M,\, p \in P(m),\, t \in T
\end{aligned}
\right\} \tag{5.11}$$

Equations 5.2 through 5.5 are accounting equations for capital cost, market revenues, operating costs, and transportation costs with the ρ, o, τ in Equations 5.3 through 5.5 being the appropriate revenue/cost coefficients. Equation 5.3, together with Equation 5.11, models market m as having piecewise constant demand curves (Gilless and Buongiorno, 1985; Lebow et al., 2003). Equation 5.6 keeps track of capacity. The initial capacities $\text{Cap}_{l,c,0}$ are given data and subsequent capacities computed from additions. Inequality (Equation 5.7) limits the wood flows of every flow type from each district to an assumed amount given in each period. This model has a simple notion of processing. Each facility is assumed to have a recipe that dictates the inputs and outputs of the facility. A constraint (Equation 5.8) limits the amount of processing to the available capacity. Equation 5.9 limits the relative proportions of each type of forest wood flow (WF) and intermediate product wood flow (CF) to be within certain proportions of recipe inputs. The two types of equations given in Equation 5.10 indicate that the inputs, which may be in different units (tons, m^3, cords), adjusted using conversion factors $\beta_{f,c}$, have to add up to the total recipe amount in operating level units. The outputs of the recipe are in fixed proportions $\alpha_{f,c}$. What is represented here is actually a very simple recipe structure where a single recipe at the facility produces fixed proportion of outputs with certain allowable inputs. Outputs of the recipe can be intermediate products that go to other facilities (CF) and/or products that go to final markets (MF). Much more complicated recipe structures are possible but to specify these realistically we need to have specific capacity situations in mind. As a very simple example, think of a sawmill, which consumes various types and classes of round wood and produces lumber, chips, shavings, sawdust, and bark. The operating level at the sawmill is expressed in terms of millions of board feet (nominal) of lumber output. The two equations given in Equation 5.11 specify that flows to a market m are sold at one of a variety of prices with the amount that can be sold at a given price range p limited by lower and upper bounds. This enables the use of price endogenous demand models (Lebow et al., 2003).

For simplicity, capacity in this model appears to last forever. Freidenfelds (1981) discusses a variety of ways of modeling capacity, which is either time limited, diminishes over time, or is retired. However, this raises the issue of time horizon. Typically, time scales for industrial investment are 20 years or less. The value of the capacity at the end of the horizon is typically not zero. How to value that capacity is an important question.

The key feature of the model involves trade-offs, the most obvious of which is the trade-off between revenues and costs, which is the classic supply/demand trade-off. Another trade-off is the one between capacity costs and transportation costs, with low capacity costs associated with a small number of large facilities and low transportation costs associated with many smaller facilities. Even within transportation costs, there is a trade-off of capacity costs versus movement costs. Facilities such as wood gathering, chipping, and rail loading facilities may make movements to other facilities cheaper.

The three different types of transportation: (i) harvested material from the woodlands to processing facilities (WF$_{f,d,l,c,t}$), (ii) intermediate products from facility to facility (CF$_{f,l_1,l,c,t}$), and (iii) finished goods to markets (MF$_{m,f,l,t}$) have their own trade-offs. The first type tends to favor small, distributed facilities. The second type favors the clustering of facilities. The third type favors locating facilities close to markets.

Another trade-off is that between relatively simple, low-cost, limited purpose mills designed for a small number of products (e.g., stud mills) and more complex mills that can process multiple species to produce a wide variety of products. As indicated above, the fixed single set of recipes at each facility (Equations 5.8 through 5.10) is purely for ease of presentation. As long as each recipe has a defined utilization of overall capacity, it is straightforward to allow multiple alternative recipes by replacing the capacity index c in OL$_{l,c,t}$ by a recipe index r and then replacing the left-hand side of Equation 5.6 by a summation over recipes. This allows considerable flexibility in matching overall flows and allows flow proportions to vary over time.

There is also a modeling trade-off here in terms of the accuracy with which one needs to represent the overall markets and capital structure and the complexity of developing reasonable models to provide insight into strategic value chain alternatives of exploiting the forest resource (see Kong et al. [2012], for an example of the potential complexity of integrated system flow models). The challenge is to find sufficient detail to provide insight into the appropriate mix of capacity type, size, and location that makes sense for the forest capabilities, without becoming overwhelmed with product market, price, and recipe possibilities.

Another trade-off follows from the complementary nature of capacities. A pulp mill or a biorefinery will usually have a lower wood cost if it uses by-products from the sawmills and the ability to sell by-products improves the profitability of sawmills. Thus, the joint profitability of two mills that are linked can be more profitable than two mills that operate independently. It is important that capacities in each sector be appropriately balanced. When viewed from the perspective of a single decision-maker model, this can be easily achieved. However, when applying such model in a multisectorial environment, it is quite possible that one sector could fail to be profitable in the optimal solution in some periods and possibly throughout the entire time horizon. This does not reflect the fact that, if mill capacities have different ownerships, not all ownerships will share the overall benefits from the optimal solution. It does not recognize the transaction costs associated with by-products and their transportations.

This model can be thought of as "optimally" defining the capacity and logistics structure as a function of the anticipated time series of the available wood sup-ply, partitioned by district d and wood type f. Thus, it matters when and where wood is available (Equation 5.7). Obviously, the establishment of capital facilities is problematic if the location of the available wood of a required type is changing significantly over time. This is particularly important if we look at how to interpret wood availability. The constraint (Equation 5.7) is written as an inequality, implying that capacity and operating plans can be chosen that do not use the available wood. However, this ignores the fact that the availability of wood in future periods may depend on the extent to which wood is used in a current period. In the short term, using less than the amount $A_{f,d,t}^{\text{Res}}$ of wood type f in period t may mean that more can be available in period $t + 1$ and later. In the long term, however, using less than $A_{f,d,t}^{\text{Res}}$ in period t may mean that because of lack of regeneration, less is available in some future period $t + k$, where k is $\gg 1$. Even more complicated, in period t, the different wood types may need to be harvested in the proportions defined as available. Some of the wood may be coming from mixed wood stands where, in order to get one species/size class, it is necessary to harvest the entire stand. Even if the stands in district d are all pure stands, if the harvest is not in proportion to the available amounts, this again implies that unharvested species will be overrepresented in the near term and may not be sufficiently regenerated in the long term.

Thus, we have a fundamental problem. We optimize the installed capacity, harvest, and transportation of wood to meet the market demands using that installed capacity based on the available wood supplies. However, the optimization of capacity and logistics may imply that the available supply of some types of wood is not used in the proportions available, eventually altering the wood supply. Capacity depends on wood supply and wood supply may in turn be affected by capacity.

5.3.2   A FOREST MANAGEMENT PLANNING MODEL

The discussion above implies that capacity planning depends on the dynamics of wood supply. Understanding these dynamics requires a detailed forest management planning model that allows for an examination of alternative landscape management strategies and silvicultural programs that takes long-term issues of regeneration and forest cover management into account. Developing such models has been the realm usually referred to as strategic forest management planning. Gunn (2007) reviewed such models and emphasized that they are designed to assess strategy rather than to develop strategy. In this section, we examine the elements of such a model, using a formulation presented in Gunn (2009).

Given that such models must be used within the realm of both sustainable environmental strategies and strategies of supply chain design, it is essential that a high degree of spatial representation be directly embedded within them. Unfortunately, this has not been the normal practice. The model presented here emphasizes the representation of spatially significant ecological issues as well as the spatial and transportation issues associated with the supply of raw materials to the forest products industry. As was the case with our capacity model above, it is meant to be illustrative and does not include all of the many issues that may arise.

The form of the model presented here is referred to as Model I, in which spatial representation has been captured because stands are preserved over time (see Davis et al. [2001] or Gunn [2009]). A stand can be managed according to one of several possible prescriptions. A given prescription applied to a stand *yields* several different types of timber over time. Examples might include the standing volume of softwood trees or the standing volumes of trees greater than some target age that define old growth. Yields can be in the form of harvested commercial timber such as pine saw log volumes. Yields can also be expressed as 0–1 quantities, such as a stand being in a forested condition in terms of required forest cover in a watershed when it is "0," or satisfying a particular age class/cover type requirement in an eco-district when it is "1."

We also have the concept of *region* so that yields occur in a specific region. Each yield type will occur in one of several regions. For example, one of the yield types might be conifer saw logs and the regions of interest might correspond to *timber sheds*, areas that might be viewed as potential sources of timber for one or more demand centers. Timber sheds can correspond to the concept of a district in the capacity model. Another type of yield might be forest condition and the regions of interest might be ecodistricts, watersheds, and/or wildlife habitat zones. The mathematical model is quite flexible. A given stand can contribute to the production of a certain type of product in a timber shed, another yield type in a watershed, and yet another yield type in an ecodistrict and all these yields can vary over time depending on the prescription chosen.

An important modeling issue is defining the trade-off between forest ecological management strategies on the one hand and the economic impacts on the traditional forest products industry and their supply chains on the other. In order to model this trade-off we need some representation of the spatial aspects of the harvest, the transportation requirements associated with them, and the overall market requirements. However, because the focus is placed on forest management planning, the modeling of market demands and the transformation processes that produce products for these markets will be less detailed than is the case with capacity planning models.

In this model, the following notation is used:

Sets:

I	set of stands
Pi	set of allowable prescriptions for stand i
Y	set of yields under strategic consideration
Y^w	set of wood products ($Y^w \subset Y$)
$R(l)$	set of regions applicable for yield type l. If $l \in Y^w$, then $R(l)$ is a timber shed for wood product l.
M	set of mills (forest products demand centers)
$S(m)$	set of demand segments for mill m

Data:

A_i	area of stand i
LY_{lrt}	lower limit on yield l in region r in period t
UY_{lrt}	upper limit on yield l in region r in period t
LD_{mt}	lower limit on demand from mill m in period t
d_{smt}^{max}	upper limit on demand segment s from mill m in period t
y_{iklrt}	yield of type l produced in region r in period t if prescription k is used for stand i
∂_{lm}	conversion coefficient of log type l to meet the demands at mill m
p_{smt}	price for segment s at mill m in period t (discounted to time 0)
c_{ik}	silvicultural costs for managing stand i with prescription k (all costs discounted back to time 0)
τ_{lrmt}	unit transportation costs for logs of type l from region r to mill m in period t (discounted to time 0)

Variables:

X_{ik}	area of stand i managed using prescription k
Y_{lrt}	yield of type l produced in region r in period t
z_{lrmt}	transportation of logs of type l from region r to mill m in period t
D_{smt}	amount of demand segment s supplied to mill m in period t

D_{smt} allows us to define demand curves for each mill m using the price endogenous ideas described in Lebow et al. (2003). Because this is a strategic forest management planning model, the demand curves here need to be defined in strategic terms. The main objective is not to predict pricing, but rather to link supply chain mill capacity to the forest strategy. This requires an ability to represent the fact that some forest products may command premium prices (p_{smt}) and the extent of the market assumed to exist for these products. There may be limited ability to use certain log types (species, size) at high prices and the representation of the low price section of the demand curve can be useful to represent disposal. The mill subscript m allows representation of both product type and location. In a strategic model, we might expect these to be broadly defined, with several facilities that require the same type of wood product and located in the same general region aggregated into a single mill type. Note again that we are oversimplifying the concept of mills, recipes, and intermediate flows. More sophisticated models need to deal with these (see Martin [2013]).

$$\text{Maximize} \sum_t \sum_m \sum_{s \in S(m)} P_{smt} D_{smt} - \sum_i \sum_{i \in I k \in P_i} c_{ik} X_{ik} - \sum_t \sum_m \sum_l \sum_r \tau_{lrmt} z_{lrmt} \tag{5.12}$$

$$S.T. \qquad \sum_{k \in P_i} X_{ik} = A_i \qquad\qquad i \in I \tag{5.13}$$

$$\sum_{k \in P_i} y_{iklrt} X_{ik} = Y_{lrt} \qquad\qquad l \in L, r \in R(l), t \in T \tag{5.14}$$

$$LY_{lrt} \le Y_{lrt} \le UY_{lrt} \qquad\qquad l \in L, r \in R(l), t \in T \tag{5.15}$$

$$\sum_{m \in M} z_{lrmt} = Y_{lrt} \qquad\qquad l \in L_w, r \in R(l), t \in T \tag{5.16}$$

$$\sum_l \sum_r \partial_{lm} z_{lrmt} = \sum_{s \in S(m)} D_{smt} \qquad m \in M, t \in T \tag{5.17}$$

$$\sum_{s \in S(m)} D_{smt} \ge LD_{mt} \qquad\qquad m \in M, t \in T \tag{5.18}$$

$$0 \le D_{smt} \le d_{smt}^{\max} \qquad\qquad m \in M, s \in S(m), t \in T \tag{5.19}$$

The objective of Equation 5.12 is to maximize the net present value of all cash flows. These include the discounted revenues associated with the delivery of logs to markets, the discounted harvesting and silvicultural costs that result from applying silvicultural prescriptions to stands, and the discounted transportation costs of delivering harvested log types from regions to mills. Equation 5.13 requires that the areas treated by all prescriptions equal the stand area. Equation 5.14 computes the yield of type l in each period in each region r where l is relevant. Equation 5.15 constrains this yield to lie within bounds chosen by the user over time. For those yields that correspond to wood products, Equation 5.16 requires that the sum of the amount transported from the region to mill centers equals the amount produced in each period. Lumber products are converted to market demands via conversion coefficients in Equation 5.17 and the total of these products delivered to market m must equal the amount sold in that market summed over the price ranges in that market. Market minimum demand is defined by the user in Equation 5.18 and the amount that can be sold at each price is limited in Equation 5.19.

An important aspect of any forest management planning model is how silvicultural prescriptions are modeled. In Model I, the decision to allocate stand i to prescription k implies an entire deterministic future for the stand. In some jurisdictions (e.g., Scandinavia) prescriptions tend to be simple, amounting to three-entry forestry with some variation in the timing of the first and second entry and the final felling. In others, there can be an enormous number of prescriptions ranging from continuous cover prescriptions, particularly for hardwood, stand conversions, and natural development/clear-cut, intensive plantation/precommercial thinning/final felling. In most cases, one can create a reasonably small number of prescriptions based on a few sound forestry principles (Martin, 2013). In Scandinavia, sophisticated growth and yield models are used to create prescriptions independently of the optimization process. These prescriptions are then provided to the Model I formulation. Examples include the Heureka system (Wikström et al., 2011). The combined use of SIMO (SIMulation and Optimization for forest management planning framework) and the JLP (a linear programming package for management planning) system accomplishes the same objectives (Lappi, 2003; Nuutinen et al., 2006; Rasinmäki et al., 2009).

Note that there are strong strategic elements in the prescription generation process. This may involve simple things like forbidding stand conversions (e.g., mixed wood stands to pure softwood) or limiting age/diameter eligibilities for certain treatments. It can include creating alternate types of prescriptions that can enter into some of the yield requirements, for example, continuous cover prescriptions or prescriptions leading to old forest conditions. It can also include ecosystem-specific or location-specific (riparian zones/steep slopes) prescriptions.

The cost structure c_{ik} can be based on access policy, harvest policy, and/or silvicultural incentives. The yields Y_{irt} are both dependent on the growth models being used and the indicators chosen (or not chosen) to be represented for the biodiversity, ecosystem condition, soil and water, and multiple benefits criteria. In economic terms, choices must be made between what species and types of forest product yields are represented. How we choose to represent the yields of various classes of saw logs, studwood, pulpwood, and other biomass governs how we can optimize the harvest and its allocation. Attention needs to be paid to the breakdown of merchantable volume into classes. Looking only at diameter limits and ignoring the fact that log merchantability is usually only in specific lengths may tend to overestimate saw log volumes and underestimate pulpwood and biomass volumes. The simulation models used in Heureka (Wikström et al., 2011) or SIMO (Rasinmäki et al., 2009) can deal with such issues and can be viewed as being the first step in the strategic modeling process. Stoddard (2008) illustrates another approach where a Model III model is used for a single stand type to generate alternate prescriptions in response to various objective functions.

One feature worth noting in Equations 5.12 through 5.19 is the absence of nondeclining yield constraints, for both timber and other products. Since variation is fundamental to ecology, it does not make sense to enforce any particular level for the relevant indicators of biodiversity, ecosystem productivity, or soil and water. The Canadian Council of Forest Ministers commitments are used to monitor and assess these indicators, not necessarily to rigidly control them. What forest planners need is the ability to limit the chosen indicators to acceptable ranges. However, for timber, nondeclining yield has often been a central feature of many provincial and corporate models and that can have adverse consequences (see Gunn [2007, 2011]). Instead, the constraints in Equation 5.14 define the yields, and the bounds in Equation 5.15 provide the ability to "steer" the yields. This construct forces a change in the thought process behind sustainability. Starting with somewhat loose bounds on the yields of interest (typically a certain deviation from current yields) and having solved the model with a given set of bounds, it is possible to observe the model feasibility or otherwise, and if feasible, the actual yields achieved. The bounds can then be revised to steer the yields along a chosen strategic trajectory. Although this may appear complex, this is in fact the process used with most simulation approaches (Davis et al., 2001; Gunn, 2007). The advantage of "steerable yields" is that it forces sustainability issues to be a strategic choice of the analysis team instead of somewhat arbitrary flow constraints. Furthermore, instead of assuming that level flow defines industry sustainability, this model requires matching the output of the forest management process to a realistic estimate of the ability to use the wood products produced (Equations 5.16 through 5.18). Similar notions are also found in JLP (Lappi, 1992) and Heureka (Wikström et al., 2011).

Nondeclining yield has often been used as a form of volume control (Davis et al., 2001). As described in Ware and Clutter (1971), this has less to do with sustainability than attempting to regulate the uneven wood flows that would result from Faustmann-like policies. Other forms of regulation include area control or control on the basis of mill supply. Ontario's recent Forest Management Manual (Ontario Ministry of Natural Resources, 2009) does not refer to nondeclining yields.

Another feature of Equations 5.12 through 5.19 is that yields of a given type can be defined by region. In the traditional views of sustainability, the region for wood supply (nondeclining) would be the entire area under consideration. However, using specific regions, of either economic or ecological interest, permits the analyst to test spatially specific strategies. This is of interest for economic timber supply regions, particularly if the analysis focuses on tenure issues. Moreover, this is a requirement in the analysis of specific ecodistricts or watersheds, using indicator yields such as age structure, old growth cover, or specific wildlife habitats. The concept of spatial analysis of forest condition is fundamental to the Heureka system (Wikström et al., 2011). The regions for different types of spatial yields can be overlapping. Thus, a stand can be in a given timber shed for timber, a given watershed for watershed cover yield, a given site capability for growth modeling, a given ecodistrict for ecosystem condition modeling, and so on.

A third feature is the consideration of supply chain issues in terms of the timber supply. This can be found both in the ability to manage regional timber supply using the steerable yield constraints and, more specifically, through Equations 5.16 through 5.19. In particular, the left-hand side of Equation 5.16 corresponds to wood product substitution to meet the needs of mill m while Equation 5.18 imposes a minimum demand requirement at mill m. Thus, by defining a certain number of "timber shed" regions, it is possible to model both transportation costs, and product substitution in meeting market demand. The time subscript in Equations 5.16 through 5.19 implies the ability to examine strategies of establishing or closing certain types of mill capacity in certain locations at certain points of time. Only a very simple representation of market possibilities is shown in Equations 5.17 through 5.19. More complex network flows are possible, such as those in the capacity model, Equations 5.8 through 5.11 (see Martin [2013]). An open question is the extent to which the details in the capacity model should be used here.

The concepts upon which this formulation is based are not particularly new. Many aspects of the spatial representation of stands and yields can be found in SPECTRUM (Greer and Meneghin, 2000). The use of demand curves can be found there and more significantly in the price endogenous linear programming system (PELPS) model (Lebow et al., 2003). The use of the steering constraints instead of nondeclining yields can be found in the JLP code (Lappi, 1992). Gunn and Rai (1987) and Barros and Weintraub (1982) both used mill sector demands instead of nondeclining yield to define allowable harvests.

Although the model constructed is not new, it has not been widely used. The emphasis here is the need to go beyond simple nondeclining yield as a definition of sustainability in the strategic model. If ecological and economic sustainability are important strategic concerns, they require spatial specification and should not be separated with the spatial issues of the supply chain. A model similar to that shown here makes it possible to assess these spatial issues directly as part of the strategy analysis process. The spatial detail in landforms and markets may lead to somewhat large models. Gunn (2009) discusses ways to reduce model size. Martin (2013) has demonstrated, in the context of Crown land in the province of Nova Scotia, that such a model can be effectively implemented. However, in this work, we are less interested in model complexity issues than the strategic issues underlying the model.

Note that the discussion here is based on the Model I formulation, rather that the Model II formulation that underlies Remsoft's Woodstock™, the most widely used forest management modeling environment. There are two reasons for this. One is to focus on sustainable prescriptions that represent good forestry. In Model II formulations, prescriptions are computed as part of the optimization process. Without considerable care as to eligibility limits, the optimization can produce prescriptions driven more by the time horizon than sound forest management. Second, as spatial detail is imposed in order to deal with ecological and economic issues, Model II formulations can grow dramatically in the numbers of variables and constraints and in model building and solution times. Gunn (2009) and Martin (2013) discuss this in more detail.*

As in the capacity planning case, there is again an issue of "whose strategy." Tenure and ownership issues have always been important and will become even more so as we move to new tenure situations (Ontario Ministry of Natural Resources, 2011), or community forestry orientations (Quebec Ministry of Natural Resources, 2012) (see also Nelson [2008] and Wyatt [2008]). What we will inevitably find is that the global optimal solution may not satisfy the needs of any particular land manager. There is also the temporal problem. Even with a single land owner, it is completely

---

* It would be a mistake to interpret these observations as a criticism of Woodstock, a mature, well-supported modeling system with a very wide user base. We do believe that there are ways to modify how Woodstock is typically used so as to deal with many of issues raised, but a discussion of these is technical and beyond the scope of this chapter.

possible to have situations where early periods are profitable and later periods not. In social terms, this is the intergenerational equity problem (Church and Daugherty, 1999; Daugherty, 1991). Even in terms of corporate finance, objective function coefficients are all discounted back to time zero using some discount factor. Discounting may well encourage policies that result in high silviculture and transportation costs in later years. It may be necessary to introduce constraints to prevent solutions that (1) impose much different profitability on different tenure holders, (2) impose much different wood costs on differing mills, and/or (3) imply that future generations of landholders and mills have unsustainable cash flows. Adding such constraints is not technically difficult but they require some thought as to their strategic implications. Much of current forest management practice has been based on an implicit assumption that constant total harvest volumes imply constant economic performance. However, from the above discussions, it should be obvious that no such assumption is warranted.

The time horizon of the model is open to question. Traditional approaches have been to use long planning horizons of 100–200 years, based on the idea that being able to harvest over such long periods is a proxy for sustainability. Based on the discussions made above, this is also open to question. If the prescriptions used are inherently sustainable at the stand level, then, rather than using long time horizon models, it may make more sense, just as in the capacity planning models, to focus on ways of evaluating desirable end of horizon forest conditions, using some combination of constraints or possibly goal programming (Tamiz et al., 1998).

Finally, we see the converse of the situation with the capacity model. Optimal forest management depends on the markets for forest products and in more complex versions of this model on the capacity to process the material for these markets. The volume of wood transported to facilities/markets must equal the harvests produced by the management prescriptions over the stands in each timber shed region (Equation 5.16).

5.4 THE NEED FOR INTEGRATED FOREST VALUE CHAIN STRATEGIC DECISION SUPPORT SYSTEMS

Given the spatial dispersion of forest resources, the spatial implications of environmental strategies for forest management, and the location and interconnection issues of industrial strategy, dealing with transportation and flow through the forest value chain requires investment in logistics to support the development and implementation of both forest management and industrial capacity strategies. This may involve information systems, harvesting systems, facilities for material receiving, and transshipment and transportation.

It is tempting to think of logistics issues as the tactics that follow the strategic decisions, but the investments in logistics systems that link forests to mills, mills to mills, and mills to customers determine both the types of flows and costs that are possible in the system. Whole tree logging systems are fundamentally different from cut to length systems in terms of harvesting costs and the flexibility of wood flows. Merchandizing systems, whether physical or virtual as information systems, alter what can be transported to mills. The interconnection systems, woods trucks, open highway trucks for logs and chips, rail, and in some instances, marine transportation, require facilities and information systems can have the potential to dramatically change the procurement costs (Forsberg et al., 2005). Finding ways to mitigate the transaction costs with efficient system behavior is also necessary to fully realize this potential (Lehoux et al., 2009).

Strategic decision-making includes deciding on the type of computer network systems that will be used to collect and store information, harvesting, processing, and shipping equipment as well as the design of government regulatory and corporate decision processes that help mitigate the externalities/transaction costs that may detract from overall system efficiency. A logistics strategy is an essential part of a forest value chain strategy that governs what is possible in the integration of the forest management and industrial strategies.

Natural disturbances raise quite a different aspect of the management of these systems. Fire, insects, disease, and storms as noted above can perturb the system dramatically with the result that the postdisturbance system may be a dramatically different system. This will likely be followed by a transient period during which harvest and production decisions are dominated by salvage or mitigative harvests. Given such uncertainty, previous solutions may not only be just suboptimal, they may be irrelevant because the state of the system has changed fundamentally. This raises a new type of instability problem. If some type of installed capacity is highly dependent on access to a specific species/size class in a particular region, the disturbance may reduce its availability to such an extent that this capacity must be greatly reduced or can no longer be sustained. This in turn, will affect other plants, which depend on the outputs of the first for some of their inputs. At the forest level, a disturbance can radically alter transportation patterns (see Broman et al. [2009]). It may not be the mills that are closest to the disturbance that are most affected. As they compete for wood from neighboring areas, these effects may ripple across the entire landscape. Davis et al. (2001, pp. 238–243) discuss how difficult it is to assess the management of multiple species habitat in the best of circumstances; when a disturbance occurs it can radically affect some critical species.

Integer programming capacity planning models and linear programming forest management planning models provide basic capabilities to examine a forest value chain strategy. The capacity model presented here for illustration is quite similar to models found in other supply chain settings although applications in the forest products sector are rare. The forest management model formulation is somewhat different than those commonly used in Canada, but well within the technical capacity widely available. However, as we have pointed out, it is possible to raise questions about the assumptions underlying these models and thus how these capabilities might be increased.

One issue is that capacity planning and forest management planning are encapsulated in two separate models. This is not just a technical matter of whether or not one is able to solve large integrated models of forest management and capacity planning. There are substantial time scale differences. Strategic forest management planning usually addresses a 150- to 200-year planning horizon while industrial strategic planning focuses on 20- to 25-year planning horizons. Forest growth modeling usually uses 5- to 10-year time periods, while capacity planning modeling is usually carried out on an annual basis. Even more fundamentally, decision-makers are different, with one being the "land manager" and the other the "capacity planner," who have different business objectives. What mechanism can we use to link these two models to explore such trade-offs?

5.4.1 ROLE OF LOGISTICS

Logistics is fundamental to both the capacity model (Equations 5.1 through 5.11) and the forest management planning model (Equations 5.12 through 5.19). Both involve trade-offs between the value of the forest products in the market and the cost of accessing, harvesting, and transforming fiber into products and delivering them to the market. In fact, it can be more subtle than that. Excellent logistics, in the form of information systems and transportation management systems, can help ensure that logs are allocated to their highest value use rather than to mills to which they can most easily be transported. In addition to the information systems, strategic commitments in terms of multimodal facilities and in terms of collaborative processes (Lehoux et al., 2009) can dramatically alter the cost structure.

As discussed in Forsberg et al. (2005) and Kong et al. (2012), these logistics models can be too complicated to include directly in strategic capacity planning or forest management models. However, experimenting with the logistics models can help develop appropriate cost structures and thus open up new possibilities in both models.

5.4.2 MULTIPLE MODELS VERSUS SEPARATE MODELS

To this point, we have discussed two separate models, a capacity model in Equations 5.1 through 5.11 and a forest management planning model in Equations 5.12 through 5.19 (and possibly a third logistics model as discussed above). In the capacity model, the optimal installed capacity and marketing of wood products depend on the wood supply. In the forest management planning model, the market for wood products defines how the forest will be managed and wood will be supplied. Would it then not make sense to combine these into a single model? The answer is not clear. The variables and constraints that define the capacity model (Equations 5.1 through 5.11) can of course be added to the forest management model (Equations 5.12 through 5.19). Equation 5.16 would then play the same role as the A_{fjd} in Equation 5.7. The constraints and variables in Equations 5.8 through 5.11 can replace the logistics and market constraints (Equations 5.17 through 5.19). The main issue is recipe and product representation detail and whether or not the economies of scale and capacity addition terms (Equations 5.2 and 5.6) should be included in the combined model. Since the nonlinear capacity cost model will normally be modeled as piecewise linear, requiring the use of integer variables, this could render a unified model too complicated to solve. As we discussed when we described the capacity model, we need enough detail to provide insight. On the other hand, too much detail can lead to long solution times and may lead to confusion rather than insight. A possible compromise might involve three models, the first two a capacity model and a forest management planning model implemented with quite detailed representations of the capacity expansion and wood flows in the capacity model, and quite detailed representation of the forest management issues in the other. If we want to look at capacity issues without the forestry issues or vice versa, the two separate models might be helpful. The third model might be a unified model with a reduced (aggregated) representation of forest management issues, ecological issues, transportation issues, and capacity production issues. Solutions to this third model might help guide the use of the other two.

Note that there are other models required in the model suite. All three models require estimates of transportation costs and mill allocation logistics. An appropriate detailed logistics model, similar to that described in Forsberg et al. (2005) or Kong et al. (2012), may be necessary to provide such estimates. Also, the representation of silvicultural prescriptions may require a strategic prescription model as discussed above.

5.4.3 DEALING WITH UNCERTAINTY

There are inevitably many sources of uncertainty that complicate strategic decision-making processes. Strategic models in any industry need to deal with uncertainty in prices and markets. These in turn can be driven by other factors such as the state of the global economy, international exchange rates, energy costs, and the availability of competing materials to name a few. Scenario planning (Schoemaker, 1995) has emerged as a way of both encouraging decision-makers to think about uncertain futures and as a way of developing some reasonable scenarios for analysis. In many ways, the estimation of important factors depends on the entrepreneurial vision of the decision-makers. One can carry out various sensitivity analyses and testing of random variations about certain base cases, but strategic vision is usually somewhat more than this. Decision-makers need support in calculating the consequences of their vision. Scenario planning decision support exercises are illustrated by Mobasheri et al. (1989) for the electric power industry. We expect that similar exercises would be useful in the forest industry.

Uncertainty concerning factors like production rates, transportation costs, production yields in the capacity models and forest growth rates, wood quality, harvest costs, and habitat availability in the forest management planning models arise in many different ways. Such uncertainty is due to a lack of process knowledge, things like moisture levels in harvested wood, uncertainty in the present forest inventory, and uncertainty concerning how the forests will grow, often due to uncertainty in site quality. Some of these uncertainties can be reduced through new technologies. Considerable work is now underway to apply new remote sensing technologies to forest inventories. Estimates of future growth require estimates of both current inventory and site quality that are often missing. However, growth processes will always be uncertain and this type of process uncertainty can often be dealt with by using rolling planning horizon processes (see Gunn [2005]). The use of robust optimization techniques may also be warranted (Bredström et al., 2011). In the end, the impact of this type of uncertainty can be assessed through sensitivity analysis and simulation studies.

A more significant type of uncertainty is that which is due to natural disturbance, which is more difficult to handle in the forest value chain. In eastern Canada, the spruce budworm has a long history of episodic destruction of the forest and in western Canada the mountain pine beetle has caused significant damage to a large area of forest. Fire has long been the dominant force on the ecology of the boreal forest and in recent years has had a very significant impact on the Montane Cordillera due largely to the extensive lodgepole pine mortality resulting from mountain pine beetle infestations. Fire has had less impact in recent years in the Atlantic Maritime region but hurricanes affect this region. Uncertainty that arises from natural disturbance differs from uncertainty concerning forest growth and yield processes. Disturbance events can dramatically reshape the large areas of forest resource landscape in a relatively short period of time (in the case of fire, virtually instantaneously). They can result in short periods of increased availability of wood due to salvage, although the uses of this wood may be limited. However, for many years, the effects of the disturbance are felt spatially due to the alteration of the age class distribution in the area directly affected but also due to the altered harvest strategies across the landscape with a combination of early reduced harvesting and later heavy harvesting due to the loss due to the disturbance.

The effects can be seen in both the capacity planning and forest management models. Natural disturbance changes both the availability of forest products (Equation 5.7) in the capacity model and changes the initial state of all the stands affected by the disturbance, thus changing the set of prescriptions available for stands going forward. Two questions thus arise: (1) How do we make capacity decisions that are relatively robust with respect to future natural disturbances? Robustness means that the value chain can still be operated in an economically feasible manner after one or more natural disturbance events. (2) How can forest management and capacity replanning provide recourse to adapt to the unknown natural disturbance events after they occur?

5.4.4 Stable Integrated Systems

The essence of the forest value chain is interconnectivity, which can occur at many levels. The widespread adoption of SFM has increased our awareness of the importance of environmental, economic, and social interactions. At a more detailed level in forest management there is a complex interaction in management of cover types, age classes, ecosystems, watersheds, and timber sheds that all have to be coordinated to achieve overall productivity. The forest products industry capacity must be capable of using the outputs of forest management and of adapting when these outputs change in quantity, type, and location over time.

Such interconnection creates a new type of problem where decision support is needed. The issue is one of system stability. Price declines and lumber market changes can affect outputs of chips, which in turn increases costs for the pulp and paper industries. Similarly, declines in pulp and paper markets reduce the demand for chips, which will have a detrimental impact on the sawmill sector. We appear to be moving to greater levels of interdependency with biorefinery and energy generation tied closely to lumber and pulp manufacture. Some new products, such as dissolving pulp (Patrick, 2011), are aimed at markets that may be quite uncorrelated with traditional paper markets. All types of markets are subject to price variability. This means that the value chain must respond to uncertain, multidimensional price signals across the many products being produced. The response to these price signals usually involves time lags. The stability question involves whether or not the overall forest product system amplifies or damps out the fluctuations in its vector of price inputs. This obviously depends on many factors that involve the ownership structure of the industry and the extent to which costs and benefits can be shared. When this is combined with uncertainties in forest inventory, forest growth, and product yields, the potential exists for very complex system dynamics.

We often respond to uncertainty in forest management by advocating the adoption of "adaptive forest management strategies." That implies a process where some agent observes that the state of the forest system differs from some desirable state and takes corrective action, somewhat akin to the process by which a thermostat controls room temperature. However, for large complex systems with time lags, most control engineers would see the concept of simple state variable control as being overly optimistic (see Gunn [2005] for a brief discussion, as well as Sandell et al. [1978] and Chen et al. [2008] for the control and interoperability issues).

This issue of multiple players becomes even more apparent within each model. In the capacity planning model, various sectors and mills within each sector often correspond to the interests of separate owners. The overall optimal plan and the wood allocations and transportation flows that correspond to that plan will most likely not be equally beneficial to all owners either during a period or over time. The transportation movements among regions, mills, and markets actually involve transactions in which there is a buyer, a seller, and a transaction price, but in the models described here, these transactions are absent and the objective involves net overall profit. It is possible to set transaction prices arbitrarily and to introduce constraints that force specific distributions of profit among the participants. However, that may not reflect desirable behavior and constrained capacity decisions and flow allocations may not reflect what could be achieved if some cooperative decision making and profit sharing could be developed (Lehoux et al., 2009).

Similarly, in the forest management planning models, there are multiple players including the Crown, large private landowners, and small private landowners. As we move to alternate tenure arrangements on Crown land, this number will increase. Adding constraints to manage cash flows among landowner groups, for wood using enterprises and over time is a possibility. However, it is not clear that this will reflect actual behavior. In particular, actual behavior for participants is usually determined one (budget year) period at a time. What mechanisms can encourage short-term behavior that is consistent with the long-term? For instance, how do we ensure that the expensive, hard to access, low-quality stands are harvested so that they can be used to regenerate good quality stands for a later harvest? These types of considerations suggest that the optimization models for capacity and forest management need to be supplemented by models that simulate multiagent behavior to enhance our ability to understand the extent to which the optimization solutions can or will be implemented.

Uncertainty raises yet more decision support questions. If uncertainty is confined to the model parameters (process, costs, yields, etc.), then longer-term strategic decisions essentially remain feasible; they just may be less than optimal and call for some small adjustments. This implies that possibly robust optimization approaches (Bredström et. al., 2011) may lead to good solutions. It is important to identify under what circumstances uncertainty implies changes in level and location of capacity and changes in silviculture programs. Alternatively, given choices of capacity and silviculture based on "known model parameters," simulation models can be developed to examine performance sensitivity when these parameters are treated as random variables.

Natural disturbances require a different approach. There are at least two issues here. First, while it may be possible to develop estimates of the mean value of loss of timber due to a disturbance (Boychuk and Martell, 1996), it can be quite difficult to model this in most forest management models and to reflect it in the availability of raw materials for the capacity planning models. Some organizations ignore fire and other natural disturbance and adjust their allowable harvests whenever fire occurs. Others calculate allowable harvest levels and then reduce these by average removals due to natural disturbance, treating these as nature's portion of the harvest. This approach is not realistic because disturbance affects the age class structure in a different way than harvesting. Second, natural disturbance radically alters the spatial structure and can dramatically change regeneration capabilities and wood flows. In some ways, the postdisturbance forest is a different forest. Harvest prescriptions that were applicable on a stand (e.g., "clear-cut harvest 10 years from now and another 70 years from now") no longer make any sense on a stand that has been burned. This is not a situation where robust optimization applies. The essence of a correct response to a disturbance is identification of the recourse opportunities, not just for the disturbed stands but also for all other stands, wood flows, and possibly capacity utilizations. Stochastic programming (Birge and Louveaux, 1997; Boychuk and Martell, 1996) may provide the right framework, but the multistage nature of the disturbance and the difficulty of specifying scenarios does not suggest that direct modeling will be helpful. Developing some scenarios through simulation of possible future natural disturbances and assuming perfect knowledge for each scenario may provide an opportunity to learn how different the capacity and silviculture responses would be in each case. Another approach would be to combine simulation and optimization (Savage et al., 2010) in a rolling planning control process where an optimization model computes a capacity, silviculture, and transportation plan looking forward from the most recent disturbance and repeating this calculation after each disturbance. An important question would be to find ways of constraining the optimization model to increase the likelihood of flexible reaction to the disturbance if and when it occurs.

Probably the most profound source of uncertainty is in the behavior of the participants in the forest value chain. As we have seen, systems level objectives are unlikely to coincide with participants' interests without some mechanism for resolving the many externalities. This suggests that there may be some merit in attempting to simulate behavior of the many participants with various perspectives on the information each has available. The use of agent-based computational economics is emerging in response to the need to understand complex systems with externalities (Tesfatsion, 2001). Frayret (2011) reviews applications of agent-based simulations in the forest products industry. Some applications are beginning both in forest management (Schwab et al., 2009; Vahid, 2011) and in the forest products supply chain.

These comments suggest that decision support for strategic analysis in the forest value chain requires more effort at developing simulation capabilities to complement the long history of optimization. Optimization models, both in the area of capacity planning and in forest management, can provide performance bounds on a strategy. In order to deal with the multiple agents, the varying issues of planning over time, and the complex uncertainties of forest value, large-scale simulation systems appear to be necessary.

5.5 CONCLUSIONS

We have attempted to identify some challenges to the development and use of decision support systems to enhance the strategic management of forest value chains. Clearly, it does not make sense to look at forest management and industrial capacity in the value chain independently, and it is possible to develop modeling approaches to generate insight. However, there remain very many real issues that require decision support. Prominent among these is how to deal with the many players in the system. If we view the overall system models as revealing what might be possible, it is incumbent on those involved to find ways to make sure that the benefits of system optimization can be shared by all participants, at least to the extent that assuming their participation in the system is justified. We also need to find ways to encourage or regulate participant behavior to reduce the chances that individual decisions damage system-wide goals.

Our observations concerning the need for more decision support models and processes and more complex models may strike some as overkill. Why can we not just keep it simple? We hope that we have made the case that analyzing strategy in the forest value chain is neither simple nor easy. Given the importance of forests and the forest industry to Canada, its regions, and its many diverse communities, we venture to suggest that the effort to deal with the real complexity is worthwhile. The many challenges mean that the development of decision support tools for such strategic analysis in the forest sector is far from finished. It has, in many ways, only just begun.

REFERENCES

Adamowicz, W. L., G. W. Armstrong, and M. J. Messmer. 2003. Chapter 6: The economics of boreal forest management. In *Towards Sustainable Management of the Boreal Forest*, eds. P. J. Burton, C. Messier, D. W. Smith, and W. L. Adamowicz. Ottawa, Ontario, Canada: NRC Research Press.

Anthony, R. N. 1965. *Planning and Control Systems: A Framework for Analysis.* Boston, MA: Division of Research, Graduate School of Business Administration, Harvard University.

Anthony, R. N., and V. Govindarajan. 2001. *Management Control Systems,* Tenth Edition. Boston, MA: McGraw Hill.

Barros, O., and A. Weintraub. 1982. Planning for a vertically integrated forest industry. *Operations Research* 30 (6): 1168–1182.

Baumol, W. J. 1977. *Economic Theory and Operations Analysis,* 4th Edition. Hemel Hempstead: Prentice-Hall.

Birge, J. R., and F. Louveaux. 1997. *Introduction to Stochastic Programming.* New York, NY: Springer Verlag.

Blinn, C. R., and M. Kilgore. 2001. Riparian management practices: A summary of state guidelines. *Journal of Forestry* 99 (8): 11–17.

Bredström, D., P. Flisberg, and M. Rönnqvist. 2011. A new method for robustness in rolling horizon planning. *International Journal of Production Economics* 143 (1): 41–52.

Broman, H., M. Frisk, and M. Rönnqvist. 2009. Supply chain planning of harvest and transportation operations after the storm gudrun. *INFOR: Information Systems and Operational Research* 47 (3): 235–245.

Boychuk, D., and D. L. Martell. 1996. A multistage stochastic programming model for sustainable forest-level timber supply under risk of fire. *Forest Science* 42 (1): 10–26.

Canadian Council of Forest Ministers. 2003. Defining Sustainable Forest Management in Canada: Criteria and Indicators. Ottawa, Canada. http://www.ccfm.org/pdf/CI_Booklet_e.pdf, (accessed May 2016).

Canadian Council of Forest Ministers. 2008a. Measuring our Progress: Putting Sustainable Forest Management into Practice Across Canada and Beyond. Ottawa, Canada. http://www.ccfm.org/pdf/CCFM_Measuring_our_progress.pdf, (accessed May 2016)

Canadian Council of Forest Ministers. 2008b. A Vision for Canada's Forests: 2008 and Beyond. Ottawa, Canada. http://www.ccfm.org/pdf/Vision_EN.pdf.

Certification Canada. 2011. Certification Status Report. Canada-wide – SFM – Mid-year 2011. Ottawa, Canada. http://www.certificationcanada.org/english/status_intentions/status.php.

Chen, D., G. Doumeingts, and F. Vernadat. 2008. Architectures for enterprise integration and interoperability: Past, present and future. *Computers in Industry* 59 (7): 647–659.

Church, R., and P. J. Daugherty. 1999. Considering intergenerational equity in linear programming-based forest planning models with MAXMIN objective functions. *Forest Science* 45: 366–373.

Daugherty, P. J. 1991. *Credibility of Long Term Forest Planning: Dynamic Inconsistency in Linear Programming Based Forest Planning Models.* PhD Thesis, Department of Environmental Science Policy and Management, University of California, Berkeley. p. 158.

Davis, L. S., and R. H. Barrett. 1992. Spatial integration of wildlife habitat analysis with long-term forest planning over multiple-owner landscapes. In Proceedings of the Workshop on Modeling Sustainable Forest Ecosystems. Washington, DC, November 1992: pp. 18–20.

Davis, L. S., K. N. Johnson, P. S. Bettinger, and T. E. Howard. 2001. *Forest Management, Fourth Edition.* New York, NY: McGraw-Hill.

Drucker, P. E. The information executives truly need. *Harvard Business Review.* January-February 1995: 54–62. Print.

Dyer, G. D. 1979. Woodlands shaped by past hurricanes. *Forest Times.* http://novascotia.ca/natr/forestry/programs/ecosystems/juan/HP-woodlandsbypast.asp (accessed January 2013).

Forsberg, M., M. Frisk, and M. Rönnqvist. 2005. FlowOpt-a decision support tool for strategic and tactical transportation planning in forestry. *International Journal of Forest Engineering* 16 (2): 101–114.

Forest Products Association of Canada. 2010. Transforming Canada's Forest products Industry: Summary of Findings of the Future Bio-pathways project. http://www.fpac.ca/publications/Biopathways%20ENG.pdf.

Frayret, J-M. 2011. Multi-agent system applications in the forest products industry. *Journal of Science and Technology for Forest Products and Processes* 1 (2): 15–29.

Freidenfelds, J. 1981 *Capacity Expansion: Analysis of Simple Models With Applications*. New York, NY: North Holland.

Gilless, J. K., and J. Buongiorno. 1985.*PELPS: Price-endogenous Linear Programming System for Economic Modeling*. Research Division of the College of Agricultural and Life Sciences, University of Wisconsin-Madison.

Greer, K., and B. Meneghin. 2000. Spectrum: An analytical tool for building natural resource management models. In United States Department of Agriculture Forest Service General Technical Report NC, pp. 174–178.

Gunn, E. A. 2005. Sustainable Forest Management: Control, Adaptive Ecosystem Management, Hierarchical Planning. In Proceedings of the 2003 Symposium on *System Analysis in Forest Resources*, ed. M. Bevers and T. Barrett, pp. 7–14. Portland, OR: General Technical Report PNW-GTR-656.

Gunn, E. A. 2007. Models for strategic forest management. In *Handbook of Operations Research in Natural Resources*, ed. A. Weintraub, T. Bjorndal, R. Epstein, and R. Romero. New York, NY: Springer US.

Gunn, E. A. 2009. Some perspectives on strategic forest management models and the forest products supply chain. *INFOR: Information Systems and Operational Research* 47 (3): 261–272.

Gunn, E. A. 2011. Integrated mill/resource capacity planning in the Canadian forest industry. *The Journal of Science & Technology for Forest Products and Processes* 1(2): 38–44.

Gunn, E. A., and A. K. Rai. 1987. Modelling and decomposition for planning long-term forest harvesting in an integrated industry structure. *Canadian Journal of Forest Research* 17 (12): 1507–1518.

Hobbs, J. E. 1996. A transaction cost approach to supply chain management. *Supply Chain Management: An International Journal* 1 (2): 15–27.

Kant, S. 2003. Extending the boundaries of forest economics. *Forest Policy and Economics* 5 (1): 39–56.

Klibi, W, A. Martel, and A. Guitouni. 2010. The design of robust value-creating supply chain networks: A critical review. *European Journal of Operational Research* 203(2): 283–293.

Kong, J, M. Rönnqvist, and M. Frisk. 2012. Modeling an integrated market for sawlogs, pulpwood, and forest bioenergy. *Canadian Journal Forest Research* 42: 315–332.

Lappi, J. 1992. JLP: A linear programming package for management planning. *Metsäntutkimuslaitoksen tiedonantoja, The Finnish Forest Research Institute Research Papers* 414: 1–134.

Lappi, J. 2003. Software by Juha Lappi. http://www.metla.fi/products/J/ (accessed June 2013).

Lebow, P. K., H. Spelter, and P. J. Ince. 2003. *FPL-PELPS: A Price Endogenous Linear Programming System for Economic Modeling, Supplement to PELPS III, Version 1.1* (Vol. 614). Madison, WI: USDA Forest Service, Forest Products Laboratory.

Lee, P., C. Smyth, and S. Boutin. 2004. Quantitative review of riparian buffer width guidelines from Canada and the United States. *Journal of Environmental Management* 70(2): 165–180.

Lehoux, N., J. F. Audy, D. A. Sophie, and M. Rönnqvist. 2009. Issues and experiences in logistics collaboration. In *Leveraging Knowledge for Innovation in Collaborative Networks*, ed. L. M Camarinha-Matos, I. Paraskakis and H. Afsarmanesh, pp. 69–76. Boston, MA: Springer Berlin Heidelberg.

Lieberman, M. B. 1987. Market growth, economies of scale, and plant size in the chemical processing industries. *The Journal of Industrial Economics* 36 (2): 175–191.

Liu, S., R. Costanza, S. Farber, and A. Troy. 2010. Valuing ecosystem services. *Annals of the New York Academy of Sciences* 1185: 54–78.

MacLean, D. A., K. P. Beaton, K. B. Porter, W. E. MacKinnon, and M. G. Budd. 2002. Potential wood supply losses to spruce budworm in New Brunswick estimated using the Spruce Budworm Decision Support System. *The Forestry Chronicle* 78: 739–750.

Manne, A. S. 1967. *Investments for Capacity Expansion: Size, Location and Time-Phasing.* Cambridge, MA: MIT Press.

Martell, D.L. and E.A. Gunn. 2014. The development and management of forest management decision support systems in Canada. pp 48–70. In *Computer-based tools for supporting forest management: The experience and expertise world-wide,* edited by Borges, Jose G and Nordström, Eva -Maria and Garcia Gonzalo, Jordi and Hujala, Teppo and Trasobares, Antonio. Department of Forest Resource Management, Swedish University of Agricultural Sciences, Umea, Sweden. ISBN 978-91-576-9236-8.

Martin, A. 2013. A Spatial Linear Programming Model of Forest Management Strategy. Masters of Applied Science Thesis, Department of Industrial Engineering, Dalhousie University.

McGill University. 2011. Canada's ecozones. Montreal, Canada. http://canadianbiodiversity .mcgill.ca/english/ecozones/index.htm.

Melo, M. T., S. Nickel, and F. Saldanha-da-Gama. 2009. Facility location and supply chain management–A review. *European Journal of Operational Research* 196 (2): 401–412.

Mintzberg, H. 1994. *The Rise and Fall of Strategic Planning: Reconceiving the Roles for Planning, Plans, Planners.* Free Press, p. 458.

Mobasheri, F., L. H. Orren, and F. P. Sioshansi. 1989. Scenario planning at southern California Edison. *Interfaces* 19 (5): 31–44.

Montreal Process Working Group. 2009. *Criteria and Indicators for the Conservation and Sustainable Management of Temperate and Boreal Forests.* Fourth Edition. http:// www.montrealprocess.org/documents/publications/general/2009p_4.pdf (accessed June 2013).

Moore, R. D., D. L. Spittlehouse, and S. Anthony. 2005. Riparian microclimate and stream temperature response to forest harvesting: A review. *Journal of the American Water Resources Association (JAWRA)* 41 (4): 813–834.

Nelson, H. 2008. Alternative Tenure Approaches to Achieve Sustainable Forest Management: Lessons for Canada. Sustainable Forest Management Network. https://era.library .ualberta.ca/files/w6634482b, (accessed May 2016).

Nuutinen, T., J. Matala, H. Hirvelä, K. Härkönen, H. Peltola, H. Väisänen, and S. Kellomäki. 2006. Regionally optimized forest management under changing climate. *Climatic Change* 79 (3–4): 315–333.

Ontario Ministry of Natural Resources. 1994. Crown Forest Sustainability Act, 1994, S.O. 1994, Chapter 25. http://www.e-laws.gov.on.ca/html/statutes/english/elaws_statutes_94c25 _e.htm. (accessed May 2016)

Ontario Ministry of Natural Resources. 2007a. *Strategic Forest Management Model - Software Version 3.2.* Toronto, Ontario: Queen's Printer for Ontario.

Ontario Ministry of Natural Resources. 2007b. Ecological Land Classification Primer. https:// dr6j45jk9xcmk.cloudfront.net/documents/2710/264777.pdf. (accessed May 2016)

Ontario Ministry of Natural Resources. 2009. Forest Management Planning Manual (2009). http://www.mnr.gov.on.ca/en/Business/Forests/2ColumnSubPage/286583.html (accessed June 2013).

Ontario Ministry of Natural Resources. 2010. Forest Management Guide for Conserving Biodiversity at the Stand and Site Scales. https://dr6j45jk9xcmk.cloudfront.net/ documents/4816/stand-amp-site-guide.pdf (accessed June 2013).

Ontario Ministry of Natural Resources. 2011. Ontario Forest Tenure Modernization Act, 2011. http://www.ontla.on.ca/web/bills/bills_detail.do?locale=en&BillID=2454&isCurrent =false&BillStagePrintId=5037&btnSubmit=go this link also takes you to Ontario Forest Tenure Modernization Act 2011, (accessed May 2016).

Ontario Ministry of Natural Resource. 2013. Harvesting in the Boreal Forest. http://www .pellet.org/images/Harvesting_in_the_Boreal_Forest.pdf. (accessed May 2016)

Patrick, K. 2011. The dissolving pulp gold rush. Paper 360. September/October 2011: 8–12. http://www.tappi.org/Hide/The-Dissolving-Pulp-Gold-Rush.aspx (accessed June 2013).

Quebec Ministry of Natural Resources. 2012. Proposals for the Selection, Establishment and Operation of Local Forests: Public Consultation Report. http://www.mrn.gouv.qc.ca/english/publications/forest/consultation/local-forests-report.pdf. (accessed May 2016)

Rasinmäki, J., A. Mäkinen, and J. Kalliovirta. 2009. SIMO: An adaptable simulation framework for multiscale forest resource data. *Computers and Electronics in Agriculture* 66 (1): 76–84.

Raunikar, R., and J. Buongiorno. 2007. Forestry economics: Historical background and current issues. In *Handbook of Operations Research in Natural Resources*, ed. A. Weintraub, C. Romero, T. Bjørndal and R. Epstein, pp. 449–471. New York, NY: Springer US.

Rotherham, T. 2003. Canada's privately owned forest lands: Their management and economic importance. *The Forestry Chronicle* 79: 106–109.

Safranyik, L., and B. Wilson. 2007. *The Mountain Pine Beetle: A Synthesis of Biology, Management and Impacts on Lodgepole Pine*. Pacific Forestry Centre, Victoria, British Columbia: Canadian Forest Service.

Sandell, N. R., P. Varaiya, M. Athans, and M. G. Safonov. 1978. Survey of decentralized control methods for large scale systems. *IEEE Transactions on Automatic Control* 23 (2): 108–128.

Santoso, T., S. Ahmed, M. Goetschalckx, and A. Shapiro. 2005. A stochastic programming approach for supply chain network design under uncertainty. *European Journal of Operational Research* 167: 96–115.

Savage, D. W., D. L. Martell, D. L., and B. M. Wotton. 2010. Evaluation of two risk mitigation strategies for dealing with fire-related uncertainty in timber supply modelling. *Canadian Journal of Forest Research* 40 (6): 1136–1154.

Schwab, O., T. Maness, G. Bull, and D. Roberts. 2009. Modelling the effect of changing market conditions on mountain pine beetle salvage harvesting and structural changes in the British Columbia forest products industry. *Canadian Journal of Forest Research* 39 (10): 1806–1820.

Schoemaker, P. J. 1995. Scenario planning: A tool for strategic thinking. *Sloan Management Review* 36: 25–25.

Strathers, R. J., T. P. Rollerson, and S. J. Mitchell. 1994. Windthrow handbook for British Columbia forests. BC Min. For., Research Branch, Victoria, BC. Working Paper 9401.

St-Hilaire, A., G. Morin, N. El-Jabi, D. Caissie. 2000. Water temperature modeling in a small forested stream: Implication of forest canopy and soil temperature. *Canadian Journal of Civil Engineering* 27: 1095–1108.

Stewart, B., and P. Neily. 2008. A Procedural Guide for Ecological Landscape Analysis: An Ecosystem Based Approach to Landscape Level Planning in Nova Scotia, Report for 2008-2. Nova Scotia Department of Natural Resources. http://novascotia.ca/natr//library/forestry/reports/Procedural%20Guide%20For%20Ecological%20Landscape%20Analysis.pdf, (accessed May 2016).

Stoddard, M., 2008. Long-Term Spatial Forest Management Planning Using the Mean-Value of Disturbance. Masters of Applied Science Thesis, Department of Industrial Engineering, Dalhousie University.

Tamiz, M., D. Jones, and C. Romero. 1998. Goal programming for decision making: An overview of the current state-of-the-art. *European Journal of Operational Research* 111 (3): 569–581.

Tesfatsion, L. 2001. Introduction to the special issue on agent-based computational economics. *Journal of Economic Dynamics and Control* 25 (3–4): 281–293.

Vahid, S. 2011. An Agent-Based Supply Chain Model for Strategic Analysis in Forestry. PhD Thesis, Department of Forestry, University of British Columbia.

Vila, D., Beauregard, R., and Martel, A. 2009. The strategic design of forest industry supply chains. *INFOR: Information Systems and Operational Research* 47 (3): 185–202.

Volney, W. J. A., and R. A. Fleming. 2000. Climate change and impacts of boreal forest insects. *Agriculture, Ecosystems and Environment* 82 (1): 283–294.

Ware, G., and J. Clutter. 1971. A mathematical programming system for the management of industrial forests. *Forest Science* (17): 428–445.

Wikström, P., L. Edenius, B. Elfving, L. O. Eriksson, T. Lämås, J. Sonesson, and F. Klintebäck. 2011. The Heureka forestry decision support system: An overview. *Mathematical and Computational Forestry & Natural-Resource Sciences (MCFNS)* 3 (2): 87–95.

Wyatt, S. W. S. 2008. First Nations, forest lands, and "aboriginal forestry" in Canada: From exclusion to co-management and beyond. *Canadian Journal of Forest Research* 38(2): 171–180.

Section 2

Forest Value Chain
Sustainability, Market Research,
Collaboration, and Agility

6 The Meaning and Means of Environmental Sustainability for the Forest Sector

Drivers and Responses in a More Responsible World

Justin G. Bull and Robert A. Kozak

CONTENTS

6.1 PREFACE

In the time between the preparation of the first and second drafts of this chapter—a period of 2 weeks—the following developments occurred in the business world: UPS, a major logistics company, created the position of chief sustainability officer (CSO); McDonalds, the fast food giant, announced sustainable procurement policies for all of its food and packaging products; Anheuser-Busch InBev, the world's largest brewer, announced water-use reduction targets for all of its facilities; and the largest American retailers and fashion designers announced the Sustainable Apparel Coalition, an effort to index the sustainability of consumer products.

Exactly what is happening here? Sustainability is widespread, and it behooves us to pause and consider what this actually means, both in broad terms, and more specifically, for the forest products sector.

To begin, we must ask, why consider sustainability? Why is it a concern for the private sector and for value chains? Indeed, it is tempting to consider sustainability only as an environmental issue, a policy issue, or a marketing priority. Throughout this chapter, however, it will hopefully become clear that sustainability is a vitally important concept to a wide variety of actors for a variety of reasons.

At the broadest level, there is an increasing awareness of the environmental and social impacts of business, especially in the wake of continued global population growth and concurrent increases in our collective levels of consumption. In a recent survey, 88% of consumers suggested that they want a company to achieve their business goals, while trying to improve society and the environment (Epstein-Reeves, 2010). Retaining, or gaining, the attention of these consumers requires action by the private sector. Beyond issues of market share, companies are also beginning to include sustainability in their corporate decision making. They see an opportunity to create "shared value," distinct from shareholder value in that corporate activities should be able to provide benefits to both society and the environment. They also see the chance to mitigate risk, reduce pollution, and/or develop new markets for environmentally friendly goods and services. While the measurement of financial value associated with sustainability is contentious, the disagreement lies not in the existence of value but where exactly the value is created.

6.2 INTRODUCTION

Very simply defined, sustainability is the idea that things in the future should be better than they are today. It is a ubiquitous concept, used and abused by academics, environmentalists, the private sector, and politicians alike. It is so pervasive that its meaning is lost in a flurry of rhetoric and marketing. Despite this, sustainability is an idea that will be fundamental in the organization and evolution of businesses and supply chains in the coming decades, and forestry, even with a comparatively strong environmental track record, is by no means immune. To corporate decision-makers, sustainability is no longer a question of *why*, but *how*.

In business, sustainability is being driven by a wide variety of actors and influences. Investors are interested as they see risks and opportunities in the environmental impacts of business. Advocates, such as environmental nongovernmental organizations (ENGOs), are working to influence the private sector to ensure better environmental outcomes. Governments are becoming increasingly stringent in their environmental regulations. Business-to-business (B2B) relationships are changing the way supply chain partners interact; the exchange of environmental information is becoming the norm rather than the exception, especially with major retailers. Consumers prefer products and companies that respect and improve the environment, although they are generally not interested in paying more. Finally, climate change, a unique global challenge, has also given carbon and energy use a prominent place in our daily lives.

The corporate world is responding in a similarly diverse fashion. Many efforts at being more sustainable are win–win solutions, improving environmental performance while reducing costs. Other solutions are initially win–lose wherein environmental performance improves, but financial performance does not. This represents a strategic decision to anticipate or develop markets for more sustainable goods, with the hope that these investments will pay off. Corporations are also partnering with ENGOs to add legitimacy and credibility to their environmental claims and efforts. They are engaging in a wide variety of reporting schemes, with the Global Reporting Initiative (GRI) and Carbon Disclosure Project (CDP) standing out as prominent examples of voluntary disclosure of environmental data. Sectors that have a shared reputation, such as forestry, have developed industry codes of conduct to ward off the perception that environmental performance is only as good as the worst offender. Tools for measuring environmental impacts, such as life cycle assessment (LCA), are gaining traction, as more and more data become available and software packages make conducting an LCA an affordable exercise in environmental management. Formalized certification schemes have emerged as a form of nonstate governance. Evolving over time, and requiring significant costs to implement and maintain, these schemes are used to verify and improve the sources and supply chains of goods such as seafood or forest products.

In forestry, certification has been the most significant response to the sustainability imperative. Indeed, forestry has been more progressive than any other sector in developing transparent, multistakeholder standards to certify the sustainability of products. Other approaches, such as LCA, are also commonplace, but can misrepresent the actual footprint of forestry. In this chapter, two illustrative examples are provided related to what appears to be, at the outset, a seemingly uncomplicated supply chain scenario—pulp and paper manufacturing. In one instance, when LCA is used to compare the footprint of paper media with digital media, the latter has invariably been portrayed as environmentally preferable. But flaws in the LCA methodology, major gaps in data describing digital media, and the inconsistent application of LCA methods have generally been ignored. Further, LCA methods have not grappled with how to measure the land-use implications of forestry, or the embodied carbon in any forest product. This simple example is provided in order to convey the need to consider other approaches—like industrial ecology—to address the complex and multifaceted environmental dimensions and implications of forestry value chains.

In 1970, the economist Milton Friedman famously said (Friedman, 1970), "it is the social responsibility of business to increase its profits." The role of business has since changed. Business can no longer relentlessly pursue profits without considering the environmental and social implications. A confluence of drivers has shaped this evolution, including increasing population (and consumption) levels, the emergence of climate change as a global environmental issue, and the 2008 economic crises which has shaken the corporate sector and undermined public trust in private industry. In 2010, the Nobel Laureate Joseph Stiglitz revisited Friedman's narrow doctrine, suggesting that "we should think about how we can create a global economic architecture which works better, for more people, in a more sustainable way" (United Nations, 2009).

This chapter describes the origins, evolution, and future directions of business and sustainability. Our focus is on environmental sustainability, not social or economic sustainability. While these latter two issues are critical, economic sustainability is well understood and already considered thoroughly by business and its profit motive. Social sustainability—which considers human rights and the concept of intergenerational equity—is equally important, but sufficiently distinct from environmental sustainability to make its inclusion here inappropriate.

We divide this chapter on environmental sustainability into five sections. To begin, we discuss what exactly "sustainability" means. This is followed by a review of drivers of sustainability, from advocacy campaigns to the regulatory initiatives of governments. We then consider responses to sustainability, such as certification schemes, environmental reporting, and corporate–nongovernmental organization (NGO) partnerships. Two case studies are presented: one on the supply chain for carbon emissions of a magazine; and another that explores the relative environmental impacts of digital and paper media. We conclude with a brief discussion of the implications for value chains in the forest products sector of Canada.

6.3   DEFINITIONS AND DIMENSIONS OF SUSTAINABILITY

6.3.1   Academic Definitions of Sustainability

A founding definition of sustainability comes from the Brundtland Commission, which described sustainability—more specifically, sustainable development— as meeting "the needs of the present without compromising the ability of future generations to meet their own needs" (United Nations, 1987). This definition evolved, with the proverbial concept of sustainability as a "three-legged stool" gaining popular adoption. Here, an ideal outcome considers economic, social, and environmental concerns together, with sustainability occurring at the intersection of these three values (Figure 6.1a).

A variation of this definition comes from the discipline of ecological economics. Ecological economics is an effort to promote understanding between ecologists and economists. It differs from neoclassical economic approaches in that neoclassical models are defined by achieving an equilibrium and final state that is independent of the path taken. Growth in the neoclassical sense is a function of accumulated savings, capital investment, and technological progress. Consequently, growth is theoretically exponential and limitless (Seager, 2008). By contrast, ecological economics models are subject to an exogenous limit: the carrying capacity of the planet. Economic models are embedded in the environment, and these models explicitly address the interdependence of human economies and natural ecosystems over time and space. Thus, sustainability is conveyed as "embedded circles" where social and economic concerns are bounded by the broader context of the environment (Figure 6.1b). In this view, society and economic systems are subsets of the environment, limited by a finite amount of natural capital.

While these definitions are conceptual, elegant, and inclusive, they do not necessarily offer the specificity required by different actors, such as governments, consumers, and corporations. In fact, it can be said that the traditional academic approach of creating distinct and isolated disciplines and reducing questions to testable hypotheses is ill suited to ask questions and provide answers about sustainability. What type of academic is best suited to address issues of sustainability—an engineer, an ecologist, an economist, a sociologist, or a political scientist? Each would approach sustainability in very different ways, none of which are more correct or better than the others.

Seager (2008) suggests that academics are not unaware of this mismatch between traditional research structures and urgent research questions. To that end, two new academic disciplines are emerging as a means of grappling with the complex and multifaceted nature of sustainability: industrial ecology and ecosystem health.

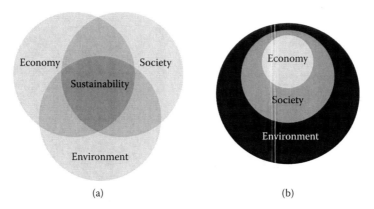

(a)   (b)

FIGURE 6.1   (a) The "three-legged stool" approach and (b) the "embedded circles" approach.

Seager, in identifying this mismatch between definitions and research capacity, follows upon the work of Mebratu (1998, p. 493), who noted that definitions of sustainability, "focus on specific elements while failing to capture the whole spectrum." Efforts to develop a theory of sustainability have led to concepts and definitions that benefit specific groups and interests, but do not reflect holistic thinking on the subject which have long been embedded in traditional beliefs and practices (Mebratu, 1998).

Industrial ecology is focused on the interactions between ecological and industrial systems. It is founded on the idea that much can be learned from natural systems to improve the environmental footprint of industrial systems. Given that industrial systems have a relatively brief history—at least compared to ecosystems—looking at the structures and successes of nature can provide insights into how to devise more effective and sustainable systems (Seager, 2008).

Figure 6.2 shows three types of systems commonly discussed in the discipline of industrial ecology. The first, Type I, is an open loop, with energy and materials

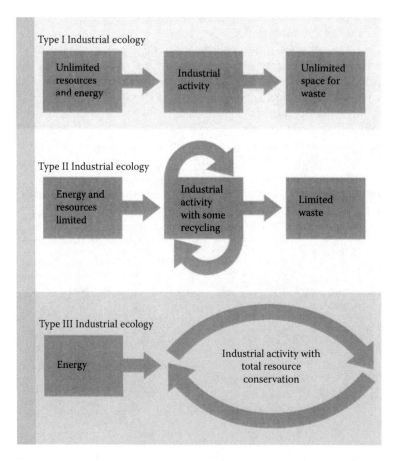

FIGURE 6.2 Industrial ecology types. (Adapted from Krones, J., *MIT Undergrad. Res. J.*, 15, 2007.)

flowing in, being processed, and flowing out. It assumes an unlimited availability of resources and an unlimited ability for the planet to absorb the impacts of industrial activities. Early industrial systems were similar to this model. Type II describes most existing industrial systems as they occur today. There are limited supplies of materials and energy, and industry cannot pollute freely without incurring costs. Some industries might be a "weak" Type II, with minimal resource conservation and waste reduction efforts, while others could be considered a "strong" Type II, where both economic and environmental motives exist for managing resource use and waste streams. A completely closed-loop Type III system is impossible for industry, as *total* resource conservation is, in essence, unachievable. However, it is something that industry can aspire to. Ecological systems are Type III systems, as nature has only one input (solar energy), and waste does not exist as every material is eventually reused by the system (Graedel and Allenby, 1995; Erkman, 1997).

Industrial ecology is more than simply envying the extreme efficiency of nature. Kalundborg, Denmark provides an example of industrial ecology that has garnered substantial academic inquiry (see Ehrenfeld and Gertler [1997] for a detailed account). A variety of industrial actors evolved—without any master plan—to integrate their waste and energy flows and become more environmentally benign (see Figure 6.3). A coal plant sells steam to a variety of users, including local households. A pharmaceutical factory installed a two-mile steam pipe rather than replacing their existing boilers, an investment that paid for itself in 2 years. The coal plant was also required by regulators to install sulfur dioxide scrubbers at a cost of $US 115 million. Fortunately, the coal plant was able to

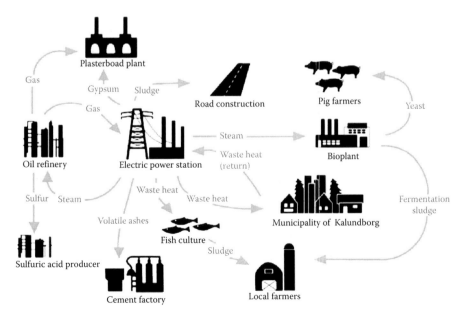

FIGURE 6.3 Industrial symbiosis in Kalundborg, Denmark. (Adapted from Ehrenfeld, J. and Gertler, N., *J. Ind. Ecol.*, 1(1), 67–79, 1997.)

sell gypsum, a by-product of the scrubbing process, to a local plasterboard plant. All of these connections between industrial actors minimize the waste of energy and materials. This symbiosis was not mandated by regulations, although some regulatory imperatives helped foster connections that would not have otherwise existed. Further, and perhaps more importantly, these decisions made a good deal of economic sense.

While industrial ecology is now emerging as a *bona fide* area of inquiry, ecosystem health is much less formalized and the most nascent of the academic disciplines focused on sustainability. It advances the goal of assessing the health of an entire ecosystem, not just the health of individual organisms or species. Further, it suggests that human health is dependent on healthy ecosystems, and that the interaction between these two systems is vital. Like ecological economics, it considers a healthy ecosystem to be a necessary factor of production in a functioning society and economy. There are three key concepts that have shaped research in ecosystem health, and thus, our conception of sustainability: vigor—a measure of total activity; organization—a measure of the quality and diversity of interactions between different parts of the system; and resilience—the ability of the system to recover from injury (Costanza and Mageau, 1999; Seager, 2008).

6.3.2 CORPORATE DEFINITIONS OF SUSTAINABILITY

The corporate world has offered its own versions of sustainability, with corporate social responsibility (CSR)* initiatives being the most commonly employed tools for achieving corporate sustainability objectives. CSR requires a firm to think beyond legal and profit imperatives by considering its impacts on society and the environment (van Marrewijk, 2003). Auld et al. (2008a) describe CSR efforts as either win–win solutions or win–lose solutions. Win–win solutions involve internal changes that are socially and/or environmentally beneficial, but also serve to increase profit. In other words, this is the "low hanging fruit" of CSR. Win–lose solutions are more complex, as a change that provides environmental or social benefits may decrease profits. As a result, the structures that influence profits (such as government policy and consumer preferences) would need to be adapted to reward the new, but initially less profitable, solution. If no such changes take place, the win–lose solution eventually fails.

Firms can be proactive in trying to develop new markets that reward what are initially win–lose efforts, or they can anticipate changes in government, consumer, or civil society behavior that will eventually result in win–lose solutions becoming profitable. For example, in the past decade, Toyota invested heavily in the development of fuel-efficient compact cars, while Ford initially remained focused on less efficient, but more profitable, sport utility vehicles (SUVs).

When win–win or win–lose solutions fail to generate profits in the long term, CSR efforts can result in a competitive disadvantage, and become marginalized rather than transformative in influencing firm behavior (Auld et al., 2008a). There is

---

* CSR is now more commonly referred to as corporate responsibility (CR).

also a danger that CSR efforts might bias our collective conception of sustainability. Does CSR mean that a company will transform to become truly more sustainable? Or does a company engage in CSR in the hopes that some of its industrial practices can be viewed—without much corporate effort—as being sustainable? This is an important caveat to consider throughout our analysis of business and sustainability. Another important issue is the idea that "you can't make an omelet without breaking eggs." Businesses will always have an impact on the environment, and it is prudent to remember this when considering some of the more utopian ideals surrounding definitions of sustainability.

6.3.3 SPECTRUM OF SUSTAINABILITY

Sustainability manifests differently within academic and corporate contexts. But in many ways, the multitudes of definitions complement one another, offering guidance at practical and abstract levels. A common element is that sustainability is a function of both time and space. For example, the time dimension of sustainability is revealed when describing the use of a reusable mug instead of a disposable cup, or altering an entire economy to rely on the production of services rather than goods, or in reducing consumption as a means of achieving intergenerational equity. The space dimension refers to the fact that sustainability initiatives can occur at the scale of a single consumer or at the scale of a factory or supply chain. When defining sustainability, we must embrace diversity, rather than struggle to find a unifying articulation. Seager (2008, p. 447) grapples with this tension between inclusivity and specificity and arrives at the following:

> Sustainability might best be defined as an ethical concept that things should be better in the future than they are at present. Like other ethical concepts such as fairness or justice, sustainability is best interpreted conceptually rather than technically.

The robustness of this definition allows for different actors to consider sustainability relative to their own interests. Figure 6.4 illustrates a spectrum of sustainability. On the left-hand side is a version of sustainability that is static. It is defined by the maintenance of the status quo, which, to some actors (such as law enforcement), would be a sustainable outcome. Moving right, there is steady-state sustainability, which is defined by reliability. Here, the ability for a system to sustain the same function over the long term is prioritized. Further right is dynamic sustainability, where the ability of a system to adapt, regrow, or evolve in light of changing circumstances is prioritized. To the far right is episodic sustainability, which is, in essence, a version of Schumpeter's "creative destruction" (Schumpeter, 1942). In this view, sustainability is defined as an evolving system passing through multiple states.

It is clear that sustainability has a range of meanings and interpretations. It should come as no surprise that a wide variety of factors motivate the adoption of sustainability practices. The next section reviews these drivers and common responses.

Sustainability spectrum

		Static	Steady state	Dynamic	Episodic
Ecology		Individual survival strategies	Predation, disease, natural selection	Reproduction, recovery, regrowth	Extinction, mutation, evolution
Industry		Law enforcement and protection services	Failure and risk analysis, quality management	Strategic planning, disaster or crisis response	Creative destruction, innovation, R&D
Function		Preserve existing function	Maximize existing function	Temporary loss of function	Function evolves
Responce to injury		Prevent or minimize loss	Replacement of loss to enhance functionality	Rapid recovery from injury	Injury changes path, resulting in new functions

FIGURE 6.4 The sustainability spectrum. (Adapted from Seager, T.P., *Bus. Strategy Environ.*, 17(7), 444–453, 2008.)

6.4 DRIVERS OF AND RESPONSES TO SUSTAINABILITY

A variety of pressures have driven the corporate world to consider environmental sustainability as part of their business strategies and practices. Each actor along any given supply chain is motivated by their own definition of sustainability and self-interests. That being said, many common themes are emerging within this burgeoning domain and in the first part of this section we consider the main "drivers" of corporate sustainability: governments, investors, eco-efficiency, the marketplace, advocates, and climate change. While we do not review these drivers in comprehensive detail, we do provide a broad overview and illustrative examples.

In the second part of this section, we consider responses to sustainability in the form of CSR practices, basing our analysis on a framework developed by Auld et al. (2008a) that categorizes CSR innovations. These seven categories exhaustively capture all of the ways in which a corporation might respond to the sustainability imperative: individual firm efforts; individual firm and individual NGO agreements; public–private partnerships; information-based approaches; environmental management systems (EMSs); industry association codes of conduct; and nonstate market-driven (NSMD) governance in the form of private-sector hard laws.

6.4.1 DRIVERS OF SUSTAINABILITY

6.4.1.1 Governments

Governments motivate sustainability by a variety of means and for a variety of reasons. Governments are, at some level, "stewards" of the environment and must be responsive to the demands of voters who, at times, prioritize environmental issues.

They are lobbied by environmental advocates and corporations, attempting to strike a policy balance between conservation and the need for economic growth. How, when, and why a government pushes for corporate environmental sustainability is dependent on the definition of sustainability employed and the environmental issue at hand. Governments have a range of options, from attempts to force specific behaviors (by banning a substance or setting an emission standard) to changing incentives (through subsidies or other policy instruments) or by correcting a lack of information (by producing impact assessments, encouraging product labeling, or providing rewards and recognition). There are a wide variety of tools available, each with strengths and weakness, and a daunting body of accompanying literature.

Governments have motivated environmental sustainability in the forest sector in a variety of ways. In British Columbia, emitters are required to pay a tax on their carbon emissions. This can be a significant cost for major emitters such as pulp and paper mills. The provincial government has committed to becoming carbon neutral, providing opportunities for forest companies to sell credits to the Pacific Carbon Trust. British Columbia is also part of the Western Climate Initiative, which could evolve into a major market for trading carbon emissions in North America. The federal government in Canada has subsidized the retrofitting of pulp and paper facilities to produce bioenergy and biochemicals as a means of participating in the new so-called "green economy." In Europe, governments have implemented sustainable procurement policies, ensuring that the government only buys forest products from legally verified suppliers.

6.4.1.2 Investors

Investors—those who provide capital to companies or hold corporate stock—have been a key driver of corporate sustainability. But to characterize all investors as drivers of sustainability would be untrue. Many, if not most, are primarily focused on profits and quarterly returns. However, large institutional investors, such as pension funds, have taken a different approach, adopting environmental, social, and governance (ESG) metrics in their investment strategies.

The United Nations Environment Programme's (UNEP) Finance Initiative is a group of over 70 investors that have adopted "Principles for Responsible Investing." The investors describe their motives for considering ESG issues as threefold (UNEP Finance Initiative, 2006). First and foremost, ESG variables can help investors make more money. Their hypothesis is that, over time, integrating thorough and systematic reviews of ESG issues will result in better financial performance. Second, some investors consider ethical values paramount in their investment decisions, ignoring any marginally positive or negative impacts to financial performance. Finally, it has been suggested that channeling investment flows in accordance with ESG issues will help to support long-term sustainable development.

The authors of the UNEP report dismiss the notion that investors are unfamiliar with the financial impacts of environmental or social risk, offering as evidence the financial liabilities associated with asbestos in building materials in the 1980s and 1990s, or the costs faced by Union Carbide following the Bhopal* explosion.

---

* In December 1984, an explosion at a Union Carbide pesticide plant released toxic gas over the city of Bhopal, India. Estimated fatalities were between 2,500 and 5,000, with up to 200,000 injured.

Speaking to agnostic or skeptical investors who think ESG issues have a tenuous connection with financial performance, the authors suggest the following (United Nations Environmental Programme (UNEP) Finance Initiative, 2006, p. 4):

> Investors who are not sure what to do would do well to refer to Pascal's wager concerning the existence of God: make believe as if the hypothesis is true, thus at little or no cost avoiding the worst-case scenario.

Another example of investors considering environmental sustainability comes from the CDP. The CDP is a nonprofit supported by 534 investors with assets totaling $US 64 trillion. The CDP administers an annual survey to over 4700 corporations, but focuses its efforts on the so-called "Global 500," a group of companies that collectively account for 11% of annual greenhouse gas emissions (Carbon Disclosure Project, 2010). Of these top 500 companies, 82% responded to the survey in 2010, the highest rate since the inaugural survey in 2003. In their annual report for 2010, the CDP identifies several important trends. They see the demand for primary corporate climate change data to be growing, with platforms like Bloomberg and Google Finance now offering carbon data alongside financial data. Previously unresponsive sectors, such as shipping and transportation, have now begun to share carbon data. CDP also sees a shift in corporate perceptions of climate change, with more respondents in 2010 identifying climate change as an opportunity (9 out of 10) than a risk (8 out of 10).

6.4.1.3   Eco-efficiency

Eco-efficiency is arguably the most tangible reason for a corporation to consider sustainability. At its core, eco-efficiency involves "producing and delivering goods while simultaneously reducing the ecological impact and use of resources." (Molina-Azorín et al., 2009, p. 1082). From this perspective, the generation of pollution is seen as inefficient. However, the financial benefits of eco-efficiency are not always immediate or evident, and its pursuit could render a firm temporarily or permanently uncompetitive.

Ambec and Lanoie (2008) identify ways in which eco-efficiency can help businesses to improve their bottom lines, for instance, by increasing their levels of risk management and external stakeholder engagement or by decreasing their costs of materials, energy, and services. They also identify three opportunities afforded by eco-efficiency to directly increase revenue streams: through better access to certain markets, through product differentiation, and through selling pollution-control technology. This variety of opportunities is demonstrative of how eco-efficiency can be implemented in a manner that makes good fiscal sense to businesses. Eco-efficiency can also relate to attempts to reduce the impacts of the transportation infrastructure required to deliver a product to market. This is most prevalent in the food industry where "food miles" are used to measure the environmental impact of delivering food to market (Engelhaupt, 2008).

Based on a review of 32 case studies, Molina-Azorín et al. (2009) present a three-stage process that relates environmental management to financial performance. Functional specialization is the first stage and involves a company reacting to the pressures of environmental regulation. Internal integration, the second stage, involves managers developing and monitoring corporate objectives on environmental management. The third stage is external integration. Here, a firm incorporates environmental performance into its overall business strategy. Many of the opportunities identified above by Ambec and Lanoie (2008) result from this three-stage process. In pursuing environmental goals, new markets of consumers who prioritize ecological issues open up and firms can become leaders in their sector. As a result, they may develop technologies or processes that put them in an advantageous position, or they may be able to patent and resell their innovations, thereby gaining first mover advantage. Molina-Azorín et al. (2009) find that firms who achieve a positive relationship between environmental management and financial performance are able to invest more in their environmental strategies, resulting in a positive feedback loop.

6.4.1.4 Marketplace

There are primarily two market forces that can drive corporate sustainability: B2B markets and consumer markets.

B2B markets influence corporate sustainability when one company seeks suppliers that have integrated environmental thinking into their business model. For example, the Home Depot implemented a policy where the purchase of certified wood is prioritized (Home Depot, 2011). Walmart has undertaken similar responsible purchasing policies and has developed a widely recognized supplier assessment mandatory for anyone who intends to sell to their goods at a Walmart store (Walmart, 2011). Providing information on carbon emissions, water usage, and plans to increase eco-efficiency, among other things, is required. Walmart is also in the process of requiring its largest suppliers to conduct similar assessments of their own suppliers. As a result, Walmart's environmental priorities are trickling up the supply chain, influencing companies who might not even sell directly to Walmart. Given the purchasing power of Walmart, these efforts hold enormous potential.

Consumers can also drive corporate sustainability. Price being equal, consumers generally prefer to purchase a "green" product (Manget et al., 2009), with some willing to pay a premium for more sustainable products (Ottman, 2011). The rapid rise of organic food reflects an increasing consumer awareness of environmental issues. Major consumer products manufacturers, such as Coca-Cola, are beginning to promote their green initiatives and are, for example, promising to make bottles that contain at least 30% plant-based materials (Houpt, 2011). Information and communications technology companies are offering mobile phones with plastic casing made from recycled water bottles (German, 2009). Hygienic paper products with only recycled content have been developed and are capturing market share (Jacob et al., 2005). There are similar examples from virtually every sector and product type.

There is a risk that corporate efforts to sate demand for green products offer little environmental benefit or, even worse, are misleading. This "greenwashing" is not surprising given the relative immaturity of markets for environmentally responsible goods. In the future, however, it is less likely that companies will be able to mislead consumers. The Internet and social media are creating increasingly savvy consumers, and facilitate the rapid dissemination of embarrassing information on corporate practices. Green marketing experts suggest that, in this transparent information age, CSR efforts should be authentic or not attempted at all (Ottman, 2011).

6.4.1.5 Advocates

Advocates are individuals or organizations who attempt to influence corporate behavior, with ENGOs serving as the most prominent example. Other advocates might include local communities who are impacted by a firm's activities or academics whose research leads them to believe that a firm's behavior could and should be improved. Different advocates will exert influence by different means, at different levels, and on different environmental issues.

If there is a corporation creating an environmental impact, there is likely an ENGO advocating for its amelioration. They do so by a variety of means: connecting directly with a firm to express their concerns and perhaps offer solutions; lobbying relevant government agencies to suggest regulatory changes that address the situation; or communicating with the public to raise awareness. Some ENGOs have a more adversarial and combative relationship with the corporate world, publicly shaming a company by releasing details of an environmental harm. Other ENGOs, in contrast, work directly with corporations to help improve their environmental performance, lending authenticity and credibility to CSR efforts. ENGOs themselves admit that working with the private sector is a potent opportunity to promote sustainability. Speaking about his relationship with the Coca-Cola Company, the CEO of World Wildlife Fund (WWF) Canada stated the following (Houpt, 2011):

> We could spend 50 years lobbying 75 national governments to change the regulatory framework[s] … or these folks at Coke could make a decision … and the whole global supply chain changes overnight.

Local communities affected by a company's environmental footprint can similarly advocate for change. How and where they advocate will depend on the particular issue, their capacity, and the responsiveness of government to community concerns. Like ENGOs, what is important from the perspective of CSR is that advocacy is likely to occur for any number of issues, at many levels, and can be combative or cooperative. How a corporation responds to advocacy is perhaps indicative of how seriously it has integrated CSR practices into its business strategy. Those that have taken the steps to measure and manage their environmental footprint, and have done so in a methodologically rigorous and transparent way, will be better positioned to engage with advocates.

6.4.1.6 Climate Change

Climate change and, by extension, carbon, deserve particular attention as a driver of sustainability. Climate change is a unique environmental issue because it is global in scale and has dire implications (IPCC, 2007). It is unique politically because so many government bodies at the international, national, provincial, and municipal levels have put forth regulations and set targets with the goal of reducing carbon emissions. Corporations, perhaps seeing the writing on the wall, have embraced carbon as an environmental issue (SwissRe, 2007). Given that carbon emissions are nearly analogous with energy use, reducing carbon often equates with reducing energy costs. Reducing emissions is, therefore, a classic win–win type CSR initiative.

Carbon simply being ubiquitous does not make it a driver of environmental sustainability. The real driver is the likelihood of carbon as a priced commodity. Governments, advised by scientists and advocates that carbon emissions should be reduced, have devised schemes to put a price on carbon (Garnaut, 2008). In the European Union, the European Trading System governs the buying and selling of emission allowances (Europa, 2005). International agreements have tried, thus far failing, to emulate this system. Regional approaches, such as the Western Climate Initiative or the Regional Greenhouse Gas Initiative, have taken steps to price carbon (RGGI, 2009; Western Climate Initiative, 2009). Some jurisdictions, like British Columbia, have imposed a carbon tax. All of these efforts result in one obvious conclusion: corporations anticipate having to pay for carbon. The prospect of a new expense embedded in all business transactions has been remarkably effective in driving businesses to consider their environmental footprints.

The monetization of carbon is what economists consider internalizing a cost. Economic systems generally fail at internalizing environmental costs, and carbon is the first attempt at a global scale. It should not be a surprise that reaching an international consensus on how to price carbon is proving to be an elusive goal. The lack of action at the international level has not stopped corporations from learning how to measure, manage, and reduce their emissions, and these efforts are reaping major benefits. Measuring the carbon footprint of a large multinational corporation is a technical feat, requiring expertise, scientific knowledge, and an evolved corporate culture. However, now more than ever, companies have an increased capacity and willingness to begin measuring this aspect of their environmental footprints. Furthermore, lessons learned here could be used in measuring the environmental impacts of using other pubic goods, like water.

6.4.2 RESPONSES TO SUSTAINABILITY

6.4.2.1 Individual Firm Efforts

Individual firm efforts occur when a firm independently makes a decision to become more environmentally responsible. Such efforts are not responses to regulations, but may be attempts to preempt government. Firms may uncover win–win CSR opportunities or adopt win–lose strategies that hold long-term financial potential. In general, internal firm efforts are not subject to externally imposed prescriptive requirements, and firms control the processes and policies developed. This flexibility and the fact that win–win solutions tend to be the focus of corporate attention explain why individual firm efforts represent the most prevalent and widespread manifestation of CSR.

6.4.2.2 Individual Firm and Individual NGO Agreements

Agreements refer to CSR efforts in which a firm engages with an NGO (usually an ENGO) or other stakeholders to address the environmental impact of a firm's operations. In general, an NGO will come to view an environmental issue from a different perspective than a firm. Firms benefit from these partnerships by adding legitimacy to their CSR efforts by collaborating with traditional adversaries.

An example of a firm–ENGO collaboration comes from the Environmental Defense Fund (EDF) and Walmart. Walmart, the epitome of a big-box American retailer, has integrated environmental variables into its practices for a variety of reasons. EDF decided to leverage this opportunity, seeing collaboration with Walmart as a chance to highlight what they perceive to be a sincere CSR effort, but also to promote best practices among retailers and along Walmart's massive supply chain (EDF, n.d.). In the forest sector, similar relationships exist. For example, Catalyst Paper partnered with WWF Canada when it set out to measure and reduce the carbon footprint of its paper products (WWF, n.d.). This allowed Catalyst Paper to develop a unique product—carbon neutral paper—while adding legitimacy through collaboration.

6.4.2.3 Public–Private Partnerships

Public–private partnerships are similar to firm–NGO agreements, but involve other interests, such as governments, or are made up of several firms and NGOs acting in concert. These partnerships can emerge as efforts to address standards development, to implement self-regulation, or to develop collaborative coregulation schemes. These partnerships are grounded in the idea that private–public collaboration can provide an efficient means of enforcing costly legislation.

An early example of public–private partnerships comes from the United States and the Environmental Protection Agency (EPA), which developed the 33/50 program. Implemented in 1991, the goal of the program was to reduce emissions of 17 toxic chemicals by 33% in 1993, and 50% by 1995. Rather than set strict prescriptive requirements, the EPA gave industry flexibility in meeting these goals. The program managed to achieve its targets 1 year ahead of schedule (EPA, 1994). Forestry has similar examples. The Great Bear Rainforest Agreement and the Canadian Boreal Forest Agreement both demonstrate the potency of partnerships among ENGOs, industry, and government. The former protected a large area of rainforest on the coast of British Columbia, allowing ENGOs to claim victory, while providing industry with a degree of security from ENGO criticism (Forest Ethics, n.d.). In the latter, ENGOs and industry (including traditional adversaries like Greenpeace and Weyerhaeuser) agreed on environmental conservational goals and the need to pursue sustainable forest management and provide jobs for forest-dependent communities (FPAC, 2010). This provided a détente of sorts, allowing industry to focus on economic issues without the constant risk of negative attention from ENGOs.

6.4.2.4 Information-Based Approaches

Many CSR efforts revolve around the provision of information related to a firm's behavior. Some efforts are voluntary, while others are mandatory. Government sponsored systems, such as the EPA's Toxic Release Inventory, require companies to disclose information on an extensive list of chemicals that their activities produce. This was brought about by advocates who argued that communities have the "right to know" about what corporations are doing. In Canada and the United States, like most other developed economies, major emitters of carbon are required to report on their emissions to the federal government. Similarly, food producers must report on the nutritional requirements of their products. Walmart, while describing the future of its Supplier Sustainability Index, has suggested that a carbon label might be required for all of the goods that it sells.

These examples speak to an increasing emphasis on transparency. By requiring the disclosure of information or by advocating for voluntary disclosure, transparency becomes the norm rather than the exception. Firms that do not participate risk losing market access. Other forms of information sharing also revolve around this notion of transparency. In the United States, the Lacey Act mandates that importers of forest products trace the origin of their imports. This has forced the development of sophisticated supply chain practices, such as radio tags that get attached to felled trees at harvest sites (USDA, 2011). Noncompliance with the Lacey Act can lead to embarrassment. Gibson Guitars, a high-end luthier in the United States, experienced public scrutiny when its operations were shut down by government officials for violating the Lacey Act. An investigation by the World Resources Institute set out to purchase 32 books randomly from a retailer, and found that three of them used illegal fiber in the paper (Nogueron and Hanson, 2010); a continued need for transparency and measurement along the supply chain remains. Similar legislation for conflict minerals (minerals from the Congo Basin which are being sold to fund armed conflict) is currently being considered in the United States which would require information and communications technology (ICT) manufacturers to trace the source of certain metals in their products (OpenCongress, 2010).

The GRI is an example of information sharing that deserves particular attention. The GRI has become the *de facto* standard* for corporate sustainability reporting, and is perceived by executives as second only to the ISO 14001 standard in influencing their CSR practices (Brown et al., 2009). The GRI is a multistakeholder process, involving industry, governments, scientists, and civil society. Various working groups develop specific guidelines for different sectors. Additionally, the guidelines evolve relatively quickly; the third version of the GRI is in place after only 9 years. The GRI has gained widespread support for a variety of reasons. It is a win–win solution as corporations adopt a reporting framework that meets the requirements of many social actors (notably, advocates and investors). The GRI also carries with it a strong sense of legitimacy and inclusivity and is a partner institution of the UNEP.

At the opposite end of the scale from enterprise-level GRI reporting is another information-based approach: Environmental Product Declarations (EPDs). These statements are equivalent to food nutrition labels, disclosing the energy, materials, water impacts, and waste emissions associated with a particular product. EPDs are derived from ISO standards (ISO, 2006) and can only be prepared once Product Category Rules (PCRs) have been developed using a multistakeholder process. The forestry sector has embraced EPDs, as they are increasingly being required of products used in the construction industry. To date, EPDs have been developed for specific wood products categories like western red cedar decking and bevel siding and glue laminated timbers, to name a few (FII, 2013).

---

* The GRI is technically a guideline rather than standard, as it is free to use and unverified.

6.4.2.5   Environmental Management Systems

EMSs are externally defined and imposed criteria about how a firm should approach its environmental footprint. The ISO 14001 standard, and more recently the 26001 standard, are the most widely adopted EMSs. Firms use these systems to measure and monitor their footprint as they may reveal opportunities to reduce inefficiencies and identify risks. They carry with them credibility, and the opportunity to attach a recognized logo to a product, but there are costs associated with implementation and third-party verification of EMS standards.

Firms can also employ other systems to measure their environmental footprint. A common tool, in both academia and the corporate world, is LCA. While ISO standards are used to define protocols on how to conduct an LCA, LCAs themselves are rarely verified by a third party. Increasingly, LCAs utilize data from life cycle inventory (LCI) databases that are privately owned and not subject to peer review. While these databases make the conduct of an LCA cost effective, they may also conceal uncertainties or gaps in data.

6.4.2.6   Industry Association Codes of Conduct

Industry codes of conduct are typically found in sectors where companies tend to share a collective reputation. That is, a bad image of industry impacts all firms, regardless of individual performance. Most often, these are sectors that produce primary or intermediary goods, like forestry, mining, or petrochemicals. Recently, other sectors have adopted codes of conduct: the coffee industry, which sells directly to consumers, has several schemes in place; and the textile industry in the United States has created a Sustainable Apparel Coalition with the backing of ENGOs (EDF), government bodies (EPA), and major clothing manufacturers (Levi Strauss, Nike, and Mountain Equipment Co-op, to name a few). These codes of conduct are often principles driven and administered by industry associations; they do not provide specific, prescriptive guidance on how firms should behave. Rather, concepts are defined and adherents promise to meet them, although verification and enforcement may not be as stringent as in other types of CSR innovations.

6.4.2.7   Nonstate Market-Driven Governance

NSMD governance is a type of CSR innovation defined by Auld et al. (2008a, 2008b). It differs from all of the above CSR tools in that it requires mandatory and enduring behavior (so-called "hard law"), but is not enforced by the state. Further, it differs from traditional hard law by continually adapting over time with the input of various stakeholders. NSMD governance also differs from other CSR innovations in that it is rooted in the supply chain. NSMD governance requires a firm to understand and manage the upstream implications of its activities, but also requires a firm to look downstream to understand that it is ultimately the market that drives NSMD schemes to be adopted. Finally, NSMD governance, unlike most CSR efforts, requires third-party verification. This imposes a significant cost to firms, but also ensures that NSMD schemes remain something akin to hard law.

The most obvious example of NSMD is certification. Certification schemes exist in several industries (seafood, for example, has seen recent growth in a variety of schemes). But forestry has led the way in the development of certification systems. Perhaps the most widely known scheme is the Forest Stewardship Council (FSC). Borne out of dialogue around sustainability at the Rio Environmental Summit in 1992, FSC is an independent body jointly controlled by an environmental committee and an industry committee (see Auld et al. [2008b] for a more thorough discussion of forest certification schemes). Other prominent schemes are the Sustainable Forestry Initiative (SFI, an industry-backed scheme in North America) and the Programme for the Endorsement of Forest Certification (PEFC, dominant in Europe with a significant Canadian presence and composed of a central body that allows for country-level definitions of certification).

Whether a firm chooses to participate in a forest certification scheme relates to the character of the firm (publicly or privately owned), the product they are manufacturing (magazine paper, dimensional lumber, pulp, etc.), trade dependence (major exporter or domestically oriented), ENGO pressure, supply chain procurement policies, and government support. Small operations may find compliance too costly and not worth the effort (Auld et al., 2008b). Large forestry firms, in contrast, are more vulnerable to pressure from advocates and can better bear the costs of compliance. In fact, in many instances, certification among larger supply chain actors is fast becoming the *de facto* mode of practice in the marketplace. Other firms, such as those that sell intermediary products, must comply with more than one scheme in order to meet various customers' demands. For example, a paper company might have to be SFI-certified to sell to an American client, FSC-certified to sell to a Canadian client, and PEFC-certified to sell to a European client.

6.5 CASE STUDIES IN SUSTAINABILITY

As a means of exploring how different sustainability drivers can lead to different sustainability responses, we present two case studies here. To enhance our understanding of sustainability, we take the academic and technical concepts discussed in this paper and demonstrate how they unfold in two real-world forest products scenarios: the Catalyst Paper supply chain, and the environmental implications of paper and digital media consumption.

6.5.1 CATALYST PAPER SUPPLY CHAIN

The work presented here stems from an investigation of the supply chain footprint of a major monthly American magazine that chose to remain anonymous. We worked with an engineer from Catalyst Paper to measure the carbon footprint of the magazine. But, over time, we found that context was an equally important part of the sustainability story. Catalyst Paper was able to develop a product that was uniquely carbon efficient (the phrase used is "manufactured carbon neutral"), but this had much to do with how Catalyst Paper engaged with its supply chain partners and with ENGOs.*

---

* For a more detailed overview of this case, see Bull et al. (2011).

6.5.1.1 Drivers

At first, Catalyst Paper was driven by eco-efficiency. By using less power to produce their paper through capital upgrades and using wood waste to generate power, they were able to gain an economic advantage. Catalyst Paper, a major energy user in British Columbia, is required to report its greenhouse gases and pay a carbon tax on fossil fuels used. Government regulation, therefore, played a role in the development of a more sustainable product.

The CEO of Catalyst Paper was meeting with a representative of WWF Canada when the idea emerged that Catalyst Paper's eco-efficiency strategy was carbon efficient, as well. Catalyst Paper saw an opportunity to market its product as manufactured carbon neutral, naming it "Catalyst Cooled™." This meant that the papermaking process would be powered entirely by hydroelectric power or energy from biomass. Parts of the process still unavoidably relied on fossil fuels and Catalyst Paper purchased carbon offsets accordingly. In short, Catalyst Paper had the unique opportunity to create a "carbon light" product by addressing the conspicuous impacts of the paper manufacturing process on the life cycle footprint of paper (see Table 6.1). Underlying all of these drivers was the impetus of climate change; Catalyst Paper would not have had to pay a carbon tax, report on their emissions, or explore a market opportunity in a carbon light product, were it not for the prominence of climate change.

6.5.1.2 Responses

This case study demonstrates several of the responses to sustainability that have been discussed. Catalyst Paper began with an individual firm effort toward eco-efficiency. In collaborating with WWF Canada, Catalyst Paper was able to add credibility and legitimacy to its environmental claims. The anonymous magazine, while conducting due diligence, contacted WWF Canada directly, helping Catalyst Paper secure the magazine as a customer. Catalyst Paper implemented information-sharing and

---

TABLE 6.1

Carbon Emissions for the Traditional Catalyst Paper Supply Chain

Activity	Carbon Emissions (CO_2E/ADt)[a] (kg)	Percentage of Total
Harvesting, road building, felling, and transport to sawmills	55	12
Sawmilling into dimensional and residual products	45	10
Transport of chips to mill	8	2
Paper manufacturing process	185	41
Transportation to print facility	127	28
Printing process	36	8
Total	456	100

[a] Carbon dioxide (greenhouse gas) equivalent per air-dried ton (ADT).

NSMD approaches; it submits an annual report to the CDP, and certifies its products with both SFI and FSC. All of this provides solid evidence of Catalyst Paper integrating environmental issues into its core business strategy.

The adoption of CSR innovations by Catalyst Paper also impacted its supply chain partners. Catalyst Paper uses the Washington Marine Group (WMG) to ship its products by barge in British Columbia. Before Catalyst Paper contacted them about measuring their supply chain footprint, WMG had never considered measuring their carbon emissions. However, in collaborating with Catalyst Paper, WMG began to adopt similar CSR efforts, even partnering with WWF Canada to help improve its environmental footprint. Similarly, Western Forest Products (WFP), a firm that provides pulp fiber to Catalyst Paper, had never measured carbon. But, it learned from Catalyst Paper's example and undertook similar efforts to measure its carbon footprint.

6.5.1.3 Implications for Sustainability

At first, the anonymous magazine publisher hoped that reducing the carbon footprint of its product would lead to public praise from the press and environmental advocates. Instead, the day after the first issue printed on Catalyst Cooled™ was released, a scathing op-ed in the New York Times criticized the magazine for not using recycled content.*

The decision to use recycled paper or carbon light paper is not a straightforward one. Recycled paper can be more difficult to print on, and some print runs are compromised as a result. Furthermore, according to Catalyst Paper, recycled fiber is in scarce supply; the constraint on recycled fiber is availability, not demand. And some recycling plants, such as one owned by Catalyst Paper in Arizona, are powered by coal. These plants emit more carbon dioxide per air-dried ton than plants producing paper from virgin fiber. This suggests that, like politics, many sustainability issues may be most germane at the local level.

6.5.2 Environmental Implications of Media

We have recently undertaken research assessing the relative environmental footprints of digital and paper media. Trying to compare two very different products, with different attributes, functions, and supply chains has proven to be a challenge. Members of the forest products industry catalyzed this initiative, as they felt that paper was being misrepresented when described as less "green" than digital alternatives. While the Catalyst Paper case study spoke to the drivers and responses of sustainability, the discussion here focuses on the challenges of measuring and defining sustainability. Before we discuss the specific results, we quickly review the difficulty of using LCAs as a tool for studying environmental sustainability.

6.5.2.1 Problems with LCAs

LCA is the predominant tool for studying environmental impacts over the life cycles of products. LCA's popularity can be attributed to several factors: it is

---

* This negative attention contributed to the magazine's desire to remain anonymous.

standardized (the International Organization for Standardization (ISO) and the Society of Environmental Toxicology and Chemistry (SETAC) collaborate to define and improve best practices in the conduct of LCAs); there are several competing data-bases—so-called LCI databases—that allow analysts to conduct LCAs without having to gather much in the way of data (although, in some instances, product-specific data is gathered); and LCAs provide widely accepted, quantified descriptions of a life cycle and enjoy a good reputation among a wide constituency (Gaudreault et al., 2007; NRTEE, 2012). An LCA that is based off of high-resolution primary data can make great strides in improving the life cycle of a particular product or process, and this is where the greatest opportunities lie for practitioners in the forest products sector. Figure 6.5 describes the general framework and scope of LCA processes.

LCAs are not without problems though. An extensive review of LCAs in the forest products sector (Gaudreault et al., 2007) found that LCAs are sensitive to subjective choices in the comparative assessments of products, and that a much better use is in trying to assess trade-offs and opportunities for improvement within the life cycle of an individual product or process. Furthermore, they found that many studies did not meet ISO standards for the conduct of an LCA, even though they were published in peer-reviewed journals. This reveals a gap in credibility. Forest products LCAs also suffer from a lack of explicit methodologies related to land use, a lack of clarity on how to account for carbon embodied in a forest product, and how to model the impacts of a forest product at its end-of-life.

Reap et al. (2008) suggest that problems in the conduct of LCAs are, by no means, limited to forestry. They developed a framework (Table 6.2) that describes the most pressing problems at each stage of conducting an LCA. These include subjectivity in the definition of a functional unit, the selection of study boundaries, and the

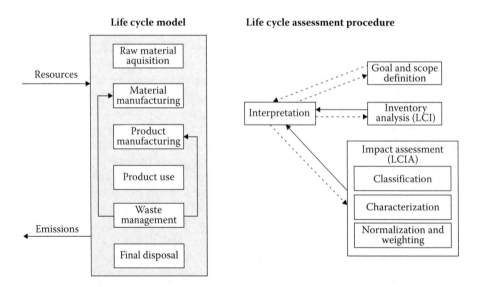

FIGURE 6.5 LCA framework and scope. (Adapted from Gaudreault, C. et al., *TAPPI J.*, 6(8), 3–10, 2007.)

TABLE 6.2
Life Cycle Assessment Problems by Phase

Phase	Problem
Goal and scope definition	**Functional unit definition**
	Boundary selection
	Social and economic impacts
	Alternative scenario considerations
Life cycle inventory analysis	Allocation
	Negligible contribution ("cutoff") criteria
	Local technical uniqueness
Life cycle impact assessment	Impact category and methodology selection
	Spatial variation
	Local environmental uniqueness
	Dynamics of the environment
	Time horizons
Life cycle interpretation	Weighting and valuation
	Uncertainty in the decision process
All	**Data availability and quality**

Source: Reap, J. et al., *Int. J. Life Cycle Assess.*, 13(5), 374–388, 2008. With permission.
Note: Bolded problems discussed in this report are identified by Reap et al. (2008).

allocation of impacts to processes. Furthermore, they see the reliance of most LCAs on third party, privately owned LCI databases as a problem; the data are generally not peer reviewed and original data sources can be obfuscated. LCAs that rely on aggregate or generalized data often overlook the influence of local variables, such as climate and geography, on the life cycles of products. As has been discussed, sustainability can be a highly localized phenomenon, and this is a critical weakness in the conduct of an LCA.

Many of the problems identified by Reap et al. (2008) manifest when we consider LCAs that compare digital and paper media (see Section 6.5.2.2). Functional unit definition, system boundaries, aggregate and unreviewed data, and a lack of consideration for localized impacts have produced comparative LCAs of questionable value.

6.5.2.2 Comparing Digital and Paper Media

Consider this: why are paper products, which can be produced from fully renewable raw materials and be fully recycled at their end of life, considered the less "green" option compared to digital media? It seems odd that something renewable and recyclable is made a pariah in a debate about sustainability. We wonder whether this is due to the tools used in measuring sustainability, a failure in communication on the part of the forest sector, or some combination of these and other factors.

We reviewed a series of seven LCAs (Table 6.3) that compared the footprint of digital and paper media. Our review was structured in accordance with the (bolded) problems identified by Reap et al. (2008) in Table 6.2. These studies generally found

TABLE 6.3

Summary of Drivers of/Responses to Sustainability Issues with Illustrative Examples

Examples	Drivers								Responses				
	Investors	Governments	Eco-efficiency	Marketplace	Advocates	Climate Change	Individual Firm Efforts	Firm–ENGO Collaborations	Public–Private Partnerships	Information-Based Approaches	Environmental Management Systems	Industry Association Codes of Conduct	Nonstate Market-Driven Governance
Kalundborg, Denmark—industrial symbiosis			X				X				X		
GHG Protocol Scope 3 Emissions[a]			X		X	X	X	X	X	X	X		
U.S. Climate Action Partnership					X	X	X	X					
EDF and Walmart Collaboration			X	X	X	X	X	X					
Coca-Cola and WWF Collaboration			X	X	X			X					
Global Reporting Initiative	X	X			X		X	X	X	X		X	
Carbon Disclosure Project	X	X			X		X			X			
Conflict Mineral Tracking		X											
Nestle Palm Oil Policy[a]			X	X	X		X	X	X	X		X	
33/50 Program		X	X		X		X				X		
Certified seafood				X	X		X					X	X
Forest sector case studies													
Canadian Boreal Forest Agreement		X		X	X	X		X	X			X	
Great Bear Rainforest Agreement		X		X	X	X		X	X			X	
Forest certification schemes				X	X		X	X		X	X		X
Home Depot—FSC purchasing policy				X	X		X			X	X		X
SFI—LEED Lawsuit[a]				X	X								X
Catalyst–WWF Collaboration				X	X		X	X		X	X	X	
Lacey Act legislation in the United States		X			X		X		X	X	X		

Note: GHG, greenhouse gas; EDF, Environmental Defense Fund; WWF, World Wildlife Fund; FSC, Forest Stewardship Council; SFI, Sustainable Forestry Initiative; LEED, Leadership in Energy and Environmental Design.

a Not explicitly mentioned in this work.

digital media to be the environmentally preferable option, and the authors expressed as much with a high degree of confidence. We found several reasons why this confidence is unwarranted. First and foremost, the footprint of manufacturing ICT products is poorly understood. Because ICT products evolve at such a rapid pace (in line with Moore's Law, expressed by the founder of Intel as the ability for circuit density to double every 18–24 months), the data used to calculate the footprint of ICT products in these LCAs were outdated and incomplete (e.g., see Deetman and Odegard [2009], Moberg et al. [2010], and Toffel and Horvath [2004]). Newer, smaller devices require increasingly pure chemicals to manufacture; a chip made in the 1990s could be produced using chemicals with impurities tolerances in the parts-per-million. Today, with Intel developing 32 nanometer chips, tolerances are in the parts-per-billion. The energy implications of these extremely pure chemicals, however, are not well understood (Plepys, 2004).

We also found that some studies (see Deetman and Odegard [2009], Enroth [2009], Gard and Keoleian [2002], Hischier and Reichart [2003], Kozak [2003], Moberg et al. [2010], and Toffel and Horvath [2004]) relied on tenuous assumptions. One study (Kozak, 2003) found that the biggest contributor of CO_2 in the life cycle of a textbook was driving a vehicle to the bookstore to make the purchase. The author neglected to consider that the textbooks being modeled were for college students at the University of Michigan. Most students at the University of Michigan live on or near campus, and several bookstores (likely vendors of college textbooks) are within walking distance. Changing this one assumption—not an unreasonable course of action—would render printed textbooks in this LCA to be the environmentally preferable option. Another study (Hischier and Reichart, 2003) recognized that using the LCA to compare such vastly different products—paper and digital media—required functional units that do not necessarily represent actual user behavior. Moberg et al. (2010) did not model the footprint of the Internet in delivering content over the Internet, because the data were simply not there. Deetman and Odegard (2009) modeled the footprint of an e-Reader, but would not share data on the physical characteristics of the device in question, due to a nondisclosure agreement. Toffel and Horvath (2004) looked at the relative footprints of having the New York Times delivered as a physical copy or to a personal data assistant (PDA). However, due to data availability issues, they relied on proxy data or estimates to describe most of the footprint of the PDA and, on some occasions, omitted measuring some components due to no data being available at all. In short, each LCA had at least one assumption or data gap that could strongly sway the study results.

User behavior also plays an important role in the footprint of digital and paper media. If a digital product is properly disposed of at an e-waste facility (as is legally required in Germany and British Columbia), then its end-of-life footprint is diminished. Unfortunately, though, most jurisdictions have no such e-waste regulations, with the majority of ICT products ending up in Africa, China, and India where they are dismantled with considerable effects on local health and the environment. Whether a newspaper is delivered to a rural home, one of only few on a delivery route, or picked up from a vendor in a busy urban area would also influence the footprint of a product. Again, we note that sustainability issues seem to be most critical at the local level.

There are some hard truths, however, for the forest products industry to contend with. Regardless of the relative environmental footprints, consumers will undoubtedly continue to adopt digital media; this is where innovation is taking place in media delivery and younger consumers (and more recently, older consumers) are embracing these new platforms. ICT devices also have massive functional redundancy. With the exception of e-readers (such as the Amazon Kindle™), ICT devices perform a huge variety of functions, depending only on software availability. This allows the footprint of an ICT device to be shared between several functions, diminishing the allocation to the paper-based function being replaced.

There is, however, a bright side for the forest products sector. When we look to industrial ecology's definition of sustainable systems, we see that forestry is very much a "strong" Type II system. The key input to paper products (wood fiber) is renewable, and the outputs of forestry are generally recyclable. ICT products can make no such claim. Unfortunately, LCAs have been predominant in determining the relative footprint of the two media types. Studies that have relied on incomplete or missing data and questionable assumptions have been prevalent, skewing discourse about the merits of paper and ICT products. Opportunities exist to use tools of CSR, industrial ecology thinking, and more effective communications with the public to better articulate and market the environmental strengths of the forest products sector.

6.6 IMPLICATIONS FOR FOREST PRODUCTS VALUE CHAINS IN CANADA

What does this this overview of sustainability mean for forest products value chains in Canada? We should begin by recognizing the unique position of the forest products sector in Canada. Beyond having a uniquely green product, Canadian forestry also exists in a country with strong regulatory regimes and a national industry that has been a world leader in the development and implementation of CSR tools. Strong regulation makes the cost of compliance with CSR innovations smaller. And a culture of innovation ensures the capacity to integrate environmental issues into business decisions. Finally, the Canadian forest products industry is fairly advanced in its thinking about how to strengthen its reputation as an environmental leader, and seize opportunities for participating in a new green economy (the focus of the Forest Product Association of Canada's Biopathways project).

However, it is not all good news. Certain forest product segments, such as newsprint, are rapidly shrinking, unlikely to return to previous levels. The operating contexts of many foreign competitors are vastly different than in Canada, with longer growing seasons, shorter rotations, cheaper labor rates, and less stringent government oversight. In other words, it is becoming increasingly difficult for commodity producers in Canada to compete in a globalized world and it is not unreasonable to assume that our share of the forest products market is shrinking. Given the growing importance of environmental issues and the urgent need to inhabit this planet in a more responsible manner, we believe that firms and industries that survive this transition will likely be those that fully integrate sustainability and CSR into their value chains. By doing so, the continued economic success of the Canadian forest products industry can be more safely assured, while offering a world-class example of environmental sustainability and responsible stewardship.

There is a need for research projects in the forest products value chain domain to identify which definition(s) of sustainability and which CSR mechanisms are most applicable. This will result in a body of work that not only has the potential to enhance economic competitiveness, but also serves to inform the systems, data, decision support tools, and knowledge needed for the forest products industry to excel at environmental sustainability.

REFERENCES

Ambec, S. and Lanoie, P. (2008). Does it pay to be green? A systematic overview. *Academy of Management Perspectives*, 22(4), 45–62.

Auld, G., Bernstein, S., and Cashore, B. (2008a). The new corporate social responsibility. *Annual Review of Environment and Resources*, 33(1), 413–435.

Auld, G., Gulbrandsen, L., and McDermott, C. (2008b). Certification schemes and the impacts on forests and forestry. *Annual Review of Environment and Resources*, 33(1), 187–211.

Brown, H.S., de Jong, M., and Lessidrenska, T. (2009). The rise of the global reporting initiative: A case of institutional entrepreneurship. *Environmental Politics*, 18(2), 82–200.

Bull, J., Kissack, G., Elliott, C., Kozak, R., and Bull, G. (2011). Carbon's potential to reshape supply chains in paper and print: A case study. *Journal of Forest Products Business Research*, 8(2), 1–8.

Carbon Disclosure Project. (2010). Carbon Disclosure Project 2010 Global 500 Report. Available at: https://www.cdproject.net/CDPResults/CDP-2010-G500.pdf (Accessed April 21, 2011).

Costanza, R. and Mageau, M. (1999). What is a healthy ecosystem? *Aquatic Ecology*, 33(1), 105–115.

Deetman, S. and Odegard, I. (2009). *Scanning Life Cycle Assessment of Printed and E-paper Documents Based on the iRex Digital Reader*. Leiden, ZH: Institute of Environmental Sciences, Leiden University.

EDF. (n.d.). Why We Work with Walmart. Environmental Defense Fund. Available at: http://business.edf.org/projects/walmart (Accessed May 25, 2011).

Ehrenfeld, J. and Gertler, N. (1997). Industrial ecology in practice: The evolution of interdependence at Kalundborg. *Journal of Industrial Ecology*, 1(1), 67–79.

Engelhaupt, E. (2008). Do food miles matter? *Environmental Science & Technology*, 42(10), 3482–3906.

Enroth, M. (2009). Environmental impact of printed and electronic teaching aids, a screening study focusing on fossil carbon dioxide emissions. *Advances in Printing and Media Technology*, 36, 1–9.

EPA. (1994). EPA Toxics Release Inventory Public Data Release Appendix A: Questions and Answers XIII 33/50 Program. Available at: http://www.mapcruzin.com/scruztri/docs/qa_3350.htm (Accessed April 21, 2011).

Epstein-Reeves, J. (2010). The Do Well Do Good Public Opinion Survey on Corporate Social Responsibility. Available at: http://dowelldogood.net/wp-content/uploads/2011/03/DWDG_CSR_Final.pdf (Accessed April 21, 2011).

Erkman, S. (1997). Industrial ecology: An historical view. *Journal of Cleaner Production*, 5(1), 1–10.

Europa. (2005). Questions & Answers on Emissions Trading and National Allocation Plans. Available at: http://europa.eu/rapid/pressReleasesAction.do?reference=MEMO/05/84&format=HTML&aged=1&language=EN&guiLanguage=en (Accessed May 25, 2011).

FII. (2013). Forestry Innovation Investment, Naturally: Wood. Environmental Product Declarations. Available at: http://www.naturallywood.com/environmental-product-declarations (Accessed June 12, 2013).

Forest Ethics. (n.d.). Landmark Great Bear Rainforest Agreement—A World-Leading Model. Forest Ethics. Available at: http://forestethics.org/great-bear-rainforest-successes (Accessed May 25, 2011).

FPAC. (2010). Canadian Forest Industry and Environmental Groups Sign World's Largest Conservation Agreement. Forest Products Association of Canada. Available at: http://www.fpac.ca/index.php/en/press-releases-full/canadian-forest-industry-and-environmental-groups-sign-worlds-largest-/ (Accessed September 21, 2010).

Friedman, M. (1970). The Social Responsibility of Business Is to Increase Its Profits. The New York Times Magazine. Available at: http://www.colorado.edu/studentgroups/libertarians/issues/friedman-soc-resp-business.html (Accessed May 25, 2011).

Gard, D. and Keoleian, G.A. (2002). Digital versus print: Energy performance in the selection and use of scholarly journals. *Journal of Industrial Ecology*, 6(2), 115–132.

Garnaut. (2008). Garnaut Climate Change Review. Available at: http://www.garnautreview.org.au/ (Accessed May 25, 2011).

Gaudreault, C., Samson, R., and Stuart, P. (2007). Life-cycle thinking in the pulp and paper industry: LCA studies and opportunities for development. *TAPPI Journal*, 6(8), 3–10.

German, K. (2009). Motorola Renew W233. Available at: http://reviews.cnet.com/cell-phones/motorola-renew-w233-t/4505-6454_7-33485041.html (Accessed May 25, 2011).

Graedel, T. and Allenby, B. (1995). *Industrial Ecology*. Englewood Cliffs, NJ: Prentice Hall.

Hischier, R. and Reichart, I. (2003). Multifunctional electronic media—Traditional media: The problem of an adequate functional unit. *International Journal of Life Cycle Assessments*, 8(4), 201–208.

Home Depot. (2011). Wood Purchasing Policy. Available at: http://corporate.homedepot.com/wps/portal/Wood_Purchasing (Accessed May 25, 2011).

Houpt, S. (2011). Beyond the Bottle: Coke Trumpets Its Green Initiatives. Available at: http://www.theglobeandmail.com/report-on-business/industry-news/marketing/adhocracy/beyond-the-bottle-coke-trumpets-its-green-initiatives/article1869437/ (Accessed April 21, 2011).

IPCC. (2007). IPCC Fourth Assessment Report. International Panel on Climate Change. Available at: http://www.ipcc.ch/publications_and_data/publications_and_data_reports.shtml (Accessed May 25, 2011).

ISO. (2006). ISO 14025. Environmental Labels and Declarations—Type III—Environmental Declarations—Principles and Procedures. Available at: http://www.iso.org/iso/catalogue_detail?csnumber=38131 (Accessed June 12, 2013).

Jacob, K., Beise, M., Blazejczak, J., Edler, D., Haum, R., Janicke, M., Löw, T., Petschow, U., and Rennings, K. (2005). *Lead Markets for Environmental Innovations*. New York, NY: Physica-Verlag HD.

Kozak, G. (2003). *Printed Scholarly Books and E-book Reading Devices: A Comparative Life Cycle Assessment of Two Book Options*. Ann Arbor, MI: Center for Sustainable Systems, University of Michigan.

Krones, J. (2007). The best of both worlds: A beginner's guide to industrial ecology. *MIT Undergraduate Research Journal*. 15(Spring Issue), 19–22.

Manget, J., Roche, C., and Munnich, F. (2009). Capturing the Green Advantage for Consumer Companies. Boston Consulting Group. Available at: http://www.bcg.com/documents/file15407.pdf (Accessed September 20, 2010).

Mebratu, D. (1998) Sustainability and sustainable development: Historical and conceptual review. *Environmental Impact Assessment Review*, 18(6), 493–520.

Moberg, Å., Johansoon, M., Finnveden, G., and Jonsson, A. (2010). Printed and tablet e-paper newspaper from an environmental perspective—A screening life cycle assessment. *Environmental Impact Assessment Review*, 30(3), 177–191.

Molina-Azorín, J.F., Claver-Cortés, E., López-Gamero, M.D., and Tari, J.J. (2009). Green management and financial performance: A literature review. *Management Decision*, 47(7), 1080–1100.

Nogueron, R. and Hanson, C. (2010). Risk Free? Paper and the Lacey Act. Available at: http://www.wri.org/stories/2010/11/risk-free-paper-and-lacey-act (Accessed April 21, 2011).

NRTEE. (2012). National Round Table on Environment and the Economy. Canada's Opportunity: Adopting Life Cycle Approaches for Sustainable Development. Ottawa, ON: National Round Table on the Environment and the Economy.

OpenCongress. (2010). H.R.4128—Conflict Minerals Trade Act. OpenCongress. Available at: http://www.opencongress.org/bill/111-h4128/show (Accessed September 21, 2010).

Ottman, J. (2011). *The new rules of green marketing: Strategies, tools and inspiration for sustainable branding*. San Francisco, CA: Berret-Koehler.

Plepys, A. (2004). The environmental impacts of electronics. Going beyond the walls of semiconductor fabs. In *Proceedings of the IEEE International Symposium on Electronics and the Environment*. Piscataway, NJ: IEEE, 159–165.

Reap, J., Roman, F., Duncan, S., and Bras, B. (2008). A survey of unresolved problems in life cycle assessment. *The International Journal of Life Cycle Assessment*, 13(5), 374–388.

Regional Greenhouse Gas Initiative (RGGI). (2009). Regional Greenhouse Gas Initiative: Home. Available at: http://www.rggi.org/home (Accessed May 25, 2011).

Schumpeter, J. (1942). *Capitalism, Socialism and Democracy*. New York, NY: Harper.

Seager, T.P. (2008). The sustainability spectrum and the sciences of sustainability. *Business Strategy and the Environment*, 17(7), 444–453.

SwissRe. (2007). The Economic Justification for Implementing Restraints on Carbon Emissions. Available at: http://www.jth.ch/thinkBank/inlinemedia/displaymedia.php?id=39 (Accessed May 25, 2011).

Toffel, M. and Horvath, A. (2004). Environmental implications of wireless technologies: News delivery and business meetings. *Environmental Science & Technology*, 38(11), 2961–2970.

United Nations. (1987). Report of the World Commission on Environment and Development: Our Common Future. [e-book]. Available at: http://www.un-documents.net/wced-ocf.htm (Accessed May 25, 2011).

United Nations. (2009). Nobel-winning economist says UN best suited to oversee global economic overhaul. UN News Service. Available at: http://www.un.org/apps/news/story.asp?NewsID=31830&Cr=financial+crisis&Cr1 (Accessed May 25, 2011).

United Nations Environmental Programme (UNEP) Finance Initiative. (2006). Show Me The Money: Linking Environmental, Social and Governance Issues to Company Value. Available at: http://www.unepfi.org/fileadmin/documents/show_me_the_money.pdf (Accessed April 21, 2011).

United States Department of Agriculture (USDA). (2011). Lacey Act. Available at: http://www.aphis.usda.gov/plant_health/lacey_act/index.shtml (Accessed May 25, 2011).

van Marrewijk, M. (2003). Concepts and definitions of CSR and corporate sustainability: Between agency and communion. *Journal of Business Ethics*, 44(2/3), 95.

Walmart. (2011). Supplier Sustainability Assessment. Available at: http://walmartstores.com/download/4055.pdf (Accessed May 25, 2011).

Western Climate Initiative. (2009). Western Climate Initiative: Home. Available at: http://www.westernclimateinitiative.org/ (Accessed May 25, 2011).

WWF. (n.d.). Climate Savers—Catalyst. Available at: http://wwf.panda.org/what_we_do/how_we_work/businesses/climate/climate_savers/partner_companies/ (Accessed May 25, 2011).

7 Overview of Wood Product Markets, Distribution, and Market Research in North America
Implications for Value Chain Optimization

Christopher Gaston and François Robichaud

CONTENTS

7.1 INTRODUCTION

It could be argued that no textbook on value chain optimization (VCO) should be compiled without a chapter dedicated to the knowledge of forest product markets and marketing. Indeed, the very word "value" suggests value to whom? Innovations in research and development of products and the value chains that produce them constantly ask the question of what do markets want, both today and in the future.

When one thinks about this, it makes complete sense to have market pull, instead of product push, as a central focus of any research activities. In reality, this is not always the case. Researchers designing their programs often focus on technologies with little consideration of the value creations for end users. Product manufacturers, including wood products producers, have placed a greater emphasis on technology and process improvements, throughput augmentations, and cost reductions rather than value creations.

A good example of a research institution that has evolved away from the product and technology push toward the market pull philosophy is FPInnovations, formerly known as Forintek Canada Corp. In 1996, with the urging wood products division of its Board of Directors, Forintek Canada Corp.* launched their Markets and Economics Group. The objective of this new research group was to bring the market emphasis into the research and development planning process and project executions in Forintek. This initiative has allowed Forintek to conduct its research focusing on the needs of its customers as well as its customers' customers, particularly in the area of new product and solutions development.

Although there were wood-use statistics for the most important Canadian and US markets, including the softwood lumber, structural panels, and other wood products, very little knowledge was known by the researchers at Forintek or industry on where these products were going. Commodity lumber and panel sales were primarily to wholesalers; from there the path was lost as to whether it went to a home builder, a home center, or an industrial highway project. This is important knowledge for researchers, as the characteristics and attributes required by the end users for Canadian wood products can vary considerably.

Figure 7.1 shows a graphical summary of the breakdown of different wood products and their market channels. In the following sections, a literature review on market research will be conducted to understand what has been done, what the market segmentations are, what product characteristics drive the customers' purchasing decisions, and how this market intelligence can support product and technology research and development. Information gathered included interviews conducted with professors, researchers, and practitioners from industry and governments in the field of market research in North America. This research, in addition to qualitative analysis, clearly indicates the impacts of market research on the Canadian forest VCO.

The results identified a number of "gaps" in the literature and existing/planned efforts, including the following:

- A lack of focus on end-use markets:
  - Residential construction
  - Repair and renovation
  - Nonresidential construction
  - Industrial

---

* Forintek Canada Corp. is today a division of FPInnovations, representing the wood products sector. The other major divisions of FPInnovations include pulp and paper/bio-products (the former Pulp and Paper Research Institute of Canada, or PAPRICAN), forest operations (the former Forest Engineering Research Institute of Canada, or FERIC), and the Canadian Fibre Centre (housed within Natural Resources Canada).

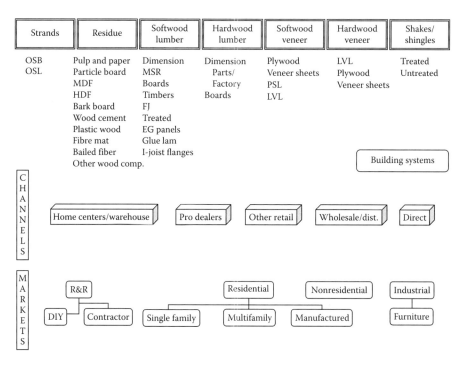

FIGURE 7.1   Wood product flows in the United States. (Reprinted from Gaston, C.W. and Fell, D., Forintek Canada Corp. Research Report, Prepared for the Canadian Forest Service, 2000.)

- A lack of focus on demand regions
- Market research has a product focus back to production rather than a product focus to the market
- Little documentation defining and describing specifiers and end uses
- Little information on marketing strategies for wood products
- Few sources of consumption and market shares for specific products
- Information on engineered wood products is limited due to the recent introduction and the proprietary nature many of the products
- Information on value-added wood products focuses on higher level development issues of the sector rather than on specific products or markets (Gaston and Fell, 2000; Gaston et al., 1999)

This chapter is focused on the Canadian wood products industry, specifically lumber, wood panels, and engineered wood products, their market research, distribution channels, the strategic changes experienced in the distribution systems and market, and the implications on wood products value chain management. The chapter is divided into six sections. In Section 7.2, we provide an overview of the current wood product market research with up-to-date product demand statistics. Then, the wood product distribution channels are described in Section 7.3 with related mega-trends in wood product end uses, focusing on North American markets and how the field has changed over the past 10–15 years and what the future trends are. The recent structural changes in the supply and demand experienced by the Canadian wood products industry, including the significantly lower housing starts in US markets after 2005/2006, the new trends of home construction and renovation markets, the emergence of new global markets, and the mountain pine beetle infestation in British Columbia, are discussed in the Section 7.4. The need for academic research in market intelligence, modeling abilities, and the policy context of wood products VCO are highlighted in Section 7.5, and finally, the concluding remarks are in Section 7.6.

7.2 AN OVERVIEW OF THE CURRENT CANADIAN WOOD PRODUCT MARKETS

Canada is a net exporter of wood products. Canadian wood products are exported predominantly to US markets with a total export value in 2005 of $25.2 billion dollars (prior to the US great recession). Among the significant export value created, commodity lumber and panels account for 70%, engineered products, including glulam and prefabricated building components, account for 3%, industrial products, such as packaging and pellets, account for 7%, and value-added products, also known as "living with wood," including furniture, doors, windows, flooring, and mill work, account for 20%. A majority of the products are made of softwood species with value-added products being made dominantly by hardwoods.

Table 7.1 shows the comparison of the dominant commodity products exported to US markets by market categories and by volumes in 2005 and 2011, before and after the recession. It indicates that the majority of Canadian commodity wood products exports are used in either housing construction or in repair and renovation. The "living with wood" export products are also largely associated with housing. This end

TABLE 7.1
US Imports of Selected Wood Products

	2005		2011	
End Use	**Lumber (bbf)**	**Structural Panels (bsf, 3/8″ basis)**	**Lumber (bbf)**	**Structural Panels (bsf, 3/8″ basis)**
New residential construction	30	24	8	7
Repair and renovation	18	8	12	6
Nonresidential construction	2	2	1	1
Industrial	13	8	12	8
Total	63	42	33	22

Source: Resource Information Systems, Inc. http://www.risiinfo.com, 2006, 2012, and 2014.

use has clearly been the most important driver for the Canadian wood product value chains.

Figure 7.2 demonstrates the historical US housing starts in the single and multi-family housing markets. The significant falling market during 2007–2013 indicates the falling demand for Canadian lumber and panel products exports to US markets. Despite the US housing market collapse, it is interesting to see that multifamily construction is increasing significantly. This has major implications on the types of wood products and systems being demanded, affecting overall wood VCO.

Figures 7.3 and 7.4 present the Canadian softwood lumber and structural panel shipments to the domestic and overseas markets over the last two decades, respectively. Significant volumes have been shipped to US markets, indicating the historical importance of US markets to Canadian softwood lumber and structural panel products. Note that while shipments to US markets have dominated historically, since 2007, the shipments to these markets have declined, while domestic and offshore shipments have increased. Today, more lumber is shipped to China and other countries by volume than to US markets. Although it is not presented here, it is worth mentioning that the export shipments to the US from Canada vary considerably by provinces. British Columbia, for example, has historically exported over 75% of its lumber products to US markets. Ontario and Quebec have more domestic consumptions than exports.

From the US market perspective, domestic production and supplies have satisfied a majority of lumber demand. Only approximately 37% of lumber was imported, dominantly from Canada (Figure 7.5). After the housing market collapse, softwood lumber imports from Canada dropped to below 30%. It should be noted, however, that US softwood lumber imports from other countries have increased significantly

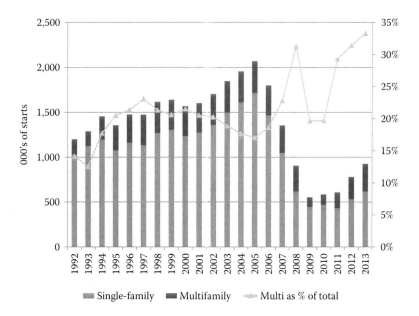

FIGURE 7.2   US housing starts. (https://www.census.gov/construction/nrc/index.html)

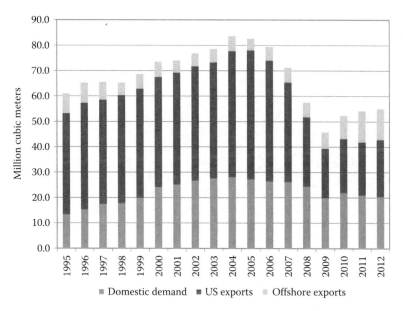

FIGURE 7.3 Canada softwood lumber shipments. (Resource Information Systems, Inc., http://www.risiinfo.com, 2006, 2012, and 2014.)

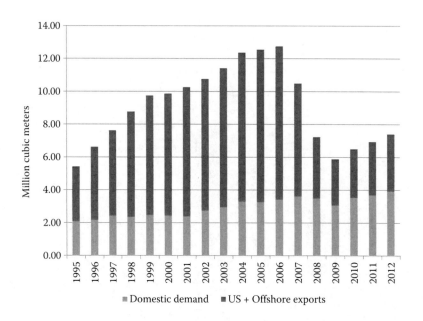

FIGURE 7.4 Canada structural panel shipments. (Resource Information Systems, Inc., http://www.risiinfo.com, 2006, 2012 and 2014.)

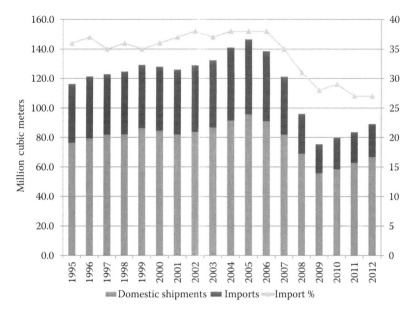

FIGURE 7.5   US softwood lumber production and imports. (Resource Information Systems, Inc., http://www.risiinfo.com, 2006, 2012, and 2014.)

by percentage, both for "living with wood" (from Chile) and for construction (from Europe, supplying imperial sizes).

It should also be noted that these figures are for softwood lumber only and dominated by the spruce–pine–fir (S–P–F) species group. Hardwood lumber imports are much smaller by volume, but nevertheless are important from value chain management and resource perspectives. In western Canada, hardwoods are dominated by aspen, poplar, and birch species, primarily used for pulp and oriented strand board (OSB) production. In eastern Canada, hardwoods include maple, birch, oak, cherry, and other species prized for their "secondary" end uses, including interior finishing and furniture. Figures 7.6 and 7.7 show the relative soft and hardwood species by growing stock in different provinces and by species, respectively. Figure 7.8 presents the relative Canadian softwood and hardwood lumber production.

Figure 7.9 indicates the historical Canadian exports by value with a breakdown of different products. Again, these values can vary considerably between provinces, with British Columbia, for example, exporting the highest amount of softwood lumber and OSB as compared to other provinces, and mainly to China and Japan.

Figures 7.10 and 7.11 show how China has emerged as an important offshore market for softwood lumber since the US recession. This has been a critical market not only because of the considerable drop in the US demand but also because of declines in the Japanese market. However, exports to China are at considerably lower unit values than those to Japan, European countries, and other countries, such as Korea, Taiwan, and the Middle East, as shown in Figure 7.12. These lower unit values are partially due to the high proportion of relatively lower grade S–P–F lumber exports (Figure 7.13).

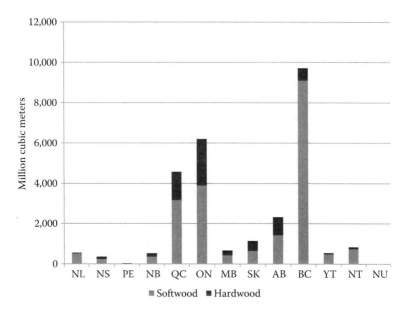

FIGURE 7.6   Canada's growing stocks of softwood and hardwood species by province (total 27.5 billion cubic meters). (The State of Canada's Forests, Annual Report.)

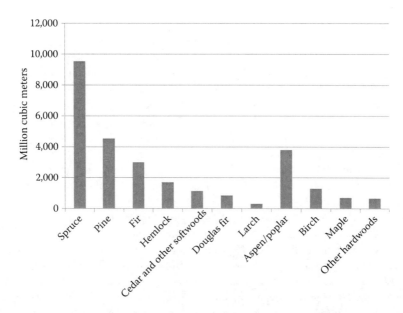

FIGURE 7.7   Canada's growing stocks by species (total 27.5 billion cubic meters). (Canadian Forest Service, 2006.)

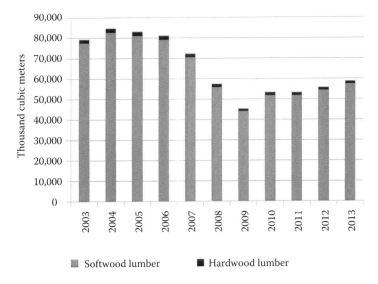

FIGURE 7.8 Canadian lumber production by species group. (Statistics Canada, http://www5.statcan.gc.ca/subject-sujet/index?lang=eng, 2014.)

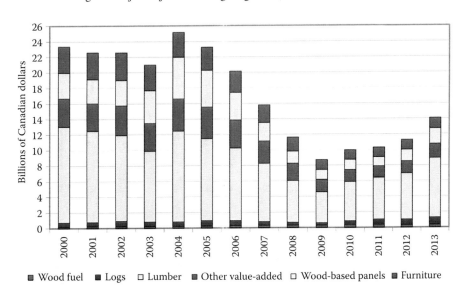

FIGURE 7.9 Canada exports of wood products. (https://www.gtis.com/gta/)

Finally, Figure 7.14 shows Canada exports of softwood logs, which have been growingly significantly since 2000 and for China since 2010.

In the United States, softwood lumber peaked at over 60 billion board feet prerecession, dropping by almost one-half in only 5 years. Similarly, the consumption of structural panels peaked at over 40 billion square feet (3/8″ basis), prerecession, dropping by almost one-half in only 5 years. The majority of the demand for lumber

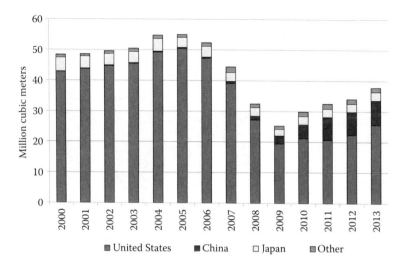

FIGURE 7.10 Canadian softwood lumber exports by destination. (Global Trade Information Services, Inc., 2014.)

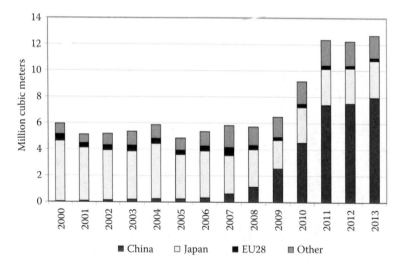

FIGURE 7.11 Canadian softwood lumber exports by non-US destination. (Global Trade Information Services, Inc., 2014.)

and structural panels is for housing, either in new construction or in repair and renovations. The latter reflects an increased share after the drop. Wood use in housing starts has been dominated by single family structures. However, the most recent data are showing a considerable rise for multifamily structures. Wood use in nonresidential construction has remained relatively low, despite the fact that the dollar value of construction is similar to residential. This shows how dominant steel and concrete are in this market, and, therefore, the opportunity for growth. For structural panels, plywood continues to lose market share to OSB. At a more micro level, this is region

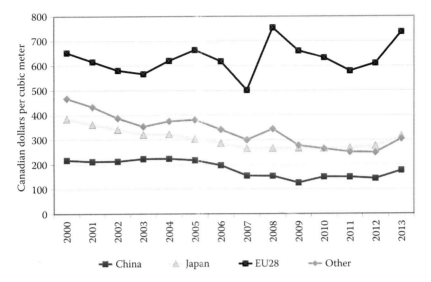

FIGURE 7.12 Canada exports (unit value) of softwood lumber by non-US destination. (Global Trade Information Services, Inc., 2014.)

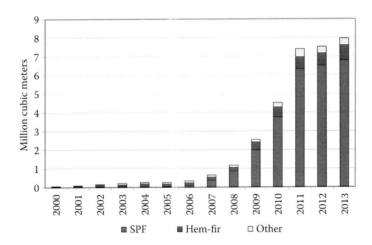

FIGURE 7.13 Canada exports of softwood lumber to China by species. (Global Trade Information Services, Inc., 2014.)

and end-use dependent. Industrial consumption of both lumber and panels has been and remains significant, even more so in recent years. There is a wide range of product grade demand here, from economy to specialty/proprietary.

Canada, by comparison, has shown modest drops in consumption in recent years compared to the United States, with a consistently higher proportion of lumber and structural panels going into repair and renovation compared to new starts. There also exists a much higher percentage of multifamily starts compared to single family in the United States.

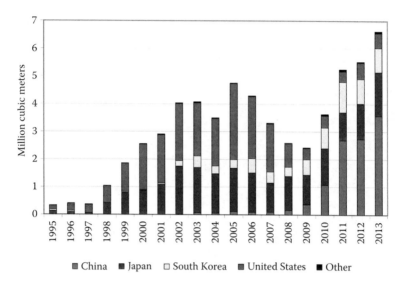

FIGURE 7.14 Canada exports of softwood logs. (Global Trade Information Services, Inc., 2014.)

To summarize the statistics presented thus far, of the $25 billion dollars of pre-US recession annual wood products exports from Canada almost half of this was commodity softwood lumber alone, mostly destined to the US markets for construction. Another roughly one-quarter was wood-based panels, again mostly for construction. The last quarter is split almost equally between higher value-adding "living with wood" products including interior finish and furniture and industrial products such as packaging, pallets, and concrete forms.

Having discussed the historical market demand and value creations of different Canadian wood products and their end uses, with each product involving a value chain of sourcing, producing, and shipping to the market to capture the market values, market research can have profound impacts on VCOs. VCO means not only to minimize the cost of production through operational and technical efficiencies but also to maximize the values by maximizing sales margins. This necessitates an investigation of the trade-offs of producing more for one end use than the others, including "fall-down" implications. For instance, it might be more beneficial to produce more "high-grade" appearance lumbers from the resource than to produce utility/economy lumbers or even biomass for energy.

To be successful in value creation, a solid definition of the value proposition is important, that is, to identify what the market needs are and what the values are, and to determine how to successfully meet the needs of the markets. Here, success means to be competitive against other competitors in products and services.

Such value proposition creation and implementation requires a strong understanding of the existing and potential distribution alternatives, a thorough knowledge of the end users and their demands, and a broad knowledge of economic, social, and environmental success metrics. These topics are discussed in the next two sections.

7.3   WOOD PRODUCT DISTRIBUTION SYSTEM

Supply chains for building materials can be complex and diverse. They are complex because many intermediaries are involved in transforming and transporting the products from manufacturers to the end users. In addition, the end users may include a variety of market segments. For instance, homebuilders, professional remodelers, homeowners pursuing do-it-yourself activities, industrial consumers such as packaging and furniture manufacturers, and nonresidential building contractors may purchase wood products. For wood products and building materials manufacturers, additional supply chain stages can be involved before the products actually reach the end users.

The most important market segments for products such as structural lumber and panels are new residential constructions and remodeling, together capturing more than 70% of structural wood products sales. Insulation products is another important market segment, accounting for nearly 60% of the market in value (Browne et al., 2009). As a result, channels through which products make their way into new residential construction and remodeling markets are of prime importance for the industry. In addition, these two segments are inextricably linked, with remodeling activities performed by homebuilders, professional remodelers, or homeowners. Note that both new construction and remodeling activities can share the same suppliers.

7.3.1   DISTRIBUTION CHANNELS

Figure 7.15 represents a simplified supply chain for homebuilding materials. From manufacturers, products are either channeled through distributors or directly to home centers, prodealers, or specialty distributors. Distributors are primarily servicing prodealers or home centers, and in some rare cases, serving directly the large home builders. Prodealers are the building material dealers selling mostly to homebuilders (Robichaud et al., 2009). Home centers sell a wide range of building materials, hardware, and home decor, while the specialty distributors sell mostly the exterior finishing products, including roofing, siding, and interior finishing products, such as drywall, insulation, doors, millwork, and floor covering (Brooks, 2011).

An important characteristic of today's distribution channels for building materials is that the lines between categories of channels are increasingly blurred. Nevertheless, manufacturers are increasingly working more directly with the intermediaries, such as prodealers, home centers, and distributors. With all the different distribution channel options, it is generally assumed that the most common flow remains from manufacturers to prodealers or home centers, then to end users or specifiers, such as homebuilders or remodelers, and finally to homeowners (Brooks, 2011).

7.3.2   FUTURE TRENDS OF DISTRIBUTION CHANNELS

Perhaps the greatest change in distribution channels over the last decade has been the increasing shares of sales and manufacturing activity among dealers. Turnkey projects, which can be defined as a combination of product and installation, are understood to add greater value by simplifying the buying process from a customer's standpoint (Brooks, 2011). Prodealers, for example, have especially taken that

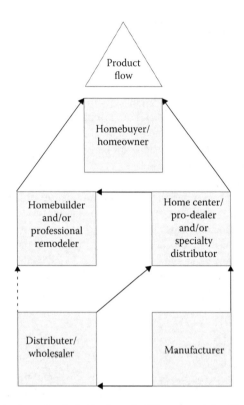

FIGURE 7.15 Typical supply chain for homebuilding materials. (From Robichaud et al., *For. Prod. J.*, 59, 11/12, 2009.)

route. It has been found that half of them in the United States actually produce roof trusses, and 20% of them do the installations (Robichaud et al., 2009). This trend also exists for wall panels and floor systems. In addition, for dealers the emphasis has shifted from buying at the lowest possible price to maximizing inventory turnaround (Brooks, 2011).

Manufacturing activity among prodealers in the United States has reached more than 60% for prehung doors. In the value-added sector, prebuilt stairs are offered by a quarter of dealers. Installed sales have been picking up based on the same logic, with entry doors, windows, and cabinetry being installed by half of the dealers according to the same source.

An increasing number of specialty distributors have emerged over the past decade in North America (Brooks, 2011). The core business of the specialty distributor is centered on exterior building materials, including roofing, siding, and windows; and interior finishing products, such as drywall and insulation. These specialty distributors generally provide an expertise in code requirements and installation techniques (Brooks, 2011).

Another important trend in distribution channels is the growth of purchasing co-ops. Co-ops play a significant role in the channel because they provide a lever for purchasing power. As co-ops move along, they will expand into additional service markets, such as house brands, human resources, and marketing (Brooks, 2011). Wholesalers have gradually lost ground to the purchasing co-ops, especially for lumber (Robichaud et al., 2009). Specialty distributors and co-ops will become increasingly important members of wood products value chains in the future. Independent dealers continue to play a very important role in the commerce of wood products, due to their local proximity to building markets. There might be further consolidation, but not at a rapid pace as experienced in past years. In the future, several macroeconomic changes are likely to impact distribution channels. These include structural changes in supply, consolidation of the supply chain, industrialization of homebuilding, the growing importance of repair and remodeling, substitution between building materials, and green building. These changes are addressed in the following sections.

7.4   STRUCTURAL CHANGES IN SUPPLY AND MARKETS

With eventual increases in housing starts, North America will likely become a net importer of many wood products, including softwood lumber. With a potential imbalance in supply and demand, which are expected to become noticeable in the near future as forecasted housing starts is increasing and allowable cuts drops in British Columbia, commodity producers are expecting price increases. Such price increases will inevitably result in changes in demand patterns and various substitutions (1) from offshore products, (2) from nonwood products, and (3) from wood-saving technologies.

These structural changes need to be considered in establishing the marketing plans and developing product mix plans to meet building material demand in the coming years. For example, as infill projects (where existing and old constructions are demolished for the establishment of new ones) become more common in urban markets, the economies of scale in building large new subdivisions are lost. Can building products and systems be designed to build efficiencies in small infill projects? With renting gaining favor and a growing rental pool, what wood products are most appropriate for these units? Will renters take care of wood floors or will owners pursue other materials to retain their investments? As the market is moving toward smaller homes and with a greater proportion of multifamily units, the usage rate of softwood lumber per housing start will decrease. Other segments within the housing market may be considered in order to attenuate this trend. For instance, wood use could be increased in outdoor living and in high-rise multifamily housing starts.

7.4.1   CONSOLIDATION OF THE SUPPLY CHAIN

For wood products suppliers, larger homebuilders bear significant meanings. Larger builders have been found to be more innovative and to be the earliest adopters of substitute materials. For instance, larger builders are more likely than smaller ones to replace solid wood with finger-jointed studs, onsite construction with prefabricated walls, and structural sheathing with nonstructural sheathing. Larger builders also leverage their strong purchasing power and actively look for a shorter supply chain.

A recent study of substitution trends in the 20 most important US markets found that large builders are open to use any industrialized building materials, including concrete and steels. For example, in flooring systems, large builders are found to use monolithic concrete slabs instead of wood (Robichaud et al., 2010). Despite these substitutions, wood-based products are still a predominant building material used by many large builders with a market share of 80%, in comparison with the smaller builders where we see a 91% market share. Foam and kraft board, alone or in combination, are another example of substitution products used by large builders with a market share of 17%, as compared to only 5% for small builders. Thus, as illustrated, substitution for nonwood products is more likely to happen as builder size increases.

As a response to homebuilders' consolidation, the consolidation among building materials dealers is more recent but has gained momentum over the past decade. An important shift within the dealers channel is the increasing importance of co-ops and buying groups as sources of supply (Robichaud et al., 2009). As prodealers and home centers consolidate, they develop more direct relationships with lumber suppliers, who in turn dedicate more resources in servicing these accounts. These customers have a great influence on the product mix (requirement of specialty grades) and on logistics and services.

7.4.2 INDUSTRIALIZATION OF HOMEBUILDING

Another change in the homebuilding industry is the industrialization of homebuilding, a change in the building process that shifts from the construction site toward factories. This process comes along with increased content in engineering and design. Industrialization happens through five channels, including prefabricated homes, structural components, building material dealers, integrated homebuilders, and integrated wood products manufacturers.

However, what does industrialization mean for wood products suppliers? It means a greater proportion of sales will be made through the industrial buyers that produce prefabricated homes within a factory before getting to the construction site. Here, maintaining quality is critical, especially with the increased use of automated processes. Automated factories demand higher quality lumber products with accurate moisture control and require a lower occurrence of wane and twist. Homebuilding industrialization may also mean more sales of specialty products with specific properties. It is possible that the demand for "proprietary grades" will increase, where customers specify various properties such as dimensions, moisture content, and finishing. The ability to master the production of these specialty products and the agility to deliver them will be important. Coordination and collaboration between primary producers and customers will help establish fruitful partnerships, especially with structural components manufacturers on the rise.

7.4.3 INCREASING IMPORTANCE OF REPAIR AND RENOVATION

Repair and renovation (R&R) markets have become increasingly important in recent years as compared to new residential constructions. A stronger remodeling market involves a strong demand for decking products. Homeowners are looking for decking materials that will last and require low maintenance. Although composite materials have made important progress in this market, the demand for products with wood appearance has come back as a more demanded attribute. Remodeling markets also require a lot of appearance products. Wood flooring, for example, has much stronger demand than the wood composite-based flooring. While wood appearance is important, product quality is essential, especially for wood products structural products in remodeling markets (Robichaud et al., 2010). This new emerging market is continuously pulling wood products manufacturers closer to the end users. This means that a strong focus will be placed on communication between the suppliers and the customers.

7.4.4 INCREASING SUBSTITUTION FOR BUILDING MATERIALS

In North America, while wood has been and remains the most used framing material in residential constructions, some shifts are evolving in material uses and construction techniques. Among the various examples of substitutions that may be found in Robichaud et al. (2010) is the shift of wall sheathing material uses from structural panels to foam sheathing in North America. Other market shifts include plastic-composites for decking, steel joists for flooring, and a growing reliance on concrete systems. With regards to construction techniques, a greater proportion of prefabricated components such as wall panels and floor decks have been used, especially in the US Northeast markets. This material substitution is being driven by builders' quest for building solutions. Builders look for specific attributes such as structural performance, durability, low cost, minimal callbacks, and ease of installation. Although wood is the preferred structural building material in North America, builders experience various business constraints and are more open to new construction alternatives. In other words, builders that were once thought to be conservative in their practices are increasingly willing to consider the adoption of new products. With the latest concerns and policies on energy efficiency, the latest trend for ideal products for homebuilders has been combining structural properties and energy performance with ease of installation.

7.4.5 GREEN BUILDING

Wood products tend to be intimately bundled with information today. This is especially acute with the greater demand for green materials. This means that wood products manufacturers have to produce and disseminate knowledge in addition to fabricating products. In the case of green buildings, such knowledge includes life cycle analysis—or part of it—such as carbon balance or energy and water content. While the production of such knowledge relies upon rigorous standards, the communication of environmental information is likely to be addressed through norms such as Environmental Products Declarations. Clearly, the product alone doesn't suffice, and suppliers need to invest time, resources, and efforts to document the environmental merits of their products.

7.5   NEED FOR RESEARCH IN WOOD PRODUCTS VCO

This section briefly describes the North American academic market research field needed to help us understand how Canadian wood products are used and how they might be used in the future. Furthermore, the trends in research topics are shown, addressing the question of the adequacy of research efforts in the area of forest products marketing for the future demands of VCO investigations. The previous section on distribution illustrated a good example of the type of information that is generated.

As mentioned in the overview of research interests and publications, an evaluation took place in the period of 1995 to 2000 as part of a "gap-analysis" done by Forintek Canada Corp. (Gaston and Fell, 2000). The academic institutes doing research related to forest products marketing were identified as follows:

- Department of Wood Science and the Centre for Advanced Wood Processing, University of British Columbia
- College of Forest Resources, Mississippi State University
- Wood Science and Forest Products, Virginia Tech
- School of Forest Resources, Pennsylvania State University
- Center for International Trade in Forest Products, University of Washington
- Forest Products Marketing Program, Louisiana Forest Products Development Center
- Forest Products Management Development Institute, University of Minnesota
- Department of Forest Products, Oregon State University
- Department of Forest Products, University of Idaho

Individual professors were identified within these universities; in addition to recording their relevant publications list, the researchers were asked to identify their main areas of research interests. These were as follows:

- Material substitution between wood and nonwood products
- Characterization of product attribute values
- Effects of risk and uncertainty on buyer behavior for wood-based products
- Development of dynamic mathematical models of market competition
- Diffusion and adoption of engineered wood-based products (including consumer perceptions)
- Co-operative marketing of wood-based products
- Using technology in the marketplace
- E-Commerce
- Marketing environmental certification
- Identifying value-added product opportunities
- Quality assurance labeling
- Quantifying various end-use markets
- Quantifying specifier perceptions of wood in nonresidential and industrial markets
- Marketing of new engineered wood products

- Decision modeling as it relates to market choices
- Quantifying interaction of wood processing, technology, and marketing
- Quantifying impact of increasingly global competition in wood products
- Life-cycle analysis of wood compared to alternative building materials
- Supply-chain management in the wood industry (Gaston and Fell, 2000)

In 2013, the publication records of these same professors were reexamined. From 1999 to the end of 2012, 1134 citations were recorded from 18 professors, with those identified at least partially related to the field of forest products marketing. Of these, 433 were peer-reviewed journal articles. This is shown over time from 1999 in Figure 7.16; note that individual professor listings changed over time due to retirements and new entrants.

The figure clearly shows a decline in output over the past decade. This has not been due to an overall decline in the number of professors, but rather a decline in the average citations per professor. This reflects a small number of new entrants over the past decade, suggesting an aging workforce that has been possibly moving away from research into administrative positions. Without the attraction of new entrants to the field of forest products marketing is worrisome for the future.

Just as worrisome from a VCO support point of view, there has been a gradual decline in the output of research papers in the "core" areas of (1) markets and products; (2) competition, globalization, modeling, and policy; (3) descriptive; and (4) supply chain, VCO, management, and e-commerce. This has been somewhat offset by a rise in the publication of papers in nonforest products market research fields, especially under the general characterization of sustainability (life cycle assessments, corporate social responsibility, etc.) and "unrelated" topics such as third-world economic development. These trends are illustrated in Figures 7.16 through 7.18.

These trends are potentially troublesome for the field of wood products VCO. Unless this growing gap of intelligence in the core areas of market research is being

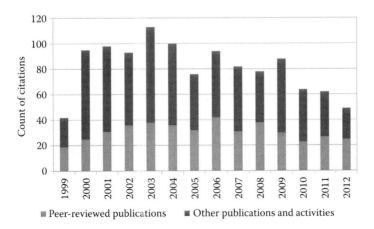

FIGURE 7.16 Number of citations from North American university professors in the field of forest products marketing.

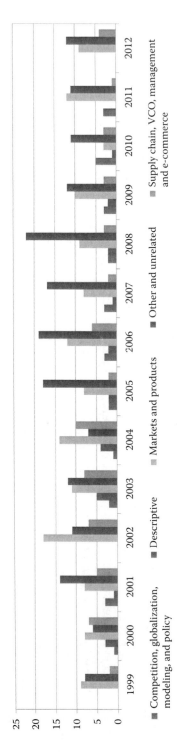

FIGURE 7.17 Peer-reviewed articles by category for North American university professors in the field of forest products marketing.

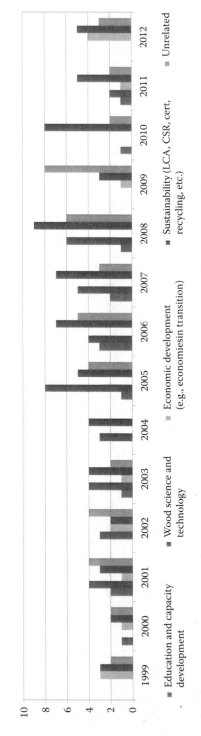

FIGURE 7.18 Peer-reviewed articles by category for North American university professors in the field of forest products marketing: "other and unrelated" category detailed.

adequately addressed elsewhere,* characterization of wood products demand will be lacking in the future. It can be argued that even today there is a lack of market research on new, nontraditional markets such as nonresidential construction (as a North American end-use example) and on emerging markets like China and India.

7.6 IMPLICATIONS FOR WOOD PRODUCTS VCO IN NORTH AMERICA

VCO implies maximizing the value of a forest resource, not just volume. This necessitates an understanding of the downstream, market side of the chain.

Understanding *where* Canadian wood products are consumed is the first part of the equation: (1) domestic demand, trade to the United States, and trade offshore (including emerging markets); (2) lumber, wood-based panels, and other engineered products and/or solutions; (3) wood use in construction versus appearance versus industrial applications; and (4) coproduct wood chips, pulp logs, and wood residues used for pulp and paper, composites, and bioproducts (including energy).

Understanding *how* wood products are consumed is the second part of the equation. Who are the product specifiers by country and by end use? What are the main demand drivers? And what are the trends and likely demands for the future? These are the important questions addressed by forest product market researchers.

In understanding the "how," there is a growing specialization of end users for wood products. Increasingly, the share of lumber processed through a factory before reaching the building site is increasing. Component and prefabricated home manufacturers, as well as manufacturers of engineered wood products such as cross-laminated timber, are demanding special characteristics. Opportunities to differentiate lumber should include dimension, finish, appearance, and moisture content. This is becoming especially relevant as the automation of manufacturing plants increase, because automated plants, that is, the structural components, require much more quality. This specialization of end users promotes specific attributes and the challenge presents to link these attributes back through the supply chain.

The same can be expected within distribution channels, where suppliers' characteristics become as important as product characteristics. There again, lumber yards need to serve specific market segments, and their supply differs if, for instance, they service production builders of entry homes or high-end remodelers. Quality is a very important attribute to building material dealers, and therefore impacts on the required supply basket can be inferred.

In the final analysis, the real challenge in the VCO process is to match these market signals (and *expected* market signals) back upstream to manufacturing optimization, to log allocation, policy, and silviculture for the future. With the focus on value, the optimization problem is getting the right log to the right manufacturing facility for the right end-use application. Resource allocation therefore considers all product-manufacturing alternatives, including lumber in various grades, coproduct chips and

---

* This may well include research done by individual companies; however this typically remains proprietary. Other nonacademic sources include governments, private institutes, and non-North American sources, all of which were beyond the scope of this summary.

residues (and whole log chipping) for pulp, wood composites, biochemicals, bioenergy, and so on. The return-to-log can be maximized if all possible "biopathways" or "value pathways"* are considered, in terms of meeting both potential domestic and export demands.

The need for understanding these growing dynamics in which the forest sector operates has never been greater. There have been considerable structural changes in the supply and demand drivers for Canadian forest products in recent years. Examples of "shocks" include the dramatic influence of the mountain pine beetle in British Columbia, and the policy shifts on the supply side, i.e. for newsprint, both of which are additionally compounded by lower demand for housing materials in the United States. These forces tend to challenge the historical areas of competitive advantage within the forestry sector, demonstrating the need to identify areas of research that will fundamentally help with transforming the industry.

VCO has the potential to be a very powerful tool for implementing industry transformation. To be successful, VCO needs to fully incorporate supply and demand drivers in its efforts to aid in industry and government's ongoing drive toward a healthy, competitive, and sustainable forest products industry. This must include a needed analysis in the fields of trade and economic impact modeling, econometrics, market research, corporate and social responsibility, policy, and business.

REFERENCES

Adair, C., and A. Schuler. (2010). *The Future of Homeownership: Implications for the Wood Products Industry*. APA—The Engineered Wood Association publication.

Brooks, G. (2011). *Scope of the Lumber and Building Materials Industry*. Prepared for the National Lumber and Building Materials Dealers Association. 179 pp.

Browne, T.R., W.G. Baumgartner, and I. Tryon. (2009). *World Insulation. Industry Study 2434*. Cleveland: OH, The Freedonia Group Inc. 421 pp.

Gaston, C.W., and D. Fell. (2000). Markets for Canadian Wood Products in the U.S. – a Gap Analysis. Forintek Canada Corp. Research Report, prepared for the Canadian Forest Service.

Gaston, C.W., G. Delcourt, and D. Cohen. (1999). Japan's Value Added Market: Wood Product Attributes and Competition: Competitor Analysis. Forintek Canada Corp. Report.

Resource Information Systems, Inc. (RISI). (2006, 2012, and 2014). http://www.risiinfo.com

Robichaud, F., P. Lavoie, and C. Gaston. (2009). Demands on lumber suppliers within the U.S. Pro-dealers channel. *Forest Products Journal*, 59(11/12): 83–92.

Robichaud, F., P. Lavoie, C. Gaston, and C. Adair. (2010). Builder perceptions of wood and non-wood products in the U.S. Top-20 metro housing areas. *Journal of Forest Products Business Research*, 07.

Statistics Canada. (2014). http://www5.statcan.gc.ca/subject-sujet/index?lang=eng

Wood Products Council. (2004). 2003 Wood Used in Non-Residential Construction, U.S. and Canada.

Wood Products Council. (2007). 2006 Wood Used in Repair and Remodelling, U.S. and Canada.

---

* For a discussion on the concept of biopathways, see the Forest Products Association of Canada website: http://www.fpac.ca/index.php/en/page/value-pathways.

8 Framework for Information and Knowledge Sharing in Collaborative Modeling of the Forest Products Value Chain

A Survey and Road Map

Riadh Azouzi and Sophie D'Amours

CONTENTS

8.1 INTRODUCTION

Canadian academic researchers, in collaboration with FPInnovations, a national forest research institute, are proposing to build a strategic research network on value chain optimization (VCO) that aims at providing the industry and policy-makers with advanced planning and decision support systems to support the design and deployment of a more competitive forest bioeconomy. Different value propositions are to be evaluated from a global value chain perspective. The research would be organized around five themes denoted as T1 to T5 as shown in Figure 8.1. Themes T1 to T4 develop a hierarchical approach to value chain optimization. Theme T5 addresses a broader comprehensive knowledge representation. The focus of this chapter is on theme T5. It stresses the need for gaining a common representation and understanding of the different components of the value chain as this will permit evaluating forest and industry strategies, supply-chain configurations, and planning approaches using simulations. More specifically, the network aims at developing and maintaining virtual business test-benches providing data sets and rich business contexts to aid the demonstration of the potential benefits of different methods for the coordination and planning of all the resources in the forest products industry. An agent-based approach has been identified as a valuable tool to facilitate the assessment of different collaborative and noncollaborative scenarios. However, the sharing of information and knowledge in a collaborative modeling context remains a key challenge. In this chapter, we look at the literature about the most significant standardization initiatives for information and knowledge sharing and agent-based modeling platforms in the forest products value chain. A road map was developed defining the collaborative knowledge–based platform to support research projects within the VCO network.

This chapter is organized as follows. Section 8.2 looks at the collaborative design of planning systems within the forest products industry. Section 8.3 presents the literature review. It begins with an overview of the components of a typical architecture for information and knowledge exchange in value chain, followed by a review of

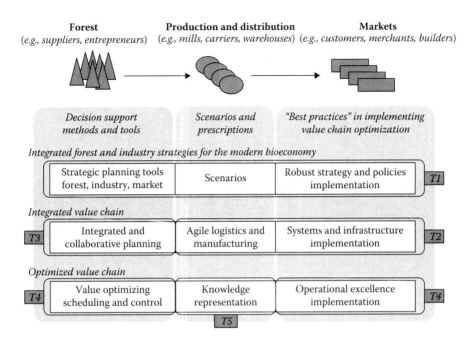

Forest   **Production and distribution**   **Markets**
(e.g., suppliers, entrepreneurs)   (e.g., mills, carriers, warehouses)   (e.g., customers, merchants, builders)

| *Decision support methods and tools* | *Scenarios and prescriptions* | *"Best practices" in implementing value chain optimization* |

Integrated forest and industry strategies for the modern bioeconomy

| Strategic planning tools forest, industry, market | Scenarios | Robust strategy and policies implementation | T1 |

Integrated value chain

| T3 | Integrated and collaborative planning | Agile logistics and manufacturing | Systems and infrastructure implementation | T2 |

Optimized value chain

| T4 | Value optimizing scheduling and control | Knowledge representation | Operational excellence implementation | T4 |

T5

FIGURE 8.1   Hierarchical framework for value chain optimization in the forest bioeconomy.

the most significant standardization initiatives, and then, the agent-based modeling (ABM) platforms in the value chains. The focus is on the standardization initiatives and ABM platforms that were specifically developed for the forestry. This section concludes with a summary of key findings. Section 8.4 presents the proposed road map. Finally, Section 8.5 presents the results of an assessment of the perspective of Canadian researchers on the required standards.

8.2   COLLABORATIVE DESIGN OF PLANNING SYSTEMS WITHIN THE FOREST PRODUCTS INDUSTRY

To transform a tree into forest products and deliver the products to the end users requires the interaction of several actors, including forest crews, manufacturers, distributors, and retailers, as shown in Figure 8.2. This network of actors must work collaboratively to ensure that the supply-chain activities of harvesting, log and biomass delivery, and the first as well as the second transformations are carried out successfully to deliver the forest products to the markets. They must constantly improve their competitiveness individually and collaboratively to maximize their profitability while respecting their social and environmental responsibilities and constraints.

In addition to the physical assets and goods flows, there are also the knowledge and information flows between the network actors, such as the product specifications, transaction information, and market intelligence. The coordination and the effective use of the information can contribute strongly to the improvement of the competitiveness of the companies involved in the value chain (D'Amours et al., 2009).

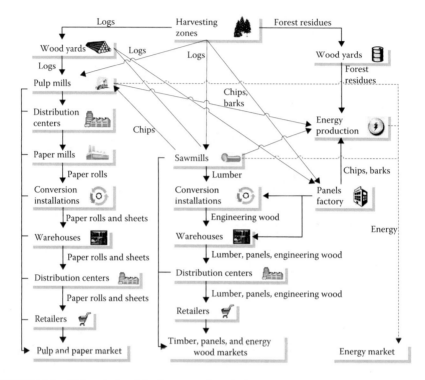

FIGURE 8.2   Value chain of the forest products industry.

Coordination is generally defined as the "process of managing dependencies among activities" (Malone and Crownston, 1994). A complete review of the different types of interdependence and coordination mechanisms can be found in Frayret et al. (2004). On the other hand, decision-making involves different levels with different information and having different impacts. Hierarchical decision-making illustrations for the forest products value chain can be found in Figure 8.3, in which rich information is passed between levels. To effectively manage and coordinate the information flows throughout the hierarchical decision-making systems is a challenging task (Rönnqvist, 2003). This challenge becomes more complex as the forest is a natural resource, the management of which is defined by socioeconomic as well as ecosystem processes and is characterized by complexity, seasonality, uncertainty, and information deficits and asymmetries (D'Amours et al., 2009; Gebetsroither et al., 2006). The fact that most of the wood comes from Crown timberlands and that the provincial as well as regional government play an important role in forest resource planning make the use of hierarchical planning and information more challenging.

From the above discussion, it is clear that in order to support this hierarchical value chain decision system, the adoption of a highly flexible framework for integrated and collaborative modeling and simulation becomes a necessity. This framework should enhance (1) the cooperation of different and multidisciplinary

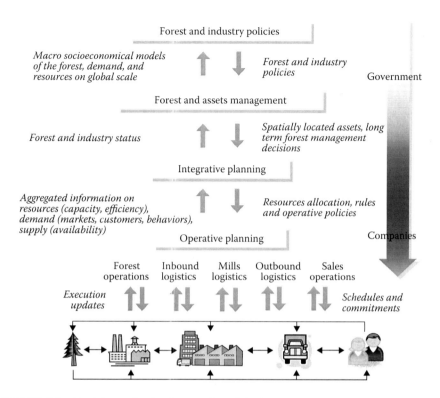

FIGURE 8.3 Decision scales for the planning of the forest products value chain.

researchers and (2) seamless information and knowledge sharing throughout the models built by these researchers (Ahmad, 2002; Liu and Kumar, 2003; Shore, 2001). To better understand what is really meant by information and knowledge sharing, it helps to visualize the value chain in a layered structure, as shown in Figure 8.4. The data resource layer (the foundation) supports the information engineering and knowledge environment layer which in turn supports the business strategies layer. Data resources are analyzed to develop information. Information exchange supports the knowledge environment (by creating new knowledge or assisting knowledge exchange). Knowledge is used in resolving complex business problems or in gaining successful business ideas.

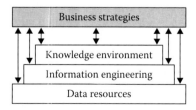

FIGURE 8.4 Visualization of the value chain as a layered structure.

8.3 LITERATURE REVIEW

This section aims at reviewing independent and industry-specific standardization initiatives for information and knowledge sharing and ABM platforms used in the value chain. Industry-independent standards are not specific to any industrial sector. Furthermore, most of the standards presented below emerged through the joint effort of members of well-known organizations. The members form a nonprofit association and the standards are based on their consensus. Such a standardization process is known as "consortium standardization" as opposed to formal, project-based, and industrial standardization (Odell, 2002). But first, we find it more convenient to take a broad view of the components involved in the typical information and knowledge exchange architecture in the value chain.

8.3.1 Overview of the Components in a Typical Architecture for Information and Knowledge Sharing in the Value Chain

Information and knowledge sharing can occur when two business partners interact. Figure 8.5 depicts the typical architecture of a business-to-business (B2B) interaction framework. Notice that the terms "framework" and "standards" are used interchangeably in the e-business literature. Basically, the interactions occur in three layers: communication, content, and business process layers (Medjahed et al., 2003). For example, in Figure 8.2, a sawmill and pulp mill need to agree on their joint business process (e.g., contracts, delivery mode). The sawmill needs also to "understand" the content of the purchase order sent by the pulp mill. Finally, there must be an agreed-upon communication protocol to exchange messages between the sawmill and pulp mill. If the communicating partners use different communication protocols, then a gateway should be used to translate messages between heterogeneous protocols. The content layer provides languages and models to describe and organize information in such a way that it can be understood and used. At this layer, companies interact through business documents that communicate a semantically complete business thought: who, what, when, where, and why (the "what" is typically the product). Business documents are composed of three types of components: core components, domain components, and business information objects. A core component is a syntax-neutral description of semantically meaningful business concepts (e.g., "date of purchase order," "sales tax," and "total amount" could be core components for parts of a purchase order). Domain components and business information objects are larger components stored in the domain library and business library, respectively. For example, a "purchase order request" business document guides how products, dates, and currencies are presented. This is achieved by schemas provided by the e-business frameworks for validating the contents of business documents (Medjahed et al., 2003; Nurmilaakso et al, 2006). The objective of interactions at this layer is to achieve a seamless integration of data formats, data models, and languages. Information translation, transformation, and integration capabilities may be needed to provide reconciliation among disparate representations, vocabularies, and semantics. Finally, the business process layer is concerned with

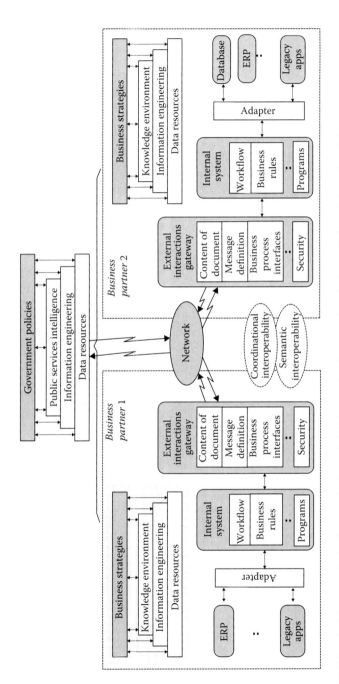

FIGURE 8.5   Architecture of a business-to-business (B2B) interaction framework and a layered view of the value chain.

the conversational interactions (i.e., joint business process) among services. The objective of interactions at this layer is to allow autonomous and heterogeneous partners to come online, advertise their terms and capabilities, and engage in peer-to-peer interactions with any other partners. Interoperability at this higher level is a challenging issue because it requires understanding of the semantics of partner business processes.

8.3.2 INDUSTRY-INDEPENDENT STANDARDIZATION INITIATIVES FOR INFORMATION AND KNOWLEDGE SHARING

In this section, the focus is on two frameworks that addressed B2B integration at both the data and process level: e-business XML (ebXML) and Rosettanet. The two can be viewed as industry-independent standardization initiatives for information and knowledge sharing.

8.3.2.1 ebXML

ebXML (www.ebxml.org) is a modular suite of specifications edited by the Organization for the Advancement of Structured Information Standards (OASIS, www.oasis-open.org) and the United Nations Center for Trade Facilitation and Electronic Business (UN/CEFACT). It provides a standard method to exchange business messages, conduct trading relationships, communicate data in common terms, and define and register business processes. ebXML does not use its own e-business vocabularies to describe business documents. Instead, ebXML documents are assembled from core components. In addition to the core components, the underlying architecture of ebXML comprises the following elements: a messaging service, registry/repository, business process specification schema, collaborative protocol profile/agreement, and core components. ebXML enables the implementation of Web Services protocols, such as Web Services Definition Language (WSDL), Universal Description, Discovery and Integration (UDDI), and Simple Object Access Protocol (SOAP). Web Services are application programming interfaces designed to support machine-to-machine interaction over a network. While ebXML serves as a starting point in some applications, the opinions on the potential of ebXML are mixed (Chituc et al., 2008).

8.3.2.2 RosettaNet

RosettaNet (www.rosettanet.org) is a nonprofit consortium of major companies (more than 400 of the world's leading information technology, electronic component, semiconductor manufacturing, and solution provider companies) working to promote open e-business. RosettaNet has published several specifications to facilitate e-business in high-technology industries. A specification consists of process definitions, message definitions, and industry-specific content. The most important part of RosettaNet is the business process specifications called Partner Interface Processes (PIPs). Currently, RosettaNet covers over a hundred PIPs and associated business documents. RosettaNet also provides a messaging specification called RosettaNet implementation framework.

8.3.3 STANDARDIZATION INITIATIVES FOR INFORMATION AND KNOWLEDGE SHARING RELATED TO THE FOREST INDUSTRY

Since the early 2000s, several standardization initiatives have been launched by recognized organizations, including forest companies, in different countries. The resulting standards were designed to facilitate electronic transactions or the implementation of system-to-system interfaces. The different actors in the forest value chain who used these standards were seeking higher efficiency and cost reductions. In this section, the most significant initiatives are reviewed.

8.3.3.1 papiNet

papiNet is a set of common electronic formats and terminology for the paper and forest products industry designed to facilitate system-to-system real-time exchange of information between buyers and sellers (www.papinet.org). It was formed in 2000 by 23 members of the Confederation of European Paper Industries (CEPI) and Graphic Communications Association (GCA). Today, the papiNet consortium includes more than 40 members (such as International Paper, StoraEnso, and Time). Several segment user groups have been formed in order to agree upon business rules, processes, and data to be used in a specific paper market segment, for instance, pulp, publication paper, fine paper, label stock, packaging, and carton board. The objective of these groups is to agree upon business rules, processes, and data to be used in a specific paper market segment, that is, a template.

papiNet provides specifications for 36 business documents and related business processes and is available in three categories: basic order fulfillment, supply-chain management, and product quality, for example, the invoice, dispatch note, stock status, stock adjust, order, and call-off. Finally, papiNet uses the ebXML messaging service as an envelope for the documents and offers a Trading Partner Agreement (TPA) template. TPA documents capture the essential information upon which trading partners must agree in order for their applications and business processes to communicate, including participation roles, communication and security protocols, and a business protocol (valid actions, sequencing rules, etc.).

8.3.3.2 ELectronic DATa or ELDAT

ELectronic DATa (ELDAT) is Germany's widely valid data interface in the wood trade, in which information is exchanged among the partners and standardized in such a way that it can be worked on by means of electronic data processing. This standard serves the exchange of product data of raw wood (logs) and contractual data between forest owners, wood consuming industries (sawmills, pulp, paper mills), wood traders, and carriers in the German wood market. The standard is conceived for file-based exchange using comma-separated text files or XML files. The data model consists of a dozen entities with a huge number of attributes, combined with an extensive controlled vocabulary. Unfortunately, no scientific reference to this standard could be found in the literature. ELDAT is a mutual initiative of the forest companies under the backing of the LWF (Technical research for silviculture and forest management institute).

8.3.3.3   Standard for Forestry Data and Communication or StanForD

StanForD (Standard for Forestry Data and Communication, www.skogforsk.se) focuses on managing bucking or merchandizing computers on board forest machines and on managing data communications in forestry (control, reporting, and monitoring logging production on harvesters and forwarders). It comprises a data standard and a file-structure standard, and also includes a Kermit-based communications protocol or data recorder to the merchandizing computer. The Forestry Research Institute of Sweden (Skogforsk) coordinated the standard with the help of manufacturers and Swedish forest enterprises. In 2010, a new version of the standard was released. It provides an XML interface and the definitions for messages of the following types: production instruction, object instruction, species group instruction, object geographical instruction, forwarding instruction, harvested production, harvesting quality control, total harvested production, forwarded production, forwarding quality control, object geographical report, and operational monitoring. Figure 8.6 illustrates the use of some of these messages. StanForD is of high interest within the Indisputable Key project (a European initiative that aims at developing methodology, technology, and knowledge to increase the utilization of production resources in the forest products value chain) since, with minimal update, it can serve to keep track of the logs throughout the supply chain.

8.3.3.4   e-Forestry Industry Data Standard

The e-Forestry Industry Data Standard (eFIDS; https://www.oasis-open.org/committees/download.php/29052/eFIDS-Description.html) is a standard designed to provide the basis for the implementation of a range of e-business applications. It has

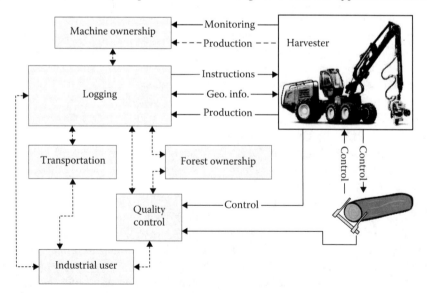

FIGURE 8.6 Illustration of how standardized messages are used in different information flows using the StanForD standard. The dashed arrows illustrate information flows that are based on other data standards such as, for example, papiNet.

been developed within the United Kingdom by a number of leading organizations within the forestry industry and there has been wide consultation on its applicability. Full use has been made of the promotional activities of government development agencies such as Scottish Enterprise as well as trade associations such as the United Kingdom Forest Products Association (UKFPA) and other forestry groups such as the Confederation of Forest Industries within the United Kingdom (ConFor, www. confor.org.uk). OASIS continues to develop eFIDS as an international standard. The schema provides an XML framework that allows a variety of trading documents and their corresponding data to be used between various parties, but it does not define the documents themselves (e.g., there is no XML element called "Delivery Note" but there is an element whose function is to contain the descriptor for a Delivery Note). The most recent version of this standard includes the following use cases: dispatch data (round timber) from supplier to buyer, weighbridge data from buyer to supplier, conventional invoicing from supplier to buyer, self-billing invoicing from buyer to supplier, dispatch/weighbridge data from third-party weighbridge to buyer/seller, any weighbridge to third party (www.forestryscotland.com).

8.3.3.5 Geographic Information Systems Data Transfer Standard

The Geographic Information Systems Data Transfer Standard (GIS DTS; www.oasis-open.org/committees/forest/charter.php) is another OASIS standard. It is meant to facilitate geographic information exchange among different systems used in the forest industry. GIS are used by the forestry industries to help manage forestry geographical information and plan complex harvest areas. However, over the last 15–20 years, several types of GIS have been adopted, with different methods of data storage being used. GIS DTS works with different types of GIS and does not require any specific data storage format, though some basic compliance to the GIS DTS is necessary. New GIS and database systems being designed for use in the forestry industry should take note of this standard, and have data models built in a compatible way.

8.3.3.6 Other Standards

The standardizations presented in this section are not exhaustive. For instance, there exist some local forest-specific standards (such as the VIOL (Timber On Line) standard Kommstandard in Scandinavia and FHPDAT (data exchange format) standard in Austria) that are not described in this section.

8.3.4 STANDARDIZATION INITIATIVES FOR INFORMATION AND KNOWLEDGE SHARING RELATED TO OTHER INDUSTRIES

Few studies on industry-specific standardization initiatives could be found in the literature. In fact, the only reviews of industry-specific standardization initiatives we found were in Nurmilaakso et al. (2006) and Chituc et al. (2008). These authors briefly reviewed several standards and attempted to analyze their interoperability aspect, limitations, and challenges. These standards include CIDX (chemical industry data exchange) for chemical industry (www.cidx.org), PIDX (petroleum industry data exchange) for petroleum industry (www.pidx.org),

AgXML (agricultural extensible markup language) for grain and grain processing (www.agxml.org), AIAG (automotive industry action group) for automotive industry (www.aiag.org), TexWeave for textile supply-chain integrated networks (www.texweave.org), Odette for automotive industry (www.odette.org), and STAR (standards for technology in automotive retail) standards for technology in automotive retail (www.starstandard.org). All these standards are made openly available to the public by consortiums formed by industry members and organizations. The general approach to development can be described using the following steps: (1) identifying the business processes that, if electronically enabled, would improve business-process efficiency and effectiveness, (2) determining the data requirements of those business processes, (3) defining and maintaining evolving XML schemas and related guidelines that support the data requirements, and (4) building commitment from participants to integrate XML-based messaging into their business processes and to provide a forum for understanding that process. In general, there are 10–40 members per consortium, and the number of documents per standard varies between 17 and 60 documents. Related business processes guidelines are also provided. Finally, all the mentioned standards are XML based; however, not all of them offer messaging, envelope, and security specifications.

8.3.5 INDUSTRY-INDEPENDENT ABM FRAMEWORKS

An agent-based model consists of a set of agents, a set of agent relationships, and a framework for simulating agent behaviors and interactions. The agents are abstractions to represent complex entities. Thus, an ABM is usually implemented as a multiagent system. In Fayret (2011), a multiagent system is presented as a technology that aims at creating general or specialized behavioral and interactional models and to implement these models in distributed and interactive computer programs, called agents. Flores-Mendez (1999) stated that "agent infrastructures deal with the following aspects: (1) ontologies: allow agents to agree about the meaning of concepts; (2) communication protocols: describe languages for agent communication; (3) communication infrastructures: specify channels for agent communication; and (4) interaction protocols: describe conventions for agent interactions." Accordingly, assuming this view of the infrastructure of a multiagent system, we can assert that standards are needed for the four aspects that the multiagent system infrastructure deals with. The research on agents, multiagent systems, and their infrastructure has been on the rise over the last two decades. Already, since the early 1990s, several authors have elaborated literature reviews about multiagent system architecture standardization (Ahmad, 2002; Flores-Mendez, 1999; Odell, 2002). Many standardization initiatives were reported by the literature, such as General Magic, knowledge-able agent-oriented system (KAoS) (www.ihmc.us/research/projects/KAoS/), holonic manufacturing systems (HMS) (hms.ifw.unihannover.de/), and control of agent-based system (CoABS) (www.objs.com/agility). The most prominent efforts may be those developed by Object Management Group (OMG) and Foundation for Intelligent Physical Agents (FIPA). Some details about these two initiatives are provided below. In general, these efforts focused on the interaction aspects of agent technology. In fact, many of the industry-independent standardization initiatives focused on key challenges facing commercial agent developers as they were bringing this technology to market. A particular interest was given to agent mobility (the ability to migrate in a self-directed way from one host platform to another). The agents communicate using languages that could hardly convey meanings; instead they convey objects with no semantics like in classical object middleware. As is shown in the following sections, the advent of the Web has forced these standardization initiatives to realign their specifications.

8.3.5.1 OMG Initiatives

OMG (www.objs.com/agent) is an international computer industry consortium that develops enterprise integration standards for modeling, middleware, and data warehousing for a wide range of industries including healthcare and manufacturing. The well knowing Unified Modeling Language (UML) is OMG's most-used specification for modeling business processes and data structures. Within OMG, there is a special interest group that aims to extend OMG's object management architecture to better support agent technology, to identify and recommend new OMG specifications in the agent area, to recommend agent-related extensions to OMG specifications, and to promote standard agent modeling techniques. As such, OMG's effort was more a bottom-up activity. Another group within OMG (OMG Ontology Working Group) is working on aligning the domain modeling activities of OMG with the Semantic Web, the extension of the Web, with related ontology development projects such as the defense advanced research projects agency (DARPA) Agent Markup Language (DARPA DAML, www.daml.org) and the IEEE Standard Upper Ontology working group (IEEE SUO, suo.ieee.org). The Semantic Web (www.semanticweb.org) is a group of methods and technologies to allow machines to understand the meaning, or "semantics," of information on the Web.

8.3.5.2 FIPA Initiatives

FIPA (www.fipa.org) is an IEEE Computer Society standards organization that promotes agent-based technology and the interoperability of its standards with other technologies. FIPA was originally formed as a Swiss-based organization in 1996 to produce software standards specifications for heterogeneous and interacting agents and agent-based systems. FIPA's effort was more a top-down activity. Its approach to multiagent systems development was based on a minimal framework for the management of agents in an open environment (Flores-Mendez, 1999). This framework is described using a reference model, which specifies the normative environment within which agents exist and operate, and an agent platform, which specifies an infrastructure for the deployment and interaction of agents. At the IEEE Computer Society, it is believed that standards for agents and agent-based systems should be moved into the wider context of software development. In short, agent technology needs to work and integrate with nonagent technologies (Flores-Mendez, 1999). To this end, FIPA has been accepted as part of the standards committees. Many of the ideas originated and developed in FIPA are now coming into sharp focus in new generations of Web/Internet technology and related specifications.

8.3.5.3 Overview, Design Concepts, Details Protocol

The Overview, Design Concepts, Details (ODD) Protocol is a standard designed as a general protocol for communicating individual-based and agent-based models. As such, it is concerned with a different perspective of ABM with regard to the OMG and FIPA standardization initiatives. It consists of a narrative description of the various elements of an ABM, contributing to a more rigorous formulation phase (Grimm et al., 2006). Each model is described using three blocks, overview, design concepts, and details, which are subdivided into seven elements: purpose, state variables and scales, process overview and scheduling, design concepts, initialization, input, and submodels. ODD has been formulated and tested by 28 authors from seven different countries, and it is gaining diffusion in ecology and in social science.

8.3.6 ABM Frameworks Related to the Forest Industry

A complete discussion of the appropriateness of agent-based modeling for the forest sector can be found in Frayret (2011). In fact, agent-based models have been used in many applications in the forest products industry. A number of frameworks were used to develop these models. This section presents a review of these frameworks.

8.3.6.1 FORAC Experimental Planning Platform

The FORAC Experimental Planning Platform (FEPP; www.forac.ulaval.ca) is composed of agents that interact with each other in order to solve the global lumber supply-chain planning problem. It addresses two relevant issues for the forest products value chain: (1) capacity to plan and coordinate operations across the supply chain and (2) capacity to analyze the dynamics and performance of different supply-chain scenarios by means of simulation (Frayret et al., 2007). From a functional point of view, the platform is made of an advanced planning and control system exploiting agent technology. Users, through the use of various dedicated graphical tools and interfaces, can develop the various supply-chain scenarios to be tested and analyzed (Figure 8.7). The configuration of the platform follows an organizational design approach that consists in the division of the supply chain into business units. In turn, this division into business units is based on the natural heterogeneity of the production process. In other words, the overall problem is split into several smaller subproblems, each of which results in the managerial problem of a single organizational unit. Consequently, every agent is modeled after a specific organizational problem (Figure 8.8). This gives each agent the ability to solve a smaller scale problem using adapted tools. The collaboration, or the cooperation, of all agents present in the supply chain solves the value creation network global problem. Many functions are shared by all agents. For example, they can interact with each other or perform tasks when necessary and by definition, they are autonomous. Moreover, by the way in which the FEPP is conceived, each agent can be represented as a modular block that can be assembled with others to model a supply chain. Several conversation protocols and agent behaviors have been implemented in order to produce optimized solutions in various situations.

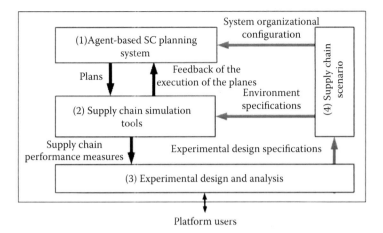

FIGURE 8.7   General overview of the FORAC Experimental Planning Platform (FEPP).

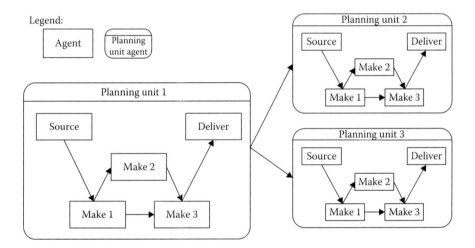

FIGURE 8.8   Components of a specific implementation of a supply-chain configuration.

8.3.6.2   Seamless Operation of Forest Industry Applications

The Seamless Operation of Forest Industry Applications (SOFIA; www.cs.jyu.fi/ai/ OntoGroup) is a logistic control platform for B2B mediation tailored to the forestry sector of Finland. It is designed to optimize logistics on the contractor site for harvesting and transportation small and medium enterprises (SMEs). It improves a contractor's order management by integrating a heterogeneous order base from different order makers. The platform is meant to serve as an integrator of information systems provided by different wood buyers and forest owner associations. The orders coming from different systems will be gathered to one integrated view allowing the contractor to apply a logistics optimization tool and decrease useless overhead in operation. The SOFIA platform design will not change existing systems of customers, but rather plug into them, and will contribute to the development of the flexible forestry service market. With this platform, the authors aim at building a strong national level network of forestry companies and organizations in order to get support and promote the new vision of flexible contracting for SMEs. Also, they intend to ensure sustainable information and communication technology infrastructure for logging and transportation SMEs. In fact, the SOFIA platform is developed on top of UBIWARE (smart semantic middleware for ubiquitous computing), a generic domain-independent middleware platform, that is meant to provide support for integration, interoperability, adaptation, communication, proactivity, self-awareness, and planning for different kinds of resources, systems, and components (Nikitin et al., 2010). UBIWARE integrates several technologies including the Semantic Web, agent technologies, ubiquitous computing, service-oriented architecture, and Web X.0 (e.g., data information and knowledge, software and services, humans, hardware, and processes). It fulfills the Global Understanding Environment (GUN) concept where various resources can be linked to the Semantic Web–based environment via adapters (or interfaces), which include (if necessary) sensors with digital output, data structuring (e.g., XML), and semantic adapter components (XML to Semantic Web). Software agents are to be assigned to each resource and are assumed to be able to monitor data coming from the adapter about the state of the resource, make decisions on behalf of the resource, and discover, request, and utilize external help if needed. Agent technologies within GUN allow mobility of service components between various platforms, decentralized service discovery, utilization of FIPA communication protocols, and multiagent integration/composition of services (Naumenko et al., 2007).

8.3.6.3 Other ABM Frameworks of Forest Industry Using Industry-Independent Platforms

Several other attempts were made by researchers to apply ABM to investigate problems related to the forest products value chain using commercial or free of charge simulation frameworks. Vanclay (1998, 2003) proposed the forest land oriented resource envisioning system (FLORES) framework for spatially explicit modeling of the human–forest interaction, at different levels of detail. This framework is intended for policy-makers and their advisors to envisage the efficacy and consequences of any initiatives (policies and incentives) to promote sustainable forestry and better land use. FLORES was built using the Simile modeling environment (www.simulistics.com). Purnomo et al. (2003) used common-pool resources and multi-agent system (CORMAS) (cormas.cirad.fr), a multiagent simulation platform specifically designed for renewable resource management systems, to examine several scenarios of sustainable forest management involving multistakeholders (central and local government, communities, nongovernmental organizations [NGOs]). Caridi et al. (2006) assessed the benefits brought by multiagent systems connected to the CPFR (Collaborative, Planning, Forecasting, and Replenishment) process where trading partners work off a common forecast. The simulations were carried out using SIMPLE++ (www.tecnomatix.com), a general-purpose system for the object-oriented, graphical, and integrated modeling, simulation, and animation of systems and business processes. More recently, in his PhD dissertation, Schwab (2008) presented CAMBIUM, an agent-based forest sector model for large-scale strategic analysis. Basically, CAMBIUM models the interdependencies and feedback loops between resource inventory dynamics and changing market conditions for finished products using an intermediate layer of autonomously interacting forest industry agents. It is used as a decision support tool for assessing the effect that changes in product demand (due to market dynamics) and in resource inventories (due to natural disturbances) can have on the structure and economic viability of individual companies and communities. Schwab used Repast_3 (repast.sourceforge.net/repast_3), a free and open source ABM toolkit. Finally, in Pérez and Dragićević (2010), the authors used aeronautical reconnaissance coverage geographic information system (ArcGIS) to test different management strategies based on an existing agent-based model for simulating mountain pine beetle tree mortality patterns in order to evaluate the influence of different forest management practices to control insect outbreak. In fact, ArcGIS was coupled with Repast (the ABM toolkit) using an extension. Because the agent-based model was enabled to include real-time GIS data feeds, it was possible to simulate and visualize lodgepole pine stands' mortality patterns unfolding at different time steps.

8.3.7 KEY FINDINGS

All the standards described above emerged from nonprofit consortiums. Figure 8.9 depicts a comparison of these standardization alternatives based on the knowledge and understanding of the authors. It attempts to position the different standards with respect to each other and to emphasize the links they have with a number of aspects organized in four quadrants: forest products value chain segments, information and knowledge sharing architecture components, planning levels or functions, ABM, and simulation. One can claim that the ideal standardization initiative should be linked to all these aspects and should correspond to the center of the circle, at equal distance from all aspects. Unfortunately, Figure 8.9 shows that most of the standards are situated close to the perimeter of the circle, meaning that they are too specialized and limited in scope. Many standards from the right-side quadrants rely on the ebXML framework to build their messaging layer. ebXML enables the transfer of documents between two or more parties. For instance, papiNet has a number of independent documents, one per message type (invoice, dispatch note, stock status, stock adjust, order, call-off) and it does not have a single, flexible document or schema to transit multiple message types. In fact, papiNet does not address semantic conflicts, it simply avoids them. Thus, each new application of the papiNet standard is much more likely to require schema amendments or extensions. For more flexibility and better connectivity and interoperability, Figure 8.9 shows that the trend is to link to the Semantic Web environment. Finally, it appears that standards makers did not tackle the ABM description and model communication. In a collaborative design context, a standard of this aspect, such as the ODD framework, would be of great help.

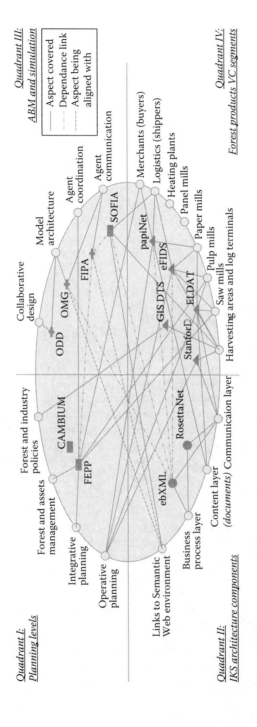

FIGURE 8.9 Comparison of different standards/frameworks.

8.4   A ROAD MAP TOWARD DEFINING CANADIAN STANDARDS FOR VALUE CHAIN MODELING

Figure 8.10 presents an overview of what could be the structure of the collaborative knowledge–based platform to support researchers within the VCO network. It links together a number of models via the Internet with one-way or two-way links. Communication agents may be used. Some models correspond to multiagent systems while others correspond to single agents. The models emphasize or could be concerned about different static and dynamic aspects of the value chain (e.g., products, machines, processes, customers, suppliers, distributors, business units, planning units), its business flows (information, goods, and finance), but also flows of the virtual artifacts used for local planning and coordination/synchronization of resource use at all operational and tactical levels. Therefore, different decision-makers operating at different stages responding to decisions made at other stages (Figure 8.3) can be simulated. What is important, here, is the interactive decision-making across various levels of authority or governance that can be portrayed and analyzed, then used to improve cross-stage and cross-sectoral coordination. Constant assumptions about the responses at other scales or at other levels (vertically or horizontally) of the value chain could be avoided. Seen from another perspective, this collaborative knowledge–based platform is a spatially explicit ABM that is used as a communication tool between researchers (knowledge resources and product/service design groups). It is a unifying translational architecture for dynamic knowledge representation removing barriers to multidisciplinary collaboration.

A basic condition for the creation of the collaborative knowledge–based platform is the development and adoption of standards approaching the ideal standard described in Section 8.3.7. In this preliminary version of the road map toward defining this standard, we highlight a range of key issues and future challenges for some of the aspects appearing in the four quadrants of Figure 8.9:

1. Regarding the aspects of "agent communication" and "links to Semantic Web environment," an automated approach to semantic interoperability needs to be developed. Semantic interoperability ensures that agents of heterogeneous behavior have a common understanding of the meanings of the requested services and data. It should be supported by a well-defined ontology (Lin et al., 2004). Ontology and semantic interoperability defines for researchers a common vocabulary to share common understanding of the

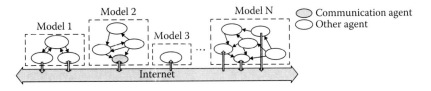

FIGURE 8.10   Illustration of the structure of a collaborative knowledge–based platform.

structure of information among people or software agents, enables reuse of domain knowledge, makes domain assumptions explicit, separates domain knowledge from operational knowledge, and analyzes domain knowledge. The approach should be supported by Web Services and it should be automated because of the dynamic and rapidly evolving environment modeled by the collaborative knowledge–based platform.

2. Regarding the aspects of "agent coordination" and "business process layer," coordinational interoperability should be achieved. Coordinational interoperability deals with the interaction between agents, operations; it imposes discipline on the interaction between the agents for preserving temporal as well as functional properties. Discipline is also needed between the researchers developing the agents. Common policies, contracts, and protocols to which the agents subject their activities are needed.

3. Regarding the aspects of "collaborative design" and "model architecture," structured thinking and notational interoperability are needed. A unified framework for systems thinking and modeling during the analysis phase would save valuable time for the researchers during problem structuring, dynamic modeling, scenario planning and modeling, implementation, and organizational learning. On the other hand, notational operability is required in order to surmount disagreements among the objects (agents, software tools, databases) on the structure, representation, or interpretation of the data. Here, a starting point could be the ODD standard. We believe that this standard could make model description and understanding much easier in the context of collaborative design.

4. Regarding the aspects in quadrant IV, all the segments of the forest products value chain should be detailed and modeled. More internal factors (manufacturing capabilities, human resources, etc.) and external factors (environmental change, macroeconomic matters, technological change, legislation, sociocultural changes, changes in the marketplace) should be included in the models. The FEPP platform appears to be very promising. Also, a standard registry describing all these models should be set up and consistently maintained. It should be very helpful to the researcher to engineer their models and scenarios. Regarding the aspects in quadrant I, links to GIS data should become a standard. GIS data should be integrated into the design process of models and solutions at all the planning levels.

These key issues and challenges should be given high priority in the academic, government, and industry research communities in the near future.

8.5   PERSPECTIVE OF CANADIAN RESEARCHERS

In this study, we examined the perspectives on decision system standardization issues from a number of Canadian researchers. Six peoples were selected based on their involvement in research on VCO including three academic researchers from three different research groups (two people from each group) throughout Canada (Québec,

Ontario, and British Columbia) and two representatives from FPInnovations. First, the participants were introduced to the context of this work and briefed about the proposed road map. Next, they were requested to respond to the following two key questions:

1. (Q1) What are the biggest challenges that they face in their efforts to collaborate with other research groups in order deliver on their mission?
2. (Q2) How can effective steps be taken against the five key issues raised in Section 8.4, notably, the automated approach to semantic interoperability, coordinational interoperability, structured thinking and notational interoperability, modeling of all the segments of the forest products value chain, and GIS data integration to give more depth to their work and contribute to not only dissemination of their results and findings but also to their models, solutions, and software components?

Four people were interviewed. In what follows, we briefly introduce the persons interviewed and provide a summary of their responses to the two questions above.

The first person interviewed was Steven Northway. He is from the Faculty of Forestry, University of British Columbia. His group developed the architecture and design of the International Forest and Forest Products trade model (IFFP) that is used to explore critical forest policy issues such as plantation development and illegal logging. The IFFP is a JAVA program/servlet that will (1) translate text files, describing the problem, into a linear programming (LP) formulation, (2) initiate the LP solver, and (3) translate the solution into HTML reports. In response to the first question, Steven asserted that there is a need for a technique that would help to build models in a much richer way, especially for strategic modeling. There is a clear need to make sure that collaborators are aware of how fine the different descriptions of the products or forests from one model to another are (granulometry issue). For example, there is a time issue when the models (generated descriptions) or the data were not generated with consistent time steps. Another example is the geographical extent, where, for instance, models built on a provincial basis may use different jurisdictions or different ecosystems. In general, the problems are well defined but there is a lack of data and the data are not detailed enough to drive the decision. Regarding the second question, Steven believes that there is no doubt about the fact that all five points are very important. They will lead to better models and better collaboration. We might need to start with the first three key issues. The first key issue (semantic interoperability) is especially important. "If you get the first point you cover a lot of things."

The second person interviewed was Francis Fournier from FPInnovations. He is involved in the transformative technologies research program developed with industry and government partners. He asserted that there is a lack of tools or decision-making solutions needed by the manufacturer for some links of the forest products value chain. At FPInnovations, they linked together their software tools in order to build an operational tool than can be used to generate more complete answers. There are many software tools already available or that are being developed. The exchange of data between software is really a challenge. For instance, the government holds inventory data in a very specific format that is highly aggregated. These data cannot be used at an operational level using their decision-making tools. Regarding the second question, Francis thinks that the different action steps brought by the road map are properly aligned with the needs of their research program. He mentioned that at FPInnovations, the different software programs are stand-alone applications. However, when two applications need to talk to each other, they do that via a bridge (a specific software component) built for them. It is clear that a standard semantic language would greatly standardize the bridges. However, Francis questions the willingness of the different parties to share their data and models.

The third person interviewed was Jonathan Gaudreault, a professor at Laval University. His research involves distributed search for supply-chain coordination using agent technology. With regards to the first question, Jonathan believes that the inconsistency in the hypotheses and the data between different groups is the main challenge. For now, in his research group, they do not attempt to integrate models by different researchers because they are aware of the huge efforts needed in order to learn about the hypotheses that were made, how to validate them, and how to modify the models in order to comply with the hypotheses. Also, the data could be inconsistent. Regarding the second question, he thinks that all the different action steps brought by the road map are interesting. However, there is a need to think about where to position the actions along the value chain as some actions could be more important than others depending on where we are in the value chain. For instance, too many research efforts were devoted to model forest resources or some wood transformation processes; however, very few attempts were made to model market or customer demand and preferences.

Finally, the fourth person interviewed was Jean-Marc Frayret, a professor at École Polytechnique de Montréal. His research involves operations and supply-chain management and distributed decision-making integration in manufacturing networks. He applies the tools of operations research and multiagent systems to design advanced planning and scheduling systems. With regards to the first question, Jean-Marc mentioned that it is relatively easy for professors to work together and exchange ideas. In fact, the professors are looking to integrate their decision tools rather than their models. However, the real challenge is not in the seamless exchange of data between models but in how we build the models to avoid gaps of variables that were not considered by any of the models. Currently, he is focused on solving specific problems for some enterprises but he definitely believes that there will be a need to standardize blocs of models and integrate models in order to build tools for large simulation. In a reply to the second question, Jean-Marc said that, in general, the proposed road map is driving the researchers in the right direction. However, we should be aware that it implies huge efforts. There are precedence relationships between the five key issues raised by the road map that should be clearly defined and strictly observed. The standardization should be made on a semantic basis. It would be very interesting if future models are built using the same semantic basis. "We can think of future agents that are enabled to use a standard semantic in their decision-making abilities and for coordination between each other."

8.6 CONCLUSIONS

We have reviewed information and knowledge sharing in the context of collaborative and agent-based modeling of the forest products value chain, including the most significant standardization initiatives and ABM platforms in the value chain. A preliminary version of a road map toward defining a standard to be used in a collaborative knowledge–based platform to support researchers within the VCO network was proposed. The research and development needs defined by this road map fit in the general trend toward Internet-based computing (also referred to as cloud computing). When asked to give their perspectives on the proposed road map, the researchers stated their support of the identified research priorities.

REFERENCES

Ahmad, H. F. 2002. Multi-agent systems: Overview of a new paradigm for distributed systems. In 7th IEEE International Symposium on High Assurance Systems Engineering (HASE'02), Tokyo, Japan, 23–25 October 2002.

Caridi, M., Cigolini, R., De Marco, D. 2006. Linking autonomous agents to CPFR to improve SCM. *Journal of Enterprise Information Management* 19(5):465–482.

Chituc, C. M., Toscano, C., Azevedo, A. 2008. Interoperability in collaborative networks: Independent and industry-specific initiatives—The case of the footwear industry. *Computers in Industry* 59:741–757.

D'Amours, S., Frayret, J. M., Gaudreault, J., LeBel, L., Martel, A. 2009. Chaînes de création de valeur. In *Manuel de foresterie*. Québec: Collectif, Les éditions MultiMondes.

Flores-Mendez, R. A. 1999. Towards a standardization of multi-agent system frameworks. *ACM Crossroads Student Magazine* 5(4):18–24. http://www.acm.org/crossroads/xrds5-4/multiagent.html, accessed 13 September 2001.

Frayret, J.-M. 2011. Multi-agent system applications in the forest products industry. *Journal of Science and Technology for Forest Products and Processes* 1(2):15–29.

Frayret, J.-M., D'Amours, S., Montreuil, B. 2004. Coordination and control in distributed and agent-based manufacturing systems. *Production Planning and Control* 15(1):1–13.

Frayret, J.-M., D'Amours, S., Rousseau, A., Harvey, S., Gaudreault, J. 2007. Agent-based supply-chain planning in the forest products industry. *International Journal on Flexible Manufacturing Systems* 19:358–391.

Gebetsroither, E., Kaufmann, A., Gigler, U., Resetarits, A. 2006. Agent-based modelling of self-organisation processes to support adaptive forest management. *Contributions to Economics* Part 4:153–172.

Grimm, V., Berger, U., Bastiansen, F., Eliassen, S., Ginot, V., Giske, J., Goss-Custard, J., Grand, T., Heinz, S. K., Huse, G., Huth, A., Jepsen, J. U., Jørgensen, C., Mooij, W. M., Muller, B., Pe'er, G., Piou, C., Railsback, S. F., Robbins, A. M., Robbins, M. M., Rossmanith, E., Ruger, N., Strand, E., Souissi, S., Stillman, R. A., Vabø, R., Visser, U., DeAngelis, D. L. 2006. A standard protocol for describing individual-based and agent-based models. *Ecological Modelling* 198:115–126.

Lin, H. K., Harding, J. A., Shahbaz, M. 2004. Manufacturing system engineering ontology for semantic interoperability across extended project teams. *International Journal of Production Research* 42(24):5099–5118.

Liu, E., Kumar, A. 2003. Leveraging information sharing to increase supply chain configurability. In 24th International Conference on Information Systems (ICIS) Proceedings. Paper 44.

Malone, T. W., Crownston, K. 1994. The interdisplinary study of coordination. *ACM Computing Surveys* 26(1):88–119.

Medjahed, B., Benatallah, B., Bouguettaya, A., Ngu, A. H. H., Elmagarmid, A. K. 2003. Business-to-business interactions: Issues and enabling technologies. *VLDB Journal* 12:59–85.

Naumenko, A., Katasonov, A., Terziyan, V. 2007. A security framework for smart ubiquitous industrial resources. In *Enterprise Interoperability II: New Challenges and Approaches*, by R. J. Gonçalves, J. P. Müller, K. Mertins, and M. Zelm (Eds), Berlin: Springer, pp. 183–194.

Nikitin, S., Terziyan, V., Lappalainen, M. 2010. SOFIA: Agent scenario for the forest industry: Tailoring UBIWARE platform towards industrial agent-driven solutions. *International Conference on Enterprise Information Systems (ICEIS)* 1:15–22.

Nurmilaakso, J. M., Kotinurmi, P., Laesvuori, H. 2006. XML-based e-business frameworks and standardization. *Computer Standards and Interfaces* 28(5):585–599.

Odell, J., 2002. *Update: Agent Standardization Efforts*. Executive Update, Vol 4. Arlington, MA: Cutter Consortium.

Purnomo, H., Yasmi, Y., Prabhu, R., Yuliani, L., Priyadi, H., Vanclay, J. K. 2003. Multi-agent simulation of alternative scenarios of collaborative forest management. *Small-Scale Forest Economics, Management and Policy* 2(2):277–292.

Pérez, L., Dragićević, S. 2010. Exploring forest management practices using an agent-based model of forest insect infestations. In Proceedings of the International Environmental Modelling and Software Society, Ottawa, Ontario, Canada, July 5–8 2010.

Rönnqvist, M., 2003. Optimization in forestry. *Mathematical Programming* 97(1–2):267–284.

Schwab, O. S., 2008. *An Agent-Based Forest Sector Modeling Approach to Analyzing the Economic Effects of Natural Disturbances*. Dissertation, University of British Columbia, Canada.

Shore, B., 2001. Information sharing in global supply chain systems. *Journal of Global Information Technology Management* 4(3):27–50.

Vanclay, J. K., 1998. FLORES: For exploring land use options in forested landscapes. *Agroforestry Forum* 9(1): 47.

Vanclay, J. K., 2003. Why model landscapes at the level of households and fields? *Small-Scale Forest Economics Management and Policy* 2(2):121–134.

9 Introduction to Agility in the Forest Product Value Chain

Jean-Marc Frayret and Nathalie Perrier

CONTENTS

9.1 DEFINITION OF AGILITY

Agility can be defined as the property of a complex system that can quickly process timely and relevant information from its volatile environment to adjust its operations, processes, or goals in order to achieve efficiently and in a sustained manner its purpose. This general definition of agility can apply to any complex systems. However, in the context of a value chain, it can be seen as either the ability to quickly and efficiently take advantage of new business opportunities or customer demands, or the ability to remain efficient and profitable in a volatile environment. Although both are not mutually exclusive, the latter does not necessarily imply taking advantage of new business opportunities.

This general definition of agility implies several key elements. First, it is an attribute, a characteristic of a system that emerges from the modus operandi of its coordinated components. Therefore, agility in a complex system is achieved thanks to, on the one hand, the concurrent design of the processes that bind the components of the system together, and, on the other hand, the design of the behavior of these components. Agility is also an attribute of a system that evolves in a volatile environment. Although this aspect of the definition does not define agility per se, it is a sine qua non condition of the system design problem. Indeed, if the environment is stable, there is no need to design an agile system. It would even be counterproductive.

Agility also involves the ability to quickly process timely and relevant information from its environment. On the one hand, this aspect of the definition requires the system to be able to sense relevant information from its environment. Such information is different for each system, which uses its own tools and processes to capture and analyze it. In complex systems, such as a value chain, the environment is more than what the system can observe. Therefore, for such a system, it is important to understand what is relevant in order to develop the tools and processes to sense what really influences the system capacity to achieve its purpose. Relevant information is not necessarily something the system can directly observe. It can also be something that is inferred from multiple observations, such as a market trend, the development of innovative technologies, or the development of new regulations. On the other hand, this aspect also requires the system to quickly process, communicate, and share information. Because complex systems are made of multiple interacting components, their performance is limited by the ability of each component to contribute to that performance. Consequently, information must flow and be processed quickly in order to achieve the most appropriate and coordinated response from each component of the system.

Next, agility implies the capacity of the system to adjust its operations, processes, or goals. In the case of a value chain, a system's response to relevant information may involve a short-term operational reaction motivated by the need, for instance, to seize the opportunity of new customer demands. In a value chain, such a reaction usually requires the ability to quickly change the operational setup of manufacturing and logistic resources. However, agility may also require a deeper adjustment of the way the system operates, by adjusting its processes as well. The constant pressure of competition usually leads to the continuous improvement of processes. However, agility requires the system to be able to transform its processes to adapt to more drastic changes in the environment, such as disruptive technological innovations and new customer expectations. For the survival of the system, it may also require an even deeper transformation of the system's goals. Adapting a system's goals has a more long-term effect on the system, as its purpose is affected, and its processes as well.

Finally, the aim of agility is the sustained success or survival of the system. Because it is an attribute that emerges from a multitude of processes that are designed and adjusted by the system itself, the concept of agility poses several questions, where answers are left outside the scope of the chapter. How do we measure agility and its impact on the system? How much agility is required? What type of agility is required?

This chapter first aims at introducing this concept and at briefly analyzing the requirements of value chain agility. This chapter also addresses some of the many improvements, methods, and techniques value chains can adopt to become agile.

The reader is referred to Gunasekaran and Yusuf (2002), Yusuf and Adeleye (2002), Sanchez and Nagi (2001), Ramesh and Devadasan (2007), Bi et al. (2008), and Vinodh et al. (2009) for recent reviews of the many facets of agility in manufacturing and supply chains. Finally, this chapter proposes a quick diagnostic of harvest and sawing operations with respect to agility.

9.2 VALUE CHAIN AGILITY REQUIREMENTS

In order to analyze the requirement of agility for an organization such as a value chain, this chapter proposes a simple framework composed of three dimensions: the organizational requirements, the information requirements, and the operational requirements, as presented in Figure 9.1. In this framework, a system implements a business model that uses and transforms resources in order to meet the needs of its customer.

Although in a value chain, the system can be considered as a collection of interacting business models, for the purpose of this analysis, we limit the definition of the value chain to a single business model. Therefore, for this business model to be agile, the system that implements that model must meet different forms of requirements. Above all, these requirements are linked to the agility objectives defined by the value chain. These objectives are briefly discussed in the following section. Section 9.2.2 then introduces information requirements. Section 9.2.3 presents organizational requirements, while finally Section 9.2.4 describes operational requirements. However, because each value chain has its own agility requirement according to the nature and volatility of its environment, these requirements cannot be described or analyzed in an exhaustive manner. This section only discusses the most general types of requirements. At the end of the section, Table 9.1 summarizes the value chain agility requirements studied in the recent literature.

9.2.1 AGILITY OBJECTIVES

If agility is a characteristic of a value chain, this characteristic is not necessarily valuable. Becoming agile is a strategic decision. It requires the value chain to

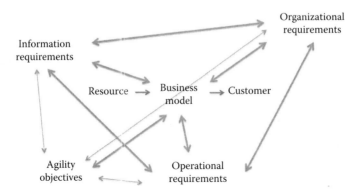

FIGURE 9.1 Framework of agility requirement analysis.

identify the limits of agility that is necessary or appropriate for the value chain to thrive in its environment.

As discussed in the introduction, agility can be limited to a value chain's operations or it can be required at the most strategic level as well. As it may involve significant investment for the value chain, the value of agility must be carefully analyzed. Simulation is an appropriate tool to analyze the value of agility, its processes, and its requirements. Among others, Gaudreault (2011) and Santa-Eulalia et al. (2012) introduce, respectively, an agent-based simulation platform and a simulation methodology, which aim to evaluate different configurations of the forest value chain

TABLE 9.1
Agile Value Chain Requirements Literature

References	Agile Initiatives to Support Value Chain Requirements
Information requirements	
Lau et al. (2003), Morel et al. (2003), Pan et al. (2003), Lü et al. (2004), Tsai and Sato (2004), Dowlatshahi and Cao (2005), Leitão (2009), Al-Tahat and Bataineh (2012)	Responsiveness to customer demand, customer responsiveness improvement, customer services improvement, responsiveness to environment changes, reduction of product design and development cycle time, product life cycle extension, cost reduction, responsiveness to fluctuating volumes, lead-time reduction, manufacturing equipment, process technology.
Organizational requirements	
Yang and Li (2002), Youssef et al. (2002), Yusuf and Adeleye (2002), Crocitto and Youssef (2003), Harrison (2003), Jiang and Fung (2003), Jin-Hai et al. (2003), Paixão and Marlow (2003), Ren et al. (2003), Ross (2003), Christopher et al. (2004), Falcioni (2004), Helo (2004), Hopp and Oyen (2004), Lee et al. (2004), Lou et al. (2004), Okazaki et al. (2004), James (2005), Lee et al. (2005), Ma and Davidrajuh (2005), Dotoli et al. (2006), Lin and Lin (2006), Lin et al. (2006), Vazquez-Bustelo and Avella (2006), Found and Harvey (2007), Papamarcos et al. (2007), Ramesh and Devadasan (2007), Wee and Yang (2007), Wikner et al. (2007), Yang et al. (2007), Yauch (2007), Zhang and Sharifi (2007), Kundu et al. (2008), Hoyer and Stanoevska-Slabeva (2009), Manenti (2009), Ren et al. (2009), Wang (2009), Xue et al. (2009), Fan (2010), Lei et al. (2011), Neves et al. (2011), Al-Tahat and Bataineh (2012), Kalbande et al. (2012), Sherehiy and Karwowski (2014)	Responsiveness to changes, responsiveness to market needs, adaptation to changing market situation, international market, global competition, total supply chain cost reduction, lead-time reduction, mass customization, increase in product variety, responsiveness to fluctuating volumes, resource management, supply chain integration, supply chain management processes, intraenterprise and interenterprise integration, cross-organizational supply chain management, distribution chain design, virtual environment, development of web-based application, adaptive production control, supplier–customer relationship management, decoupling points selection, competitive manufacturing, competitive capabilities, software development teams, organizational structure, management hierarchies, work organization, incentive compensation, business process, quality measures.

(Continued)

TABLE 9.1   *(CONTINUED)*
Agile Value Chain Requirements Literature

References	Agile Initiatives to Support Value Chain Requirements
	Operational requirements
Li et al. (2002), Chengying et al. (2003), Lim and Zhang (2003), Marapoulos et al. (2003), Moore et al. (2003), Yusuf et al. (2003), Zhou et al. (2003), Flower (2004), Harris (2004), Ip et al. (2004), Jiao et al. (2004), Le et al. (2004), Li et al. (2004), Maione and Naso (2004), Sato and Tsai (2004), Watanabe and Ane (2004), Devadasan et al. (2005), Venables (2005a, 2005b), Wang et al. (2005a, 2005b), Borissova et al. (2006), Chen et al. (2006), Feng and Yamashiro (2006), Helo (2006), Kamal (2007), Trappey et al. (2007), Fung et al. (2008), Garcia et al. (2008), Minoni and Cavalli (2008), Franzen et al. (2009), Gen et al. (2009), Yang and Yang (2010), Albuquerque et al. (2012), Al-Tahat and Bataineh (2012), Frei and Whitacre (2012), Mehrsai et al. (2014)	Responsiveness to market needs, responsiveness to customer specifications and changes, increase in product variants, organization-wide improvements, lead-time reduction, resource information, modular manufacturing equipment, manufacturing cost reduction, product quality improvement, engineering effectiveness in new product development, response flexibility, volume flexibility, changing production orders, resource utilization improvement, embedded systems development, virtual enterprise, production management, advanced planning scheduling, lot-sizing production planning, schedule-based planning, productivity improvement, shop floor management, product design and development, manufacturing cell planning, job scheduling in a virtual environment, flexible job-shop scheduling, automated guided vehicle assignment, flexible manufacturing processes, combined process, equipment and plant design, elimination of wastes, optimization.

performance. Santa-Eulalia et al. (2011) use this simulation platform to study the robustness of a forest value chain, while Cid et al. (2009) use it to study the impact of various decoupling point positions in the value chain. At the operational scale, Beaudoin et al. (2012) and Gil and Frayret (2015) use, respectively, discrete event and agent-based simulation to improve log yard operations. Farnia et al. (2013a, 2013b) use agent-based simulation to study the impact of the introduction of auctions in the Québec forest value chain. In the pulp and paper industry, Sauvageau and Frayret (2015) present an agent-based simulation study of different procurement strategies in the recycled pulp industry. Finally, in the hog fuel sector, Pledger et al. (2014) use an agent-based simulation of grinding and transportation operations in order to analyze the impact on performance of different logistic configurations and fiber moisture content.

This small sample of simulation studies demonstrates the usefulness of simulation to evaluate different strategies to implement agility. Although simulation does not tell you how to become agile, it is definitely instrumental to assess the value of agility and to identify the best configurations of agile processes. In order to design agile processes and decision structures, the value chain must understand what is required for a value chain to achieve some level of agility.

9.2.2   Information Requirements

Information requirements deal with the value chain's ability to sense, process, communicate, and share information (Lau et al., 2003). As discussed in this chapter's introduction, it is one of the key elements of agility. It can be generally described according to four dimensions. Because a value chain's purpose is to meet its customers' demand, market is the information requirement dimension of agility in which value chains have traditionally a lot of interest (Leitão, 2009). For a value chain to be profitable, it must produce what its market will buy. This implies knowing the required products mix to provide, their right price and quality, as well as the expected services. The second information requirement dimension concerns resource and supply. Resource availability and supply conditions, such as shortage and seasonality, constrain the value chain's ability to meet customer needs (Tsai and Sato, 2004). It is therefore necessary for the value chain to sense, analyze, and communicate such information in order to make the most appropriate decisions. Along the same lines, the third dimension concerns logistics, production, and operations. More specifically, it concerns manufacturing and logistics availability and equipment status, their yield, and operational cost (Al-Tahat and Bataineh, 2012). It also includes work-in-progress information and the location of material and resources. Such information is necessary for the system to control its operations and achieve its objectives. Finally, the fourth dimension concerns information quality. It is another type of information requirement as it deals with information availability, its accuracy, its levels of aggregation, and timeliness. It represents the meta-information the value chain needs about the information it uses to make decisions, in order to manage risk and decision confidence in an environment that is volatile.

In order to achieve a certain level of agility, a value chain must understand its information requirements and develop the capability to sense, process, communicate, and share this information within the value chain. It must also determine the quality of information required. In particular, the design of such a capability requires understanding the trade-off between the value and the cost of such information.

9.2.3   Organizational Requirements

Organizational requirements deal with how the value chain is organized and respond to changes. It includes the business and decision processes through which agility is implemented. Again, it can be described according to four dimensions. The first dimension deals with processes and decision structures. In order for a value chain to be agile, it must adopt processes and decision structures that lead to the most appropriate response, given the information it possesses. At a high level, this dimension concerns the value chain's business model (e.g., make-to-stock, produce-on-demand, assemble-on-demand) and how appropriate it is to deal with the level of volatility of its business environment. It also concerns its decoupling point location (i.e., the push–pull boundary) (Kundu et al., 2008), as well as its decision hierarchy and processes (i.e., how decision are made and propagated throughout the value chain) (Found and Harvey, 2007; Ma and Davidrajuh, 2005).

The second dimension deals with internal and external process integration. More specifically, this dimension deals with how elementary business processes, from forecasting demand to demand fulfillment, are integrated together within and across the organization's boundaries (Kalbande et al., 2012; Lin et al., 2006; Paixão and Marlow, 2003; Wee and Yang, 2007). It also deals with collaboration between value chain partners as well as their contractual agreements (Al-Tahat and Bataineh, 2012; Dotoli et al., 2006; Found and Harvey, 2007; Harrison, 2003; James, 2005; Lin et al., 2006, 2011; Ren et al., 2009; Sherehiy and Karwowski, 2014; Vazquez-Bustelo and Avella, 2006; Yusuf and Adeleye, 2002; Zhang and Sharifi, 2007). Because value chain agility requires a quick response to changes, interorganizational business processes must be seamless, in order to support quick information exchange (e.g., customer demand, order status, production capability), collaborative decision-making (e.g., order plan, delivery schedule), and quick contingency management to minimize the impacts of any forms of disruption. This dimension also deals with incentives and revenues (Papamarcos et al., 2007) and risk sharing across the value chain.

The third dimension is performance assessment. Along with the previous dimension, performance assessment is an important requirement that deals with how value chain partners measure success. If success is not measured in a compatible manner across the value chain, partners might be individually working in a counterproductive manner. On the contrary, if value is measured in a similar manner, partners will naturally work to support each other's performance, which, in the end, contributes to an agile value chain that has a unified perception of value. Performance assessment deals with indicators such as material recovery, quality measures, production yields, customer satisfaction, stakeholder satisfaction, and environmental impact indicators (Al-Tahat and Bataineh, 2012; Dowlatshahi and Cao, 2005). It represents what partners in the value chain must control in order to thrive to their environment.

Finally, the fourth dimension deals with business objectives. Value chain partners must not only have similar perceptions of performance assessment, they must also have compatible objectives, for instance with respect to customer service level objectives. Objective compatibility leads to compatible efforts across the value chain. In the retail industry (Simatupang and Sridharan, 2002), alignment of the value chain objectives can be achieved by managing, across the value chain, product categories as strategic business units (i.e., category management).

9.2.4 OPERATIONAL REQUIREMENTS

Finally, the third requirement concerns how operations are carried out in order to respond adequately to changes in the environment. More specifically, it deals with resource utilization optimization and operational flexibility. Three dimensions can be used to describe these requirements. The first concerns the elimination of wastes. Industrial engineering and continuous improvement methods, such as lean manufacturing and lean supply chain management, are a fundamental aspect of agility. Because timeliness is the core of agility, any form of operational waste and inefficiency leads to slower than optimal response time. It can even be argued that it is a necessary condition to becoming an agile organization. More specifically, it involves initiatives such as reducing nonquality to an acceptable level, rework, unnecessary inventory, and unnecessary handling and transportation (Al-Tahat and Bataineh, 2012).

The second dimension deals with decision optimization. Many agile value chain decision problems can be modeled and solved using optimization techniques (Chen et al., 2006; Feng and Yamashiro, 2006; Fung et al., 2008; Gen et al., 2009; Ip et al., 2004; Jiao et al., 2004; Le et al., 2004; Lim and Zhang, 2003; Maione and Naso, 2004; Marapoulos et al., 2003; Wang et al., 2005a, 2005b). Decision optimization concerns how to deliver the right product, to the right customer, at the right time for the right price. It involves all the operational (i.e., short-term), tactical (i.e., mid-term), and strategic (i.e., long-term) decisions across the value chain (Stadtler, 2005). Some of these decisions are made by individual value chain partners and only concern their own operations. Others are made in a collaborative manner and concern parts of, if not the entire value chain (Stadtler, 2009). Decision optimization aims at eliminating wastes in the form of inadequate resource utilization that arises from poor decision-making in procurement, production, distribution, and sales activities. Each industrial sector has specific constraints that must be modeled, before decisions can be optimized using computers and algorithms. The forest product industry has a vast literature in this domain. The reader is referred to Martell et al. (1998), Rönnqvist (2003), and D'Amours et al. (2008) for literature reviews on this subject.

The third dimension concerns process flexibility. Flexibility is the characteristic of a process that can operate in an efficient manner over a specific range of conditions. The ability to quickly react to changes can require a certain level of flexibility, or the ability to quickly change a process configuration, or setup, to adapt to an expected range of variability. Flexibility involves fast setup equipment (Al-Tahat and Bataineh, 2012) or the ability to manage efficiently production campaigns (Franzen et al., 2009; Garcia et al., 2008; Gen et al., 2009; Harris, 2004; Maione and Naso, 2004; Minoni and Cavalli, 2008).

9.3 ACHIEVING AGILITY

The requirements of agility, briefly overviewed in the previous section, describe only the general aspects a value chain initiative must encompass in order to achieve agility. This section deals more specifically with some of the many improvements value chains can implement to become agile.

This section covers only some of the many agile value chain initiatives. The first concerns the need to reduce perceived volatility. The second deals with the need to buffer against variability, while the third concerns the reduction of inefficiencies. Finally, the fourth deals with the need to design flexible processes and operations. Based on each of these four initiatives, Tables 9.2 through 9.5 summarize, at the end of each section, respectively, some of the strategies and technologies of agility studied in the recent literature. Agile strategies are required to meet the criteria and requirements of agility, which have been discussed in Section 9.2. Achieving agility also requires focusing on agile technology solutions, where the strategies are applied.

TABLE 9.2
Examples of Agile Value Chain Strategies and Technologies to Reduce Perceived Volatility

References	Agile Strategies	Agile Technologies
Marapoulos et al. (2003), Christopher et al. (2004), Flower (2004), Mondragon et al. (2004), Feng and Yamashiro (2006), Lin and Lin (2006), Lin et al. (2006), Chae et al. (2007), Sharma and Gao (2007), Fung et al. (2008), Yang and Yang (2010), Al-Tahat and Bataineh (2012), Huang et al. (2014)	Collaborative manufacturing, reduce total demand variance by adjusting retailers' order sizes, information-based supply chains, virtual integration, process alignment, quantitative analysis, qualitative analysis, redesign, quick introduction of new products, information processing network, organizational control, integration of information systems (IS)/information technologies (IT), people, business processes and facilities, delivery speed, product range, virtual manufacturing cell reconfiguration, process planning, combined process planning and partner selection based on core competencies, aggregate product, process and resource planning.	IS, object-oriented IS, demand forecasting, portfolio theory, computer-aided process planning (CAPP), expert system, rapid prototyping, e-manufacturing, computer-aided design (CAD), computer-aided manufacturing (CAM), computer-integrated manufacturing (CIM), operations research models, optimization, enterprise planning, web-centric codevelopment environment.

TABLE 9.3
Examples of Agile Value Chain Strategies and Technologies to Buffer against Variability

References	Agile Strategies	Agile Technologies
Jiang and Fung (2003), Moore et al. (2003), Harris (2004), Mondragon et al. (2004), Tsai and Sato (2004), Wang et al. (2005a, 2005b), Helo (2006), Kamal (2007), Wikner et al. (2007), Kundu et al. (2008), Al-Tahat and Bataineh (2012)	Decoupling points selection, make-to-order, product postponement, inventory centralization, late configuration, customized assembly based on mature component designs, virtual production, machine system design, constant costs per unit, mass-customization, setup cost, and lead-time reduction.	Knowledge-based decision support tool, assembly variant design system, operations research models, optimization, simulation, unified modeling language, virtual production systems, virtual manufacturing, virtual engineering, three-dimensional (3D) graphical simulation, communication technology, tooling strategies.

TABLE 9.4

Examples of Agile Value Chain Strategies and Technologies to Reduce Inefficiencies

References	Agile Strategies	Agile Technologies
Coronado et al. (2002), Jin et al. (2002), Youssef et al. (2002), Yusuf and Adeleye (2002), Crocitto and Youssef (2003), Harrison (2003), Jin-Hai et al. (2003), Paixão and Marlow (2003), Pan et al. (2003), Zhou et al. (2003), Helo (2004), Jiao et al. (2004), Le et al. (2004), Sato and Tsai (2004), Dowlatshahi and Cao (2005), Ma and Davidrajuh (2005), Venables (2005a, 2005b), Onuh et al. (2006), Ramesh and Devadasan (2007), Wee and Yang (2007), Yang et al. (2007), Garcia et al. (2008), Minoni and Cavalli (2008), Gen et al. (2009), Hoyer and Stanoevska-Slabeva (2009), Fan (2010), Vinodh et al. (2010), Lei et al. (2011), Vinodh and Kuttalingam (2011), Al-Tahat and Bataineh (2012), Kalbande et al. (2012)	Supply chain response-time coordination, automation of supply chain management processes, supply chain redesign and integration, rapid response manufacturing, technology integration, information resource integration, on-time resource planning, on-time distribution planning, artificial intelligence, time-based technologies, integration of management and technology, logistics, business process redesign, product life cycle reduction, product functionality improvement, JIT capacity, ramp-up time reduction, optimum-volume manufacturing, optimal design scheduling, flexible scheduling, rescheduling, modular production facilities, fast production cycle times, multiple winners, intraenterprise and interenterprise integration, core competencies, custom solutions, global value chain, virtual enterprise and IT alignment, virtual enterprise built through dynamic alliances, virtual collaboration, virtual product development, reconstruction of manufacturing system, quality control, elimination of waste, make-to-lot size.	Supply chain–wide planning and control systems, auto-ID, RFID, electronic POS data, integrated vendor–buyers inventory system, JIT, reverse engineering, rapid tooling, IS/IT systems, web-based IS, integrated IS, management IS, computer-aided engineering, integrated CAD/CAM, CAPP, CIM, total quality management, design for manufacturability, e-commerce, e-manufacturing, virtual organization, web-based application, communication technology, team building, semiconductor devices, real-time control, electronic data access, flexible scheduling system, agent-based tool, multiagent systems, integrated manufacturing system, production planning, and control system, ERP, manufacturing resources planning, rapid prototyping, advanced manufacturing and IT, reconfigurable system, flexible manufacturing systems, machine tool accessories, product configurators, advanced planning systems, multiagent-based scheduling, optimization models, decision support system (DSS).

CAD, computer-aided design; CAM, computer-aided manufacturing; CAPP, computer-aided process planning; CIM, computer-integrated manufacturing; ERP, enterprise resource planning; ID, identification; IS, information system; JIT, just-in-time; POS, point-of-sale; RFID, radio frequency identification; T, information technology.

TABLE 9.5
Examples of Agile Value Chain Strategies and Technologies to Design for Flexibility

References	Agile Strategies	Agile Technologies
Li et al. (2002), Eaton (2003), Harrison (2003), Lau et al. (2003), Lim and Zhang (2003), Morel et al. (2003), Ren et al. (2003), Ross (2003), Su and Chen (2003), Harris (2004), Lee et al. (2004), Lou et al. (2004), Maione and Naso (2004), Mondragon et al. (2004), Devadasan et al. (2005), Lee et al. (2005), Borissova et al. (2006), Chen et al. (2006), Dotoli et al. (2006), Kulturel-Konak (2007), Papamarcos et al. (2007), Zhang and Sharifi (2007), Fekri et al. (2009), Franzen et al. (2009), Leitão (2009), Manenti (2009), Xue et al. (2009), Abdul Kadir et al. (2011), Al-Tahat and Bataineh (2012), Frei and Whitacre (2012)	Supply chain infrastructure, strategic and profitable proximity sourcing, integrated e-supply chain network design, supply chain reconfiguration, supply chain planning, knowledge and new learning, lot-for-lot replenishment, information sharing, flexible manufacturing, build-to-order, supply chain design, vendor-managed inventory (VMI), product design knowledge associated with decision support, customer-oriented design and development, information-driven and integrated virtual process, participative management, design-supplier-manufacturing planning, design for quality, integrated design, group technology, production process reengineering, rapid redesign, virtual systems, intelligent manufacturing systems, business to manufacturing, enterprise integration, management integration, self-organizing assembly systems, management motivation, culture of change, cross cultural management, incentive compensation, cellular manufacturing, customer–supplier relationship, networked relationships, quality over product life, product with substantial value addition, first-time right design, flexible small-batch production, design efficiency, fixture design, design for reconfigurability, manufacturing systems reconfiguration, responsive process planning and scheduling, facility design and reconfiguration, dynamic and stochastic facility layout.	Virtual agent, agent-based supply chain system, multiagent system, virtual supply chain, data interchange, operations research models, optimization, evolutionary optimization, simulation, DSS, supply chain optimization, design of experiments, integrated design system, flexible manufacturing system, robotics, CAD/CAM, CAPP, CIM, virtual reality, virtual machine tool, virtual machining, web-based systems, hardware interactions, standard for the exchange of product model data for numerical control, holonic manufacturing systems, integrated manufacturing control system, unified modelling language, information and intelligence control, advanced manufacturing technology, IS to support product development, flexible processing technology, rapid prototyping, manufacturing resources optimization, computer-aided tool.

9.3.1 Reduce Perceived Volatility

In order to become agile, a value chain must first understand the volatile nature of its business environment. Volatility may come from different sources. Often, the first that comes to mind is market volatility. In order to better understand this form of volatility, value chains can forecast market demand and market trends (Lin and Lin, 2006). This includes the forecasting of, among others, orders, market prices, overseas markets, and exchange rates. Forecasting techniques, both quantitative, such as time series, and qualitative, such as informal judgment and subjective assessment, can be used alone or in conjunction to better understand this type of volatility. An indirect approach to reduce perceived volatility is to reduce delivery time. Although delivery time does not directly affect the volatility of the environment, it reduces the need to forecast demand further in the future. Therefore, the improvement of logistic operations, using simple techniques, or more advanced approaches such as cross-docking in distribution networks, contributes indirectly to reducing the level of perceived volatility a value chain must deal with.

Volatility can also come from the supply side of the value chain. This is particularly true in the forest product industry. Natural resource industries generally involve some level of variability with respect to resource availability and quality. This can be related to the seasonal nature of the supply or the inaccurate or lack of detailed information about the volume and quality of available resources. This can have adverse and difficult to predict effects on market prices. Such effects can be exacerbated by the global nature of supply, which can come from any part of the global economy. In order to better know supply conditions, a natural resource value chain can adopt advanced technologies to have more detailed information and state-of-the-art data processing capabilities. In the context of manufacturing value chains, supply contracts can also be implemented to stabilized supply conditions.

Finally, volatility can also come from production and logistic operations. Transformation and production processes with process control issues can lead to output variability, such as in the semiconductor industry (Sahnoun et al., 2014). Output variability can also be directly linked to input variability, such as in the lumber industry. In order to address these quality issues, quality management techniques have been developed to bring consistency to the value chain's ability to meet customer needs. Statistical process control (SPC) is a standard and systematic technique to identify out-of-control process conditions.

The need to reduce perceived volatility arises from the need to reduce the scope and extent of variability, before the value chain invests in agile technologies and processes. For instance, input material and supply classification can reduce the need to invest in a flexible product system (Gil and Frayret, 2015), while accurate demand forecasts can improve customer satisfaction. The implementation of formal management and planning methods, such as inventory management, project management, manufacturing resource planning, or stage-gate methods, can also reduce variability with respect to inventory level, production, and new product development and introduction. Information technology also offers a wide range of solutions to support these management processes, including enterprise resource planning (ERP) and production life cycle management (PLM) systems (Helo, 2004; Huang et al., 2014; Kalbande et al., 2012; Venables, 2005a, 2005b).

9.3.2 BUFFER AGAINST VARIABILITY

Once variability is reduced to an acceptable if not a minimum level, the value chain can adopt different buffering techniques to protect against its adverse effects and avoid having to deal with each individual changes.

The most common buffering technique is inventory management. A value chain can optimize its product inventory levels according to its customer demand dynamics in order to achieve high customer service with the right inventory level. Inventory management is a rich domain with many techniques and solutions for a wide range of situations and industries. Multiechelon inventory management models can be used in many different contexts in order to integrate inventory management across the value chain (Shahi and Pulkki, 2014). Inventory management can also be used as a technique to delay product differentiation in order to reduce the time needed to differentiate products according to customer needs. In other words, the management of work-in-process inventories can be optimized in order to maintain, at specific locations, undifferentiated products, subassemblies, and components that can be assembled together and finished to meet different customer demands (Harris, 2004; Wang et al., 2005a, 2005b). Because they can be used to meet different demands, these inventories are less likely to be outdated, all the while allowing the value chain to quickly customize products and reduce delivery time. These techniques lead to manufacturing strategies known as assemble-to-order and make-to-order (Al-Tahat and Bataineh, 2012; Tsai and Sato, 2004; Wikner et al., 2007). Delayed product differentiation can also be implemented as a transportation postponement strategy, in which product transportation is delayed as much as possible until customer demand is known with greater accuracy. Finally, modern inventory management techniques also exploit the concept of product consignment, according to which a product's owner remains its supplier, regardless of its location, until the final customer buys the product, or the manufacturer consumes the product (i.e., assembling or manufacturing its own products). Vendor-managed inventory (VMI) is the most common inventory management technique based on product consignment (Wee and Yang, 2007).

Another common technique to buffer against variability is time management. Time management can be defined as the management of customers' expectation with respect to time. The most common technique consists in using a delivery lead time that includes all procurement and production processes, handling, and storage times plus a buffer that can absorb unexpected variations. A lead time defines the amount of time required to meet a specific need. While the lead time allows the customer to plan its operations, it allows the supplier to deal, in a black box manner, with the variability from its operations and suppliers. Because a lead time must be competitive with respect to market expectations, value chains must select appropriately the location of the decoupling point, in other words, the push–pull boundary (Cid et al., 2009; Harris, 2004; Kundu et al., 2008). The location of the decoupling point is of strategic importance for a value chain because it defines explicitly the minimum set of operations that remains to be completed before the fulfillment of an order. A more advanced approach to manage customers' expectation with respect to time involves order promising. Order promising is a process that aims at fulfilling customer demand based on resource availability, by allocating planned capacity and on-hand inventories in advance. Order promising involves the concept of available-to-promise (i.e., available on-hand and planned inventories) and capable-to-promise (i.e., available production capacity).

9.3.3　REDUCE INEFFICIENCIES

Perceived variability is intensified by the value chain's inefficiencies. In other words, a value chain that is slower to react to change must be able to forecast its future state, including its customer needs, further ahead, as explained in the previous section. In order to reduce this adverse effect of inefficiencies, value chains can first reduce their operational inefficiencies using some forms of process quality management. Among these techniques, lean manufacturing, six sigma (6σ), and total quality management (Youssef et al., 2002) are the most common industrial engineering techniques used to reduce inefficiencies. For instance, lean manufacturing targets the elimination of different forms of wastes (e.g., inventory, rework, motion, waiting, transportation, overproduction, and processing) using different tools and techniques (Al-Tahat and Bataineh, 2012). Value stream mapping, flow charts, Ishikawa diagrams (i.e., cause-and-effect analysis), and Pareto and control charts, among other tools, are used to diagnose and identify operational wastes and all sorts of inefficiencies. Next, other tools, such as 5S (i.e., sort, set in order, shine, standardize, sustain), *poka-woke* (i.e., fail-safe), Deming wheel (i.e., plan–do–check–act), and SPC are used to reduce these inefficiencies and prevent their reoccurrence. Along this line, improved process and machine ergonomics, setup time reduction using single minute exchange of die (SMED), and fixtures can also increase efficiency (Li et al., 2002; Mondragon et al., 2004). Inefficient equipment can also be replaced or modernized in order to increase production capacity. Similarly, inefficiencies can be reduced with the design of efficient production procedures in order to use the right material with the right machine. Once inefficiencies have been reduced, or eliminated, processes can be standardized and cross-training can be used, so improved processes can be taught to new employees, in order to further reduce learning time. Total productive maintenance, with the implementation of preventive maintenance programs, can also be used to increase machine availability and to directly contribute to reducing variability from within the value chain.

Another general approach to reducing inefficiencies deals with advanced planning and scheduling. Operations planning and scheduling involves the use of advanced planning tools, which require the development of short-term operational to long-term strategic optimization models. These models aim at achieving an economically efficient use of the value chain's resources. For instance, capacitated lot-sizing optimization aims to plan which products should be produced, when, and in what quantities under capacity and set up constraints in order to meet customer demand. Operations planning also includes machine scheduling, distribution planning, transportation planning, and route planning. It is also a vast academic domain of research, although the forest products industry has yet to adopt such tools. At the value chain level, state-of-the-art collaborative processes, such as collaborative planning, forecasting, and replenishment (CPFR), allows value chain partners to jointly plan replenishment operations based on a agreed-upon demand forecast. Other more experimental approaches are being developed in order to support the collaborative planning of manufacturing operations. The reader is referred to Stadtler (2009), Frayret (2009), and Lehoux et al. (2014) for recent reviews of this subject.

Along this line, operational efficiency can be achieved with the implementation of operations control strategies, including advanced dispatching rules (e.g., minimum slack time, minimum due date) and just-in-time (JIT) control strategies (e.g., Kanban, CONWIP). A control strategy aims to control in a reactive manner the workflows and inventories in a manufacturing system modeled as a queuing system. Dispatching rules and mechanisms are used in order to prioritize operations and guide their flow throughout the system. Usually designed in order to be relatively simple to operate, some advanced control strategies can involve more complex forms of information exchange, such as the production authorization card (MacDonald and Gunn, 2010).

Yet another technique to further reduce inefficiencies is value chain process and information chain integration. Process and information chain integration aims at reducing the inefficiencies at the interface between processes. On the one hand, information chain integration aims at streamlining order and order status information exchange across the value chain. This includes point-of-sale (POS) data and inventory level information exchange. Along the same line, products and resources can also be tracked and monitored in real-time using the global positioning system (GPS) and radio frequency identification (RFID). On the other hand, process integration deals with the coordination of workflows across companies within the value chain (Yang et al., 2007). For example, new product development and introduction involves many processes in different departments and organizations, such as engineering, design, regulation/certification, marketing, cost and risk management, and prototyping. Process integration aims ultimately at creating a virtual network of organizations and their performance can be monitored and controlled as a single organizational unit. Information technology, such as ERP and PLM systems, provides functions that allow for management and exchange of different types of information across the value chain. In general, these systems can also support the management of workflow between organizations.

9.3.4 Design for Flexibility

In order to push further a value chain's agility, flexible processes, products, and organizational structures can be developed and adopted. For instance, flexible manufacturing systems using computer numerical controlled (CNC) machines, robots (Frei and Whitacre, 2012), and manufacturing execution systems (MES) can automatically coordinate the manufacturing processes of several machines to meet customized orders with minimal setup time.

Flexibility can also be achieved through the development of product platforms. A product platform involves the joint design of standard product components and processes (including equipment and fixtures), which can be used in various configurations in order to create a wide variety of products in an efficient manner. This technology allows the value chain to produce a large number of different products, while taking advantage of standardized and efficient processes.

Finally, agility can also be achieved with the design of flexible forms of organization and by developing a culture of change that allows for self-organization (Eaton, 2003). For instance, virtual network organizations based on some form of service-oriented architecture, in which partners can be assembled on-demand (D'Amours et al., 1997), can provide the organizational flexibility to quickly adapt to change at the highest level possible. Collaboration across organizations involving resource pooling and sharing (e.g., trucks, pallets, storage, R&D) can also provide some level of flexibility in which benefit and risk are shared.

9.4   AGILITY REQUIREMENTS IN THE SOFTWOOD VALUE CHAIN

This section proposes a brief overview of some of the issues in the forest value chain with respect to agility. More specifically, this section first proposes a quick diagnostic of the softwood lumber value chain. Next, we present some of the many initiatives, both academic and industrial, to improve agility in this industry. The reader is referred to Vahid et al. (2014) for a review of value chain analysis frameworks for the forest product industry. For other sectors of the forest products value chain, the reader is referred to Hughes et al. (2014), who introduce a review of wood pellet supply chain management approaches, and Mansoornejad et al. (2011), who address the need for design flexibility in the forest biorefinery sector.

9.4.1   QUICK DIAGNOSTIC

The softwood value chain includes all operations from harvesting, to log delivery at the mills, to sawing, drying, and planing, to lumber distribution and secondary application transformations, as illustrated in Figure 9.2. It also includes by-products transformation and distribution such as slash grinding and sawdust transportation to heating and electricity generation plants. Coproducts in the forms of wood chips are transformed into other products, such as pulp and paper, biomass solid fuel, or biorefined products. This sector of the forest value chain currently lives in a profound transition from being almost exclusively dedicated to paper production, to the production of bioproducts and biofuel.

The softwood value chain has two main sources of variability and several characteristics that increase its complexity. On the one hand, being a natural resource industry, this value chain deals with input material variability. Unlike other forest product industry sectors based on plantation, the Canadian softwood value chain must transform logs of different sizes, shapes, and species from mixed natural

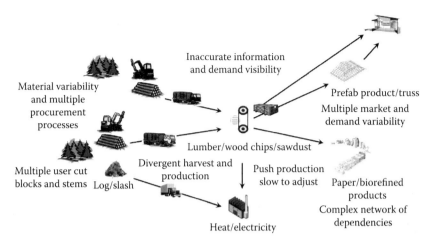

FIGURE 9.2   Softwood value chain.

forests. Consequently, before logs are delivered to sawmills, their characteristics are rather poorly, or inaccurately, known, although forest companies have some information regarding the roadside volume and log length.

On the other end of the softwood lumber value chain, and like most value chains in any sectors, it deals with customer demand variability, mostly related to the construction sector. Because forest product transformation is a divergent process that produces many coproducts simultaneously, it must also deal with uncoordinated demand processes. In other words, demand variability for these coproducts can lead to unsold inventories for products with low demand. The reader is referred to Marier et al. (2014) for a study of the Sales and Operations Planning process in the lumber industry. Concerning forest operations, harvesting is planned in advance in order to give sufficient time to build the required forest roads and camps for the harvest teams. On public land, these plans must be approved with the community and cannot be changed easily. Therefore, the output of forest operations generally lacks short-term flexibility, although weekly operations can be adapted with respect to log size, for instance. In some provinces, such as Québec, auctions create potential volume flexibility, as forest companies can increase or decrease log procurement according to their own ability to generate profit. However, it also increases the complexity of the procurement process.

These basic sources of variability require flexibility with respect to sawing operations. For instance, sawing equipment in the Canadian softwood lumber industry is capable of scanning input logs and work-in-process to react to any log variation in order to prioritize the production of specific lumber dimensions. This agility allows sawing to adapt to material variability. However, sawmills are not configured in order to efficiently produce lumber for different markets, which have different size and quality requirements resulting in out-of-specification products (Fournier, 2011). In other words, this technology is not yet utilized in order to adapt product mix and product prioritization according to customer demand. Furthermore, the industry does not have the capability to analyze whether it can produce specific orders (Fournier, 2011). Along the same line, operations coordination and planning in sawmills, including sawing, drying, and planing, are not yet supported by decision support tools.

9.4.2 Agility Requirements on Harvesting

Harvesting operations produce several outputs simultaneously, including logs of various sizes and species and postharvest slash, the value of which is increasing on the bioenergy market (Ralevic et al., 2010). The output of harvest operations is the first constraint the softwood value chain must deal with because everything else depends upon the quantity, quality, and mix of log deliveries. It is therefore critical to have accurate information about this resource in order to make the best harvest planning decisions and deliver the right logs to the right mill at the right time. This information includes species, stem dimension and shape, and forest conditions. This information can be captured using light detection and ranging (LiDAR) technology, or less accurately, using aerial pictures. The respective values of these different technologies must be assessed with respect to how they are implemented, and for what purpose. For example, Shorthouse (2012) studies the possibility of using enhanced preharvest inventory information in order to predict green lumber value recovery. Another important aspect is information sampling, both spatially and in time. Data can be measured in a continuous manner or it can be measured at specific locations and dates. If tree growth does not require a continuous measurement of their size and shape, the status of harvest equipment, their location, roadside inventories, on-board log volume, and length can change quickly. For instance, real-time on-board volume and length information can be used in order to optimize log yard operations and enable hot logging (i.e., direct from the truck to sawing). Along this line, forest information can be specific to each stem or log or it can be specific at a more aggregated level, such as a pile or an inventory location.

Value chain agility also requires accurate information of harvest operations and transportation performance, yield, and cost. It also requires near real-time machine availability, status, and reliability in order to improve planning and operations coordination.

From an organizational standpoint, processes and decision structure should be designed in order to achieve the value chain agility objectives. For instance, cut-to-length operations require early bucking decisions. However, the postponement of bucking decisions allows for the late differentiation of stems into logs of length that match customer demand. Similarly, the postponement of transportation decisions also allows for on-demand deliveries. This strategy can be implemented using transhipment facilities in the forest, the function of which would be to centralize on-demand deliveries to mills from a single facility. Furthermore, procurement processes based on auctions are theoretically more flexible than timber licenses, especially if volume flexibility is an issue. However, the adoption of such a process requires a cultural change in the industry. An even greater cultural change is the adoption of the make-to-order philosophy, according to which logs and stems are cut-to-order and sawn-to-order. This philosophy is already used in northern European countries. It concerns both harvest operations and sawing. Finally, another organizational requirement for agility involves internal and external process and information chain integration. Harvest operations involve different actors including forest companies, entrepreneurs, forest cooperatives, and governments. This implies both organizational processes and information systems should be specifically designed and coordinated in order to eliminate all sources of inefficiencies. For instance, on-board systems for entrepreneurs and drivers can be connected to forest product information systems in order to send back and forth demand information, order status, inventory levels, and so on. Such technology would be necessary for procurement strategies based on log consignment in a vendor-managed inventory philosophy.

Finally, from an operational standpoint, the achievement of agility in this industry requires the streamlining of operations, and the elimination of wastes, such as unnecessary inventory, decay, and truck line-ups. One way to eliminate inefficiencies is to optimize planning decisions in order to cut the right log with the right stem, and transport it to the right mill at the right time, with the best transportation mode. Forest operations planning optimization has been thoroughly studied in the academic community. Many different aspects have been studied from bucking optimization (Chauhan et al., 2009, 2011; Dems et al., 2013) to annual tactical planning (Beaudoin et al., 2007, 2008), to forest road design (Meignan et al., 2012, 2013), and forest camps location (Jena et al., 2012), and to auction configuration (Farnia et al., 2013a, 2013b). Similarly, transportation, including truck scheduling (El Hachemi et al., 2013), and backhauling and coalition formation among forest companies have been investigated (Audy et al., 2012). Concerning by-products of forest operations, similar challenges must be addressed in order to improve efficiency. For instance, hog fuel grinding operation configuration and transportation fleet size must be optimized in order to meet demand (Pledger et al., 2014). The reader is referred to Martell et al. (1998), Rönnqvist (2003), and D'Amours et al. (2008) for thorough reviews on the subject of operations research applications and challenges in the forest product industry.

9.4.3 AGILITY REQUIREMENTS ON SAWMILL OPERATIONS

Sawmill operations produce simultaneously several coproducts (i.e., lumber of various sizes; wood chips) and by-products (i.e., bark, sawdust) from logs, using a rather complicated network of equipment (i.e., head rig, canter, edger, trimmer, sorter) and conveyors. Sawmills also generally possess several kiln driers of various sizes and locations in the yard for air drying, as well as a planing facility, where dry rough lumber is planed. It is an industry that is driven by production campaigns to produce lumber for different markets, although its equipment cannot be quickly adapted from one campaign to the other (Fournier, 2011). In order to increase agility, a sawmill must have accurate information about log availability (both volume and quality) and on-hand lumber inventory. Log sorting, by improving the level of information on log availability, can reduce the need for flexibility and enable ribbon feeding (i.e., continuous feeding of logs of a similar shape). A sawmill should also be able to use roadside and on-board log inventory information in order to optimize yard operations and enable hot logging.

The ability to accurately forecast demand for each market application is also necessary to produce the right inventory to react quickly to customer demand. Along the same line, in order to be able to know whether an order can be met on time, production yield, cost, and machine availability and reliability must also be known with a certain level of accuracy.

From an organizational standpoint, the processes and decision structures must be designed to provide the required level of agility. For instance, the position of the decoupling point between the push and pull production strategies within the sawmill processes must be identified with respect to the desired response time objectives (Cid et al., 2009). Next, these processes must also be coordinated within the sawmill operations (Gaudreault et al., 2010) and with the procurement and the sales and distribution functions. In particular, if a forest product company possesses several sawmills, orders must be allocated to the right sawmill with respect to production and transportation capacity and cost. This information must therefore be exchanged regularly between the different production sites and the sales and distribution function. This includes the capability to manage what-if scenarios in order to quickly analyze the most profitable way to meet a customer order. This requires yet another drastic cultural change from a pure make-to-stock strategy, in which product sales are based solely on on-hand inventory and yield forecast, to a customer-driven strategy, such as saw-to-order, dry-to-order, or plan-to-order. This requires knowing the lumber recovery factor of each log type in order to select which logs should be transformed to meet a specific mix of orders. The adoption of a customer-driven production strategy requires advanced simulation tools to analyze the best strategy in a specific context and with specific agility objectives.

Finally, from an operational standpoint, the path towards agility requires the elimination of wastes and inefficiencies (Gunn et al., 2013). The basic principles of lean manufacturing must be adapted to lumber production. This includes setup time reduction (i.e., SMED). Green lumber production involves two types of setup. The first type of setup concerns the automatic positioning of the logs and work-in-process into the equipment feeder (i.e., scan-and-set process). It is dome using scanner information about the shape of the lumber pieces to transform. Although it is almost instantaneous, the need for such a setup can be reduced if the logs to process have a similar shape. This allows for the reduction of the minimum distance between the pieces to process using ribbon feeding. The second type of setup concerns the need to change priority lists in each piece of equipment in the sawmill in order to produce lumber for a different market. The reduction of this type of setup is particularly critical for a campaign-driven industry, which must optimize its batch sizes according to the order list and the target market (Fournier, 2011). The elimination of waste also includes the minimization of out-of-spec products. Along the same lime, log yard operations, including log handling, sorting, and lumber handling, must be optimized in order to reduce unnecessary travel time and the time to empty the bins where specific sizes of lumber are accumulated during production.

Another important constraint to become an agile value chain deals with kiln size. Different species and lumber thickness must be kiln-dried. However, because they require different drying programs, they cannot be mixed together during the generally lengthy drying process. Consequently, large kiln batch sizes constrain sawmills' ability to deliver small volumes of lumber, because a kiln must be completely filled before drying can start.

Finally, in order to become agile, the lumber production industry must develop and adopt decision support tools to optimize production campaigns (Zanjani et al., 2009, 2010) and coordinate operations throughout the entire process (Gaudreault et al., 2010) in order to quickly adapt to demand mix and material variability, and take advantage of opportunities.

9.5 CONCLUSIONS

This chapter has proposed a definition of the concept of agility as being the property of a complex system that can quickly process timely and relevant information from its volatile environment to adjust its operations in order to achieve efficiently its purpose. The information, organizational, and operational requirements of agility were quickly analyzed in order to better understand the general processes and techniques a value chain must adopt to become agile. In particular, the needs to (1) reduce the value chain's perceived variability, (2) protect against variability, (3) reduce inefficiencies, and (4) design and adopt new flexible processes and technologies were briefly overviewed.

Next, the status of the softwood value chain was quickly analyzed with respect to its ability to be agile. Hence, the softwood value chain is slowly evolving towards agility. However, much of this transformation remains to be done. In order to successfully achieve this transformation, it must first adopt a customer-driven philosophy, and define its own agility objectives. This requires a profound cultural change in an industry that is designed for the purpose of producing commodity products.

This effort toward agility also requires the development of new tools and processes that specifically address the needs of this value chain. In particular, as mentioned in the introduction, tools must be developed in order to better understand how agility can contribute to companies' prosperity, as well as to measure and assess the level and the type of agility that is required. Because agility is an emerging characteristic of a complex system, simulation tools in general and agent-based simulation in particular (see Frayret [2011] for a general overview of multiagent technology in the forest product industry) appear to be appropriate tools to use. Such a technology is still relatively new compared to traditional operation research-based tools. Furthermore, because of its ability to create swarm intelligence, multiagent technology can contribute to building interconnected, reactive, and proactive systems across the value chain. Such technology would allow operations to be coordinated and adapted to the individual environment of each business unit of the forest product value chain.

REFERENCES

Abdul Kadir, A., Xu, X., Hämmerle, E. 2011. Virtual machine tools and virtual machining—A technological review. *Robotics and Computer-Integrated Manufacturing*, 27(3), 494–508.

Albuquerque, C. O., Antonino, P. O., Nakagawa, E. Y. 2012. An investigation into agile methods in embedded systems development. Proceedings of the 12th International Conference on Computational Science and Its Applications, June 18–21, Salvador de Bahia, Brazil.

Al-Tahat, M. D., Bataineh, K. M. 2012. Statistical analyses and modeling of the implementation of agile manufacturing tactics in industrial firms. *Mathematical Problems in Engineering*, 2012, 1–23.

Audy, J.-F., D'Amours, S., Rönnqvist, M. 2012. An empirical study on coalition formation and cost/savings allocation. *International Journal of Production Economics*, 136(1), 13–27.

Beaudoin, D., Frayret, J.-M., LeBel, L. 2008. Hierarchical forest management with anticipation: An application to tactical-operational planning integration. *Canadian Journal of Forest Research*, 38(8), 2198–2211.

Beaudoin, D., LeBel, L., Frayret, J.-M. 2007. Tactical supply chain planning in the forest products industry through optimization and scenario-based analysis. *Canadian Journal of Forest Research*, 37(1), 128–140.

Beaudoin, D., LeBel, L., Soussi, M. A. 2012. Discrete event simulation to improve log yard operations. *INFOR*, 50(4), 175–185.

Bi, Z. M., Lang, S. Y. T., Shen, W., and Wang, L. 2008. Reconfigurable manufacturing systems: the state of the art. *International Journal of Production Research*, 46(4), 967–992.

Borissova, A., Fairweather, M., Goltz, G. E. 2006. Combinatorial process and plant design for agile manufacture. *Research in Engineering Design*, 17(1), 1–12.

Chae, H., Choi, Y., Kim, K. 2007. Component-based modeling of enterprise architectures for collaborative manufacturing. *International Journal of Advanced Manufacturing Technology*, 34(5–6), 605–616.

Chauhan, S. S., Frayret, J.-M., Lebel, L. 2009. Multi-commodity supply network planning in the forest supply chain. *European Journal of Operational Research*, 196(2), 688–696.

Chauhan, S. S., Frayret, J.-M., LeBel, L. 2011. Supply network planning in the forest supply chain with bucking decisions anticipation. *Annals of Operations Research*, 190(1), 93–115.

Chen, L., Macwan, A., Li, S. 2006. Model-based rapid redesign using decomposition patterns. *Journal of Mechanical Design*, 129(3), 283–294.

Chengying, L., Xiankui, W., Yuchen, H., 2003. Research on manufacturing resource modeling based on the 0-0 method. *Journal of Materials Processing Technology*, 139, 40–43.

Christopher, M., Lowson, R., Peck, H. 2004. Creating agile supply chains in the fashion industry. *International Journal of Retail & Distribution Management*, 32(8), 367–376.

Cid Y. F., Frayret, J.-M., Léger, F., Rousseau, A. 2009. Agent-based simulation and analysis of demand-driven production strategies in the lumber industry. *International Journal of Production Research*, 47(22), 6295–6319.

Coronado, A. E., Sarhadi, M., Millar, C. 2002. Defining a framework for information systems requirements for agile manufacturing. *International Journal of Production Economics*, 75(1–2), 57–68.

Crocitto, M., Youssef, M. 2003. The human side of organizational agility. *Industrial Management & Data Systems*, 103(6), 388–397.

D'Amours, S., Montreuil, B., Soumis, F. 1997. Price-based planning and scheduling of multiproduct orders in symbiotic manufacturing networks. *European Journal of Operational Research*, 96(1), 148–166.

D'Amours, S., Ronnqvist, M., Weintraub, A. 2008. Using operational research for supply chain planning in the forest products industry. *INFOR*, 46(4), 265–281.

Dems A., Frayret J.-M., Rousseau, L.-M., 2013. Effects of different cut-to-length harvesting structures on the economic value of a wood procurement planning problem. *Annals of Operations Research*, 232, 65–86.

Devadasan, S. R., Goshteeswaran, S., Gokulachandran, J. 2005. Design for quality in agile manufacturing environment through modified orthogonal array-based experimentation. *Journal of Manufacturing Technology Management*, 16(6), 576–597.

Dotoli, M., Fanti, M. P., Meloni, C., Zhou, M. 2006. Design and optimization of integrated e-supply chain for agile and environmentally conscious manufacturing. *IEEE Transactions on Systems, Man and Cybernetics, Part A: Systems and Humans*, 36(1), 62–75.

Dowlatshahi, S., Cao, Q. 2005. The relationships among virtual enterprise, information technology, and business performance in agile manufacturing: An industry perspective. *European Journal of Operational Research*, 174(2), 835–860.

Eaton, M. 2003. Running the lean agile marathon. *Manufacturing Engineer*, 82(6), 14–17.

El Hachemi, N., Gendreau, M., Rousseau, L.-M. 2013. A heuristic to solve the synchronized log-truck scheduling problem. *Computers & Operations Research*, 40(3), 666–673.

Falcioni, J. G. 2004. The agile engineer. *Mechanical Engineering*, 126(9), 4.

Fan, Q. 2010. Research on the operations strategy of Chinese manufacturing enterprises under the background of international subcontracting. Proceedings of the IEEE International Conference on Industrial Engineering and Engineering Management, October 29–31, Xiamen, China.

Farnia, F., Frayret, J.-M., Beaudry, C., Lebel, L. 2013a. Time-based combinatorial auction for timber allocation and delivery coordination. *Forest Policy and Economics*, 50(1), 143–152.

Farnia, F., Frayret, J.-M., LeBel, L., Beaudry, C. 2013b. Multiple-round timber auction design and simulation. *International Journal of Production Economics*, 146(1), 129–141.

Fekri, R., Aliahmadi, A., Fathian, M. 2009. Predicting a model for agile NPD process with fuzzy cognitive map. *International Journal of Advanced Manufacturing Technology*, 41(11–12), 1240–1260.

Feng, D. Z., Yamashiro, M. 2006. A pragmatic approach for optimal selection of plant specific process plans in a virtual enterprise. *Journal of Materials Processing Technology*, 173(2), 194–200.

Flower, S. 2004. Balancing innovation in an agile world. *Manufacturing Engineer*, 83(5), 40–43.

Found, P., Harvey, R. 2007. Leading the lean enterprise. *Engineering Management*, 17(1), 40–43.

Fournier, F. 2011. Agile issues: Eastern Softwood and Hardwood Industry. VCO—Workshop on Agile Manufacturing and Logistics for the Forest Value Chain, February, 2011, Montréal, Canada.

Franzen, V., Kwiatkowski, L., Martins, P. A. F., Tekkaya, A. E. 2009. Single point incremental forming of PVC. *Journal of Materials Processing Technology*, 209(1), 462–469.

Frayret, J.-M. 2009. A multidisciplinary review of collaborative supply chain planning. Proceedings of the IEEE International Conference on Systems, Man, and Cybernetics, October 11–14, San Antonio, TX.

Frayret, J.-M. 2011. Multi-agent system applications in the forest products industry. *Journal of Science and Technology for Forest Products and Processes*, 1(2), 15–29.

Frei, R., Whitacre, J. 2012. Degeneracy and networked buffering: principles for supporting emergent evolvability in agile manufacturing systems. *Natural Computing*, 11(3), 417–430.

Fung, R. Y. K., Liang, F., Jiang, Z. 2008. A multi-stage methodology for virtual cell formation oriented agile manufacturing. *International Journal of Advanced Manufacturing Technology*, 36(7–8), 798–810.

Garcia, M. E., Valero, S., Argente, E., Giret, A., Julian, V. 2008. A fast method to achieve flexible production programming systems. *IEEE Transactions on Systems, Man and Cybernetics, Part C: Applications and Reviews*, 38(2), 242–252.

Gaudreault, J. 2011. Evaluating supply chain performance using multi-agent simulation. 2011. VCO—Workshop on Agile Manufacturing and Logistics for the Forest Value Chain, February, 2011, Montréal, Canada.

Gaudreault, J., Forget, P., Frayret, J.-M., Rousseau, A., D'Amours, S. 2010. Distributed operations planning for the softwood lumber supply chain: Optimization and coordination. *International Journal of Industrial Engineering*, 17(3), 168–189.

Gen, M., Lin, L., Zhang, H. 2009. Evolutionary techniques for optimization problems in integrated manufacturing system: State-of-the-art-survey. *Computers & Industrial Engineering*, 56(3), 779–808.

Gil, A., Frayret, J.-M. 2015. Log classification in the hardwood timber industry: Method and value analysis. *International Journal of Production Research,* DOI:10.1080/00207543 .2015.1106607.

Gunasekaran, A., Yusuf, Y. Y. 2002. Agile manufacturing: A taxonomy of strategic and techno-logical imperatives. *International Journal of Production Research*, 40(6), 1357–1385.

Gunn, E. A., MacDonald, C., Saatadayar, S., Sohrabi, P. 2013. Sawmill manufacturing: mov-ing towards lean. *Value Chain Optimization Network Webinar*, April 3, 2013. Available online: http://www.reseauvco.ca, accessed on 18 April 2016.

Harris, A. 2004. Reaping the rewards of agile thinking. *Manufacturing Engineer*, 83(6), 24–27.

Harrison, A. 2003. Customised, integrated and agile supply chains. *Manufacturing Engineer*, 82(3), 28–31.

Helo, P. 2004. Managing agility and productivity in the electronic industry. *Industrial Management & Data Systems*, 104(7), 567–577.

Helo, P.T. 2006. Agile production management: An analysis of capacity decisions and order-fulfillment time. *International Journal of Agile Systems and Management*, 1(1), 2–10.

Hopp, W. J., Oyen. M. P. 2004. Agile workforce evaluation: A framework for cross-training and coordination. *IIE Transactions*, 36(10), 919–940.

Hoyer, V., Stanoevska-Slabeva, K. 2009. IT impacts on operation-level agility in service industries. Proceedings of the European Conference on Information Systems, June 8–10, Verona, Italy.

Huang, P.-Y., Pan, S. L., Ouyang. T. H. 2014. Developing information processing capability for operational agility: Implications from a Chinese manufacturer. *European Journal of Information Systems*, 23(4), 462–480.

Hughes, N. M., Shahi, C., Pulkki, R. 2014. A review of the wood pellet value chain, modern value/supply chain management approaches, and value/supply chain models. *Journal of Renewable Energy*, 2014, 1–14.

Ip, W. H., Yung, K. L., Wang, D. 2004. A branch and bound algorithm for sub-contractor selection in agile manufacturing environment. *International Journal of Production Economics*, 87(2), 195–205.

James, T. 2005. Stepping back from lean. *Manufacturing Engineer*, 84(1), 16–21.

Jena, S. D., Cordeau, J.-F., Gendron, B. 2012. Modeling and solving a logging camp location problem. *Annals of Operations Research*, 232(1), 151–177.

Jiang, D. Z., Fung, R. Y. K. 2003. An adaptive agile manufacturing control infrastructure based on TOPNs-CS modeling. *International Journal of Advanced Manufacturing Technology*, 22(3), 191–215.

Jiao, L. M., Khoo, L. P., Chen, C. H. 2004. An intelligent concurrent design task planner for manufacturing systems. *International Journal of Advanced Manufacturing Technology*, 23(9–10), 672–681.

Jin, J. W., Ni, Y. R., Xi, J. T., Fan, F. Y., Jin, Y. 2002. Research on the application of tech-nology-oriented rapid response manufacturing in a distributed network environment. *Journal of Materials Processing Technology*, 129(1–3), 579–583.

Jin-Hai, L., Anderson, A. R., Harrison, R. T. 2003. The evolution of agile manufacturing. *Business Process Management Journal*, 9(2), 170–189.

Kalbande, D., Shah, C., Nigam, A., Kothawade, P. 2012. Integrating ERP to accelerate busi-ness process agility: A case study and critical research review in Indian pharmaceutical industry. Proceedings of the International Conference on Communication, Information and Computing Technology, October 19–20, Mumbai, India.

Kamal, M. 2007. Agile manufacturing of a micro-embossed case by a two-step electro-magnetic forming process. *Journal of Materials Processing Technology*, 190(1–3), 41–50.

Kulturel-Konak, S. 2007. Approaches to uncertainties in facility layout problems: Perspectives at the beginning of the 21st century. *Journal of Intelligent Manufacturing*, 18(2), 273–284.

Kundu, S., McKay, A., de Pennington, A. 2008. Selection of decoupling points in supply chains using a knowledge-based approach. *Journal of Engineering Manufacture*, 222(11), 1529–1549.

Lau, H. C. W., Wong, C. W. Y., Pun, K. F., Chin, K. S. 2003. Virtual agent modeling of an agile supply chain infrastructure. *Management Decision*, 41(7), 625–634.

Le, V. T., Gunn, B. M., Nahavandi, S. 2004. MRP-production planning in agile manufacturing. Proceedings of the IEEE International Conference on Intelligent Systems, June 22–24, Varna, Bulgaria.

Lee, S. M., Harrison, R., West, A. A. 2005. A component-based control system for agile manufacturing. *Journal of Engineering Manufacture*, 219(1), 123–35.

Lee, C. K. M., Lau, H. C. W., Yu, K. M., Fung, R. Y. K. 2004. Development of a dynamic data interchange scheme to support product design in agile manufacturing. *International Journal of Production Economics*, 87(3), 295–308.

Lehoux, N., D'Amours, S., Langevin, A. 2014. Inter-firm collaborations and supply chain coordination: Review of key elements and case study. *Production Planning and Control*, 25(10), 858–872.

Lei, D., Lian, L., Xiao, S. 2011. The system construction and research review of agile manufacturing. Proceedings of the International Conference on Management and Service Science, August 12–14, Wuhan, China.

Leitão, P. 2009. Agent-based distributed manufacturing control: A state-of-the-art survey. *Engineering Applications of Artificial Intelligence*, 22(7), 979–991.

Li, P. G., Li, S. X., Rao, Y. Q. 2004. Coalition formation and its application in planning for agile manufacturing cell. *International Journal of Advanced Manufacturing Technology*, 24(3–4), 298–305.

Li, W., Li, P., Rong, Y. 2002. Case-based agile fixture design. *Journal of Materials Processing Technology*, 128(1), 7–18.

Lim, M. K., Zhang, Z. 2003. A multi-agent based manufacturing control strategy for responsive manufacturing. *Journal of Materials Processing Technology*, 139(1–3), 379–384.

Lin, C-T., Chiu, H., Tseng, Y.-H. 2006. Agility evaluation using fuzzy logic. *International Journal of Production Economics*, 101(2), 353–368.

Lin, C., Lin, Y.-T. 2006. Mitigating the bullwhip effect by reducing demand variance in the supply chain. *International Journal of Advanced Manufacturing Technology*, 28(3–4), 328–336.

Lou, P., Zhou, Z. D., Chen, Y.-P., Ai, W. 2004. Study on multi-agent-based agile supply chain management. *International Journal of Advanced Manufacturing Technology*, 23(3–4), 197–203.

Lü, B., Li, Z., Liu, K. 2004. Study on integrated infrastructure for agile manufacturing systems. *Journal of Integrated Design and Process Science*, 8(4), 99–105.

Ma, H., Davidrajuh, R. 2005. An iterative approach for distribution chain design in agile virtual environment. *Industrial Management & Data Systems*, 105(6), 815–834.

MacDonald, C., Gunn, E. A. 2010. A framework for analysis of production authorization card-controlled production systems. *Production and Operations Management*, 20(6), 937–948.

Maione, G., Naso, D. 2004. Modeling adaptive multi-agent manufacturing control with discrete event system formalism. *International Journal of Systems Science*, 35(10), 591–614.

Manenti, P. 2009. Profitable proximity. *Engineering & Technology*, 4(1), 75–75.

Mansoornejad, B., Pistikopoulos, N. E., Stuart, P. 2011. Incorporating flexibility design into supply chain design for forest biorefinery. *Journal of Science and Technology for Forest Products and Processes*, 1(2), 54–66.

Marapoulos, P. G., Bramall, D. G., McKay, K. R., Rogers, B., Chapman, P. 2003. An aggregate resource model for the provision of dynamic 'resource-aware' planning. *Journal of Engineering Manufacture*, 217(10), 1471–1480.

Marier, P., Bolduc, S., Ali, M. B., Gaudreault, J. 2014. S&OP network model for commodity lumber products. Research paper CIRRELT-2014-25. Québec, Canada: CIRRELT.

Martell, D. L., Gunn, E. A., Weintraub, A. 1998. Forest management challenges for operational researchers. *European Journal of Operational Research*, 104(1), 1–17.

Mehrsai, A., Thoben, K.-D., Scholz-Reiter, B. 2014. Bridging lean to agile production logistics using autonomous carriers in pull flow. *International Journal of Production Research*, 52(16), 4711–4730.

Meignan, D., Frayret, J.-M., Pesant, G. 2013. Interactive planning system for forest road location. Research paper CIRRELT-2013-37. Québec, Canada: CIRRELT.

Meignan, D., Frayret, J.-M., Pesant, G., Blouin, M. 2012. An automated heuristic approach to forest road location. *Canadian Journal of Forest Research*, 42(12), 2130–2141.

Minoni, U., Cavalli, F. 2008. Surface quality control device for on-line applications. *Measurement*, 41(7), 774–778.

Mondragon, A. E. C., Lyons, A. C., Kehoe, D. F. 2004. Assessing the value of information systems in supporting agility in high tech manufacturing enterprises. *International Journal of Operations & Production Management*, 24(12), 1219–1246.

Moore, P. R., Pu, J., Ng, H. C., Wong, C. B., Chong, S. K., Chen, X., Adolfsson, J., Olofsgard, P., Lundgren, J.-O. 2003. Virtual engineering: An integrated approach to agile manufacturing machinery design and control. *Mechatronics*, 13(10), 1105–1121.

Morel, G., Panetto, H., Zaremba, M., Mayer, F. 2003. Manufacturing enterprise control and management system engineering. *Annual Reviews in Control*, 27(2), 199–209.

Neves, F. T., Correia, A. M. R., Rosa, V. N., de Castro Neto, M. 2011. Knowledge creation and sharing in software development teams using agile methodologies: Key insights affecting their adoption. Proceedings of the Conference on Information Systems and Technologies, June 15–18, Chaves, Portugal.

Okazaki, Y., Mishima, N., Ashida, K. 2004. Microfactory-concept, history, and developments. *Journal of Manufacturing Science & Engineering*, 126(4), 837–844.

Onuh, S., Bennett, N., Hughes, V. 2006. Reverse engineering and rapid tooling as enablers of agile manufacturing. *International Journal of Agile Systems and Management*, 1(1), 60–72.

Paixão, A. C., Marlow, P. B. 2003. Fourth generation ports—A question of agility? *International Journal of Physical Distribution & Logistics Management*, 33(4), 355–376.

Pan, P. Y., Cheng, K., Harrison, D. K., 2003. A web-based agile system for rolling bearing design. *Integrated Manufacturing Systems*, 14(6), 518–529.

Papamarcos, S. D., Latshaw, C., Watson, G. W. 2007. Individualism-collectivism and incentive system design as predictive of productivity in a simulated cellular manufacturing environment. *International Journal of Cross Cultural Management*, 7(2), 253–265.

Pledger, S., Ghafghazi, S., Bull, G. 2014. Analyzing forest biomass supply chains using agent based simulation modeling. Value Chain Optimization Network Webinar, September 24, 2014. Available online: http://www.reseauvco.ca, accessed on 18 April 2016.

Ralevic, P., Ryans, M., Cormier, D. 2010. Assessing forest biomass for bioenergy: Operational challenges and cost considerations. *Forestry Chronicle*, 86(1), 43–50.

Ramesh, G., Devadasan, S. R. 2007. Literature review on the agile manufacturing criteria. *Journal of Manufacturing Technology Management*, 18(2), 182–201.

Ren, J., Yusuf, Y. Y., Burns, N. D. 2003. The effects of agile attributes on competitive priorities: A neural network approach. *Integrated Manufacturing Systems*, 14(6), 489–497.

Ren, J., Yusuf, Y. Y., Burns, N. D. 2009. A decision-support framework for agile enterprise partnering. *International Journal of Advanced Manufacturing Technology*, 41(1–2), 180–192.

Rönnqvist, M. 2003. Optimization in forestry. *Mathematical Programming*, 97(1–2), 267–284.

Ross, A. 2003. Creating agile supply. *Manufacturing Engineer*, 82(6), 18–21.

Sahnoun, M., Bettayeb, B., Bassetto, S.-J., Tollenaere, M. 2014. Simulation-based optimization of sampling plans to reduce inspections while mastering the risk exposure in semiconductor manufacturing. *Journal of Intelligent Manufacturing*, 1–15, DOI 10.1007/s10845-014-0956-x.

Sanchez, L. M., Nagi, R. 2001. A review of agile manufacturing systems. *International Journal of Production Research*, 39(16), 3561–3600.

Santa-Eulalia, L. A., Aït-Kadi, D., D'Amours, S., Frayret, J.-M., Lemieux, S. 2011. Agent-based experimental investigations about the robustness of tactical planning and control policies in a softwood lumber supply chain. *Production Planning and Control*, 22(8), 782–799.

Santa-Eulalia, L. A., D'Amours, S., Frayret, J.-M. 2012. Agent-based simulations for advanced supply chain planning: The FAMASS methodological framework for requirements analysis and deployment. *International Journal of Computer Integrated Manufacturing*, 25(10), 963–980.

Sauvageau, G., Frayret, J.-M. 2015. Waste paper procurement optimization: an agent-based simulation approach. *European Journal of Operational Research*, 242(3), 987–998.

Sato, R., Tsai, T. L. 2004. Agile production planning and control with advance notification to change schedule. *International Journal of Production Research*, 42(2), 321–336.

Shahi, S., Pulkki, R. 2015. A simulation based optimization approach to integrate inventory management of a sawlog supply chain with demand uncertainty. *Canadian Journal of Forest Research*, 45(10), 1313–1326.

Sharma, R., Gao, J. X. 2007. A knowledge-based manufacturing and cost evaluation system for product design/re-design. *International Journal of Advanced Manufacturing Technology*, 33(9–10), 856–865.

Sherehiy, B., Karwowski, W. 2014. The relationship between work organization and workforce agility in small manufacturing enterprises. *International Journal of Industrial Ergonomics*, 44(3), 466–473.

Shorthouse, K. D. 2012. Pre-harvest lumber value recovery modeling: integrating sawline laser-scanning with an enhanced forest inventory, M.Sc. Thesis, Faculty of Natural Resources Management, Lakehead University, Thunder Bay, Ontario, Canada.

Simatupang, T. M. and Sridharan, R. 2002. The collaborative supply chain. *International Journal of Logistics Management*, 13(1), 15–30.

Su, D., Chen, X. 2003. Network support for integrated design. *Integrated Manufacturing Systems*, 14(6), 537–546.

Stadtler, H. 2005. Supply chain management and advanced planning—Basics, overview and challenges. *European Journal of Operational Research*, 163(3), 575–588.

Stadtler, H. 2009. A framework for collaborative planning and state-of-the-art. *OR Spectrum*, 31(1), 5–30.

Trappey, A. J. C., Lin, G. Y. P., Ku, C. C., Ho, P.-S. 2007. Design and analysis of a rule-based knowledge system supporting intelligent dispatching and its application in the TFT-LCD industry. *International Journal of Advanced Manufacturing Technology*, 35(3–4), 385–393.

Tsai, T., Sato, R. 2004. A UML model of agile production planning and control system. *Computers in Industry*, 53(2), 133–52.

Vahid S., Lehoux N., de Santa-Eulalia L. A., D'Amours S., Frayret J.-M., Venkatadri U. 2014. Supply chain modelling frameworks for forest products industry: A systematic literature review. CIRRELT working paper. Québec, Canada: CIRRELT.

Vazquez-Bustelo, D., Avella, L. 2006. Agile manufacturing: Industrial case studies in Spain. *Technovation*, 26(10), 1147–1161.

Venables, M. 2005a. Small is beautiful. *IEE Review*, 51(3), 26–27.

Venables, M. 2005b. When size does matter. *Manufacturing Engineer*, 84(2), 6–7.

Vinodh, S., Kuttalingam, D. 2011. Computer-aided design and engineering as enablers of agile manufacturing. A case study in an Indian manufacturing organization. *Journal of Manufacturing Technology Management*, 22(3), 405–418.

Vinodh, S., Sundararaj, G., Devadasan, S. R. 2009. Total agile design system model via literature exploration. *Industrial Management and Data Systems*, 109(4), 570–588.

Vinodh, S., Sundararaj, G., Devadasan, S. R., Kuttalingam, D., Rajanayagam, D. 2010. Achieving agility in manufacturing through finite element mould analysis: An application-oriented research. *Journal of Manufacturing Technology Management*, 21(5), 604–623.

Wang, W.-P. 2009. Toward developing agility evaluation of mass customization systems using 2-tuple linguistic computing. *Expert Systems with Applications*, 36(2), 3439–3447.

Wang, A., Koc, B., Nagi, R. 2005a. Complex assembly variant design in agile manufacturing. Part I: System architecture and assembly modeling methodology. *IIE Transactions*, 37(1), 1–15.

Wang, A., Koc, B., Nagi, R. 2005b. Complex assembly variant design in agile manufacturing. Part II: Assembly variant design methodology. *IIE Transactions*, 37(1), 17–33.

Watanabe, C., Ane, B. K. 2004. Constructing a virtuous cycle of manufacturing agility: Concurrent roles of modularity in improving agility and reducing lead time. *Technovation*, 24(7), 573–583.

Wee, H. M., Yang, P. C. 2007. Mutual beneficial pricing strategy of an integrated vendor-buyers inventory system. *International Journal of Advanced Manufacturing Technology*, 34(1), 179–187.

Wikner, J., Naim, M. M., Rudberg, M. 2007. Exploiting the order book for mass customized manufacturing control systems with capacity limitations. *IEEE Transactions on Engineering Management*, 54(1), 145–155.

Xue, F., Sanderson, A. C., Graves. R. J. 2009. Multiobjective evolutionary decision support for design-supplier-manufacturing planning. *IEEE Transactions on Systems, Man and Cybernetics, Part A: Systems and Humans*, 39(2), 309–320.

Yang, S. L., Li, T. F. 2002. Agility evaluation of mass customization product manufacturing. *Journal of Materials Processing Technology*, 129(1–3), 640–644.

Yang, W., Li, L., Ma, S. 2007. Coordinating supply chain response-time. *International Journal of Advanced Manufacturing Technology*, 31(9–10), 1034–1043.

Yang, Y., Yang, S. 2010. Research of agile manufacturing strategy based on the competitive strategic situation cognition. *International Conference on System Science, Engineering Design and Manufacturing Informatization*, November 12–14, Yichang, China.

Yauch, C. A. 2007. Team-based work and work system balance in the context of agile manufacturing. *Applied Ergonomics*, 38(1), 19–27.

Youssef, M. A., Zubair, M., Sawyer, G., Whaley, G. L. 2002. Testing the impact of integrating TQM and DFM on the ability of small to medium size firms to respond to their customer needs. *Total Quality Management*, 13(3), 301–313.

Yusuf, Y. Y., Adeleye, E. O. 2002. A comparative study of lean and agile manufacturing with a related survey of current practices in the UK. *International Journal of Production Research*, 40(7), 4545–4562.

Yusuf, Y. Y., Adeleye, E. O., Sivayoganathan, K. 2003. Volume flexibility: the agile manufacturing conundrum. *Management Decision*, 41(7), 613–624.

Zanjani, M. K., Ait-Kadi, D., Nourelfath, M. 2010. Robust production planning in a manufacturing environment with random yield: A case in sawmill production planning. *European Journal of Operational Research*, 201(3), 882–891.

Zanjani, M. K., Nourelfath, M., Ait-Kadi, D. 2009. A multi-stage stochastic programming approach for production planning with uncertainty in the quality of raw materials and demand. *International Journal of Production Research*, 48(16), 4701–4723.

Zhang, Z. D., Sharifi, H. 2007. Towards theory building in agile manufacturing strategy: a taxonomical approach. *IEEE Transactions on Engineering Management*, 54(2), 351–370.

Zhou, Z. D., Wang, H. H., Chen, Y. P., Ai, W., Ong, S. K., Fuh, J. Y. H., Nee, A. Y. C., 2003. A multi-agent-based agile scheduling model for a virtual manufacturing environment. *International Journal of Advanced Manufacturing Technology*, 21(12), 980–894.

Section 3

Forest Value Chain Tactical Planning and Wood Flows

10 Tactical Planning and Decision Support Systems in the Forest-Based Value Creation Network

Jean-François Audy, Azadeh Mobtaker,
Mustapha Ouhimmou, Alexandra
Marques, and Mikael Rönnqvist

CONTENTS

10.1   INTRODUCTION

Forests worldwide create environmental, social, and economic value. Focusing on the latter, the value of exports in forest products was estimated at US$231 billion in 2012 (FAO, 2014a), while the formal forest sector employs some 13.2 million people across the world (FAO, 2014b). For a major forested country such as Canada, the forest sector contributed to 1% of the nominal GDP and provided 200,000 direct jobs in 2013 (NRC, 2014). Similar figures are found in other countries where the forest industry is important. In Sweden (www.skogsindustrierna.se), the forest industry sector represents 2.5% of the gross domestic product (GDP), the number of direct jobs in 2013 was 55,000 (175,000 indirect jobs), revenues are about US$25 billion, and the export value is US$15 billion. In Portugal, the gross value added (GVA) of forest-based companies in 2012 was worth 1.746 million euros (about 1.2% of the national GVA), corresponding to 9.1% of the total exports and 1.7% of total employment (www.aiff.org). To create this value, the forest products industry is organized in a complex industrial system known as a value chain, starting from the forest and continuing to the delivery of products to end customers (markets) as well as recapturing the value (or disposal) of a product at the end of its use/life span. Planning such an extended industrial system, accounting for its distributed and dynamic nature, constitutes a challenging task. In past years, research in supply/value chain management has contributed to major improvements in the forest sector as well as in other industrial sectors. Among the most important outcomes are the advanced planning methods embedded in decision support systems (DSS) that are often modules of the overall business system of a company [i.e., enterprise resource planning (ERP) system]. This chapter aims to provide a broad overview of a number of planning methods and DSS for tactical decisions (i.e., mid-term decisions ranging from a couple of months to a few years) in the forest-based value creation network (FVCN) since the 1990s that have been published in the literature. The solution methodologies and decision-making frameworks behind these methods/DSS are discussed. The motivation is to furnish readers with an easy-to-read and pedagogical summary on what has been done worldwide, highlighting the most successful DSS developments by reporting their most significant applications and benefits, present trends and gaps in planning methods/DSS, and future research directions and links for further reading. As such, an exhaustive literature review is beyond the scope of this chapter, but throughout the chapter, we have identified a number of reviews focusing on specific value chains within the extended FVCN. Also, although there are many commercial software programs that have been developed and utilized, their methodology and models are not known in detail and are hence not included.

The chapter is organized as follows. In Section 10.2, the five main value chains of the FVCN are introduced. Then, Section 10.3 discusses the main planning problems encountered in the FVCN and presents a generic mathematical model to illustrate typical tactical decisions. Section 10.4 reviews a number of planning methods and DSS in each of the five main value chains and also reviews methods/DSS spanning over two or more value chains. A discussion about the gaps and trends in planning method/DSS development, the issues and challenges for their implementation, and future research directions are presented in Section 10.5. Concluding remarks end the chapter in Section 10.6.

10.2   VALUE CHAINS IN THE FVCN

The transformation of raw materials from the forest into finished products involves several consecutive activities performed by a number of private and public organizations. The mixture and number of the involved organizations vary according to several country-to-entity features such as forestland ownership structure, level of vertical business integration, business models and practices in place, and so on. This complex set of entities that work together to perform the transformation activities via different types of relationships to create economic, environmental, and social values is known as a *value chain* or a *value creation network* (D'Amours et al. 2011). Thus, the FVCN could be illustrated according to its five main value chains (Figure 10.1). Four value chains produce sets of finished products sold over different market channels, that is, from left to right in Figure 10.1: biorefinery value chains; pulp and paper products value chains; lumber, panel, and engineered wood products value chains; and bioenergy value chains. All of these value chains are linked to a fifth value chain, the forest value chain (top of Figure 10.1), for their procurement, which also comes from flows in various raw materials (including by-products) between some of the value chains. To a certain extent, all these raw material flow links lead to interdependent value chains in constant adjustment to sustain the raw material flow equilibrium at the FVCN level. A description of each of these five value chains is provided in Sections 10.4.1 through 10.4.5, respectively.

10.3   PLANNING OF THE VALUE CHAINS IN THE FVCN

10.3.1   VALUE CHAIN PLANNING MATRIX

A supply chain can be subdivided into four main processes consisting of substantially different planning tasks (Fleischmann et al., 2008). Procurement involves the operations directed toward providing the raw material and resources necessary for production. Production is the next process in which the raw materials are converted into intermediary and/or finished products. Thereafter, distribution includes the logistics taking place to move the products either to companies further processing the product (e.g., value-added mills) or to ship for sales to distribution centers, and then to retailers. The sales process deals with all demand planning issues including customer or market selection, pricing strategy, forecasting, and order-promising policies. The planning within each process is typically managed according to three time-perspective planning horizons: strategic (long-term planning), tactical (mid-term planning), and operational (short-term planning). Strategic planning is related to the design and structure of the value chain while operational planning is related to the scheduling instructions for the execution of the operations in the value chain. Serving as a bridge between the strategic and operational level, tactical planning addresses the definition of rules and policies through a global analysis of the value chain, needed for guiding day-to-day operations. Often, the tactical planning horizon covers a full seasonal cycle and the decisions seek to balance demand forecast and facilities' capacities to avoid shortage and excess. In the FVCN, the tactical decisions play a key role in meeting

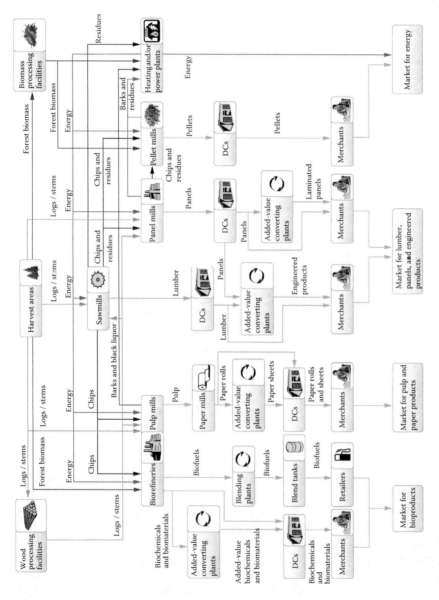

FIGURE 10.1 Five main value chains composing the forest-based value creation network.

the need to plan in advance and to address seasonal aspects such as the impacts of weather conditions on the operations such as thaws affecting transportation, frozen ground constraining harvesting blocks, forest fires affecting procurement, seasonal demand for lumber, and seasonal variation of biomass moisture content.

Fleischmann et al. (2008) present the typical planning problems in a supply chain using the form of a two-dimensional matrix structured according to the main processes along the supply chain (i.e., procurement, production, distribution, and sales) and the planning horizons (i.e., strategic, tactical, operational). At each intersection of these two dimensions, a number of planning problems, with associated decisions, are reported. A planning matrix for the forest value chain, lumber value chain, and pulp and paper value chain has been proposed by Rönnqvist (2003), Singer and Donosco (2007), and Carlsson et al. (2009), respectively. However, it is worth noting that depending on the country-to-company specificities and business context, some of the planning problems could be shifted up or down in the planning horizon, removed or added, combined or separated, and so on. In Tables 10.1 through 10.5, we present a nonexhaustive list of references addressing tactical level planning problems in a value chain of the FVCN and indicate (using a X mark) which of the main process(es) along the given value chain they cover. It should be noted that we considered the transportation decisions within the distribution process.

Different characteristics of the FVCN increase the complexity when it comes to planning. First, we must consider the divergent nature of the material flow where a different mix of products can be obtained from the harvesting of a single standing tree and where not all products have a demand. In addition, different markets ask for various quality attributes (e.g., dryness, moisture content, National Lumber Grades Authority's standards) and different dimensions, which lead to a manifold product basket. Second, the intrinsic variability of natural raw material characteristics, the diversity of orographic conditions in which the procurement operations need to be conducted in the forestland sites, and the external and not-controlled environment highly subject to changing weather conditions all affect the availability of the raw material and performance of forest operations. Some of the characteristics of the raw materials, such as the moisture content, also change over time depending on the storage duration and conditions. Thus, sources of uncertainty are introduced in the very early stage of the FVCN, requiring planning strategies to handle such uncertainties. One way to deal with those cases is to consider business and anticipation decisions in the modeling of the planning problem. Third, raw material can be used to fulfill demand of several value chains. In some contexts (e.g., pulpwood shortage is pulling saw-wood or high energy price on the market is increasing price paid for any wood quality), there is a competition for the raw material among and within the value chains (e.g., Kong et al. [2012] study the market interactions between the pulpwood and forest fuel biomass). Such competition changes the wood flow equilibrium in the FVCN, thus leading to temporary or even permanent restructuring of some value chains. Fourth, the usual wide geographic spread of the units involved in the FVCN, starting with the forest areas for supply in raw material to the international markets to sell final products, requires efficient management of transportation and inventory. Fifth, as mentioned by Marier et al. (2014), there are very different planning problems to be solved in each manufacturing facility. For instance, a softwood lumber sawmill involves a production process where one input leads to several outputs (one-to-many in the sawing and finishing) and also a one-to-one batch process (drying). At energy-producing units, there may be many-to-one as the demand (output) is for energy only and several assortments can be used as input. Sixth, there are typically very large volumes often transported with multimodal transportation options, including road, railway, and maritime transport. Seventh, there are many stakeholders involved in the value chains, for example, governments, companies, First Nations, carriers, entrepreneurs, and local communities (including hunters, campers, etc.). Each of these groups has its own objectives and agendas. Hence, there is a need to include multiobjective modeling as well as shared use of forest resources in many cases when several stakeholders are integrated.

10.3.2 VALUE CHAIN PLANNING SUPPORT

The complexity of the tactical planning problems and the economic importance of their decisions have motivated research on computer-based planning support for several decades. Several techniques such as optimization, simulation, and hybrids of them (e.g., simulation and optimization combination, see Marques et al. [2014a]) can be found in the literature. For operational research (OR) techniques, the literature reports the use of linear, integer, mixed-integer, and nonlinear models. The solution method in use depends on the type of model used, required solution time, and includes dynamic programming and linear programming (LP) methods, branch and bound methods, column generation, multicriteria decision-making, heuristics, and metaheuristic approaches.

To allow decision-makers (DMs) to benefit from this computer-based planning support, DSS embedding the planning methods have been developed and deployed in the industry. To the best of the authors' knowledge, the earliest applications in the forest sector can be traced from the 1950s (see review by Bare et al. [1984]). At the present time, the contribution of the DSS on the improvement of the quality and transparency of decision-making in natural resources management is well established (Reynolds et al., 2007). As an example, the wiki page of the Forest DSS Community of Practice (www.forestDSS.org) reports 62 DSS for forest management developed in over 23 countries, covering a broad range of forest ecosystems, management goals, and organizational frameworks.

These DSS can either be focused on one specific problem or an attempt to combine more, either at the same planning level or from two consecutive planning levels in order to avoid suboptimization (Rönnqvist, 2003). In this context, Marques et al. (2014a, b) propose distinguishing between a fully integrated planning problem and a decoupled planning problem with the anticipation of related decisions. The fully integrated planning problem considers simultaneously various interrelated *business decision variables*. This means that obtaining the problem result ends the decision-making process and choices made in respect to each of the single decision variables will then be implemented in the course of processes that are often conducted separately. Even if this model is tractable, all decisions are not often implemented in practice. In contrast, the decoupled planning problem has a main set of business decisions but also includes other *anticipation variables* in order to anticipate the impact on/from other related planning problems. The anticipation variables improve the quality of the results of the main problem as the impact of the business decisions can be described in the model. The outcome of such problems ends the decision-making process but only for the business decisions. A new decision-making process will be conducted for the secondary problem, which will then provide the best choice to be implemented.

In a literature review on DSS in the transportation domain, Zak (2010) reports two definitions of transportation DSS that could be generalized to all DSS addressing any planning problem along a value chain in the FVCN. The first definition gives a broader meaning to DSS by including all computer-based tools supporting the decision-making processes in transportation. Thus, all information management systems, data analysis methods, and spreadsheets applied to solve transportation decision problems can be designated as transportation DSS according to this first definition. The second definition gives a narrower meaning to transportation DSS: it is "(…) an interactive computer-based system that supports the DM in solving a complex (…) transportation decision problem. (…) a [ideal] role of a 'computer-based assistant' that provides the DM [with] specific transportation-focused information, enhances his/her knowledge of a certain transportation decision problem and amplifies the DM's skills in solving the considered transportation decision problems." Therefore, a DSS must manage the information required for planning, execute the planning technique (e.g., the solution method set to address the planning problem), and display the arising plans using graphical user interfaces and, common in forest DSS, spatial maps. Moreover, to enable flexibility, the planning technique must allow solving several instances of different characteristics of a given planning problem that, of course, represent decision(s) to be made, in practice, by a DM. A DSS may even present the comparison among the results of the different instances in graphical user interfaces. In Section 10.4, we discuss a number of DSS that fall into the second definition by Zak (2010) and that address a tactical level planning problem in a value chain of the FVCN.

10.3.3 GENERIC MATHEMATICAL MODEL FOR TACTICAL PLANNING

To illustrate the typical decisions to be made in tactical planning of a value chain in the FVCN, we present a general mathematical model. This model assumes a vertically integrated company that manages a forest-to-customer value chain or a value chain where all members coordinate their operations toward a common objective. Also, we stress that the model is only one example of many possibilities depending on the required level of detail.

We allow for flows between manufacturing plants and a combination of direct flows from manufacturing plants to customers directly or via distribution centers. This model is a general LP model with some network structure. As we have process descriptions with general input/output values, it is not a network flow model. It is also a divergent value chain, that is, the number of products increases through the chain.

In this formulation, manufacturing mills represent any forest products manufacturing plant such as a sawmill, pulp and paper mill, lumber and engineered wood mill, and biorefinery and bioenergy mill.

Consider the following sets, parameters, and variables:

Sets and Indices

$s \in S$: set of suppliers
$m \in M$: set of manufacturing mills
$d \in D$: set of distribution centers
$c \in C$: set of customers
$t \in T$: set of periods
$r \in R$: set of recipes used in manufacturing mills
$p \in P$: set of products (our definition of products includes raw material, semifinished products, coproducts, and finished products)

Parameters

c_{pst}^{pur}: purchasing cost per unit of product p from supplier s in period t
c_r^{rec}: production cost for each activity level when using recipe r
c_{pij}^{tr}: transportation cost of each unit of product p from node i to node j
c_{np}^{inv}: inventory holding cost of product p at node $n \in \{S \cup M \cup D\}$
b_{pst}^{pro}: procurement capacity of supplier s for product p in period t
b_{pmt}^{pc}: production capacity of manufacturing mill m for product p in period t
b_{mt}^{pm}: production capacity of manufacturing mill m in terms of available machine hours at period t
b_{dp}^{s}: storage capacity of product p at the distribution center d
d_r^{rec}: machine hours that processing recipe r takes, on a unit activity level
f_{rp}^{in}: the quantity of product p consumed when using recipe r on a unit activity level (activity level can be interpreted as how many times a standard recipe is used)
f_{rp}^{out}: the quantity of product p produced when using recipe r on a unit activity level
d_{cpt}: demand quantity of product p by customer c at period t

Decision Variables

X_{pijt}: flow of product p from node i to node j at period t
Y_{rmt}: activity level of recipe r at manufacturing mill m at period t
I_{pnt}: inventory level of product p at node n at the end of period t ($n \in \{S \cup M \cup D\}$)

Objective Function

$$Min \; z = \sum_{s \in S}\sum_{m \in M}\sum_{p \in P}\sum_{t \in T} c_{pst}^{pur} X_{psmt} + \sum_{r \in R}\sum_{m \in M}\sum_{t \in T} c_r^{rec} Y_{rmt} + \sum_{i \in \{S,M,D\}}\sum_{j \in \{M,D,C\}}\sum_{p \in P}\sum_{t \in T} c_{pij}^{tr} X_{pijt} +$$

$$\sum_{p \in P}\sum_{n \in \{S \cup M \cup D\}}\sum_{t \in T} c_{np}^{inv} I_{pnt}$$

The objective is to minimize the total cost of a four-echelon value chain (suppliers, manufacturing mills, distribution centers, and customers) with respect to the constraints mentioned below. The total cost includes purchasing costs from suppliers, processing costs, transportation costs throughout the value chain, and inventory holding costs at suppliers, manufacturing mills, and distribution centers.

Constraints

Procurement capacity constraints of suppliers:

$$\sum_{m \in M} X_{psmt} \leq b_{pst}^{\text{pro}} \ \forall \ s \in S, \ p \in P, \ t \in T$$

Production capacity constraints of manufacturing mills in terms of quantity of products produced:

$$\sum_{r \in R} f_{rp}^{\text{out}} Y_{rmt} \leq b_{pmt}^{\text{pc}} \ \forall \ m \in M, \ p \in P, \ t \in T$$

Production capacity constraints at mills in terms of machine hours:

$$\sum_{r \in R} d_r^{\text{rec}} Y_{rmt} \leq b_{mt}^{\text{pm}} \ \forall m \in M, \ t \in T$$

Storage capacity constraints of distribution centers:

$$I_{pdt} \leq b_{dp}^s \ \forall \ d \in D, \ p \in P, \ t \in T$$

Customers' demand constraints (including product flows from mills directly and via distribution centers):

$$\sum_{d \in D} X_{pdct} + \sum_{m \in M} X_{pmct} = d_{cpt} \ \forall \ c \in C, \ p \in P, \ t \in T$$

Flow conservation constraints of manufacturing mills:

$$\sum_{r \in R} f_{rp}^{\text{out}} Y_{rmt} + \sum_{s \in S} X_{psmt} + \sum_{o \in M} X_{pomt} + I_{pm,t-1} = \sum_{r \in R} f_{rp}^{\text{in}} Y_{rmt} + \sum_{d \in D} X_{pmdt} + \sum_{o \in M} X_{pmot} + \sum_{c \in C} X_{pmct} -$$
$$+ I_{pmt} \ \forall m \in M, \ p \in P, \ t \in T$$

Flow conservation constraints of products at distribution centers:

$$\sum_{m \in M} X_{pmdt} + I_{pd,t-1} = \sum_{c \in C} X_{pdct} + I_{pdt} \ \forall \ d \in D, \ p \in P, t \in T$$

All decision variables must be nonnegative.

10.4 DECISION SUPPORT IN THE VALUE CHAINS OF THE FVCN

In the following sections, we describe the main decisions and planning problems arising in tactical planning in each of the aforementioned value chains in the FVCN. We also review a number of models and solution methods proposed in the literature. Furthermore, we provide an overview of existing DSS for tactical planning developed worldwide since the 1990s. These DSS could be at different development stages, that is, from a DSS proof-of-concept developed by researchers and tested on a real/realistic problem instance to an operating DSS in use by DMs in the industry or government. Each DSS is discussed according to the decision(s) made, the planning approach used, the quantitative and/or qualitative results obtained, and to what extent the DSS is implemented in the industry (e.g., used by DMs, used for consulting analysis). Finally, for each reference, we also indicate in which of the main process(es) along the value chain the planning method/DSS is used.

10.4.1 FOREST VALUE CHAIN

The forest value chain includes the entities responsible for managing forestlands, those handling forest harvesting and wood transportation up to the manufacturing mills. There are several articles that describe this value chain, for example, the review by D'Amours et al. (2008, 2011). In general terms, tactical forest planning is done by the forest manager (that may or may not be the forest owner) or by the entity that purchased the wood (still standing trees), which may be the mill or a wood contractor that intermediates the wood supply to the mills. In most cases, harvesting and forwarding operations are outsourced to small-scale local entrepreneurs that manage the manpower and own or rent the machinery. Forest harvesting operations in a cut-to-length system includes tree felling (final felling or thinning operations), tree bucking into logs of different lengths, and forwarding the logs to pick-up points (landing) adjacent to logging roads. Felling and bucking operations are done by specialized workers with manual chainsaws or mechanized harvesting systems depending on the characteristics of the stand and equipment availability. The forwarding can also be done with mechanized forwarders. Log processing and sorting can occur at the harvesting site. It involves removing the limbs and the tops of the trees and bucking them into merchantable log lengths. Each log is sorted into assortments according to grade, dimensions (length and diameter), and species. The assortments are individually piled at the roadside. Log transportation is usually outsourced to a third company that manages a fleet of log trucks and drivers. Logs may be directly transported to an industrial transformation site (see, e.g., the review on forest-to-mill transportation by Audy et al. [2012a]) or to intermediate stockyards located at strategic logistic nodes (e.g., close to the railway network). There, the logs are temporarily stored. In a full-/whole-tree harvesting system, the processes taking place at a stump in a cut-to-length system are postponed at one or several stages from the landing site to the industrial transformation site. In some regions, tower hauling is used for forwarding purposes. In very special cases, depending on road accessibility and site conditions, helicopters may be used to transport the logs.

The number and nature of the entities involved determines the way these forest operations are planned from the strategic to the operational level, across the forest value chain. Unlike strategic planning, the distinction between tactical and operational planning is sometimes narrow and greatly country- and company-specific. In some pulp and paper industries, the term tactical is not used, therefore they designate as operational planning the entire process of scheduling forest operations on a 12-month basis (e.g., Murray and Church, 1995; Epstein et al., 1999b). In any case, it is commonly acknowledged in the literature that tactical harvest planning deals with decisions of selection of the harvesting stands and scheduling harvesting across a planning horizon that may vary from 1 to 5 years, depending on the complexity of the problem and the species composition, allocating available manpower and existing harvesting machinery systems to the stands to be harvested, and determining allocation to customers (e.g., sawmills), as well as road engineering (building new roads or maintaining existing ones). Operational harvest planning relates to detailed scheduling decisions that precede and determine the real-world operations (D'Amours et al., 2008). The length of the planning periods is generally such that in tactical planning several stands can be harvested in the same period (months or years). In operational harvest planning, the harvesting of a stand covers several periods (months or days). Another difference is that tactical planning often uses aggregated demand information on assortments without spatial data whereas operational planning includes location of industries and a more detailed description of the assortments needed. Detailed discussion of tactical and operational planning problems is provided in Marques et al. (2014a, b).

The references on planning method/DSS in the forest value chain that are discussed in this section are listed in Table 10.1 with an indication of the main process(es) covered along the value chain. Please note that this is a nonexhaustive selection, aiming to capture the most relevant DSS found in the literature to support forest harvesting and/or raw material transportation-related decisions. In fact, in many DSS (e.g., Optimed, Beaudoin et al. 2007, RoadOpt, and FlowOpt) transportation and forest harvesting are jointly planned, with the goal to fulfill the demand at the mill that may encompass different types of product assortments. Few such DSS also address the production process. The DSS discussed in this section rely on LP, Integer Programming, or Mixed-Integer Programming (MIP) formulations. Binary (or continuous) decision variables state when each stand should be harvested. Integer or continuous wood flow decision variables relate to the amount of wood transported from a stand to the mill in a given period or a given product assortment. The solution methods include both exact and heuristic methods. Case-specific heuristics are used in some of the systems (e.g., FlowOpt) as a way to obtain good solutions in

TABLE 10.1
Scope along the Forest Value Chain Addressed by the Reviewed Literature

References	Procurement	Production	Transportation/ Distribution	Sales
	Main Processes along the Value Chain			
Planex (Epstein et al., 1999a)		X	X	
Optimed (Epstein et al., 1999a)		X	X	
FlowOpt (Forsberg et al., 2005)	X		X	
Carlgren et al. (2006)	X	X	X	
RoadOpt (Karlsson et al., 2006; Flisberg et al., 2014)			X	
Beaudoin et al. (2007)	X	X	X	
MaxTour (Gingras et al., 2007)			X	
Bredström et al. (2010)	X	X		
FPInterface (Favreau, 2013)	X	X	X	

short computational time. All the DSS also have in common a development tailored to a real industrial problem. Therefore, the time (months to 2–5 years) and spatial scales (group of stands to forest region) of planning are very diverse in adapting to the reality of the DM.

Optimed runs for 2–5 years divided into summer and winter seasons to support harvest and transportation planning, considering multiple types of assortments (including sawn timber and pulp logs), with the goal of maximizing the net present value of the forest management or minimizing the total harvesting costs across the planning period. Harvesting is driven by the forecasts of the demand at the mill over the planning period for different types of product assortments. The number of assortments impacts not only the price at the mill but also the harvesting cost. Harvesting is mainly constrained by the total volume available at the forest site, which is estimated by growth and yield models. Optimed also considers road network design and planning. This means that the decisions to upgrade a given road segment or to build a new one in a period are made according to when harvesting is expected to occur in the stands served by that road segment and its required accessibility conditions. DSS for tactical forest value chain planning often acknowledges the seasonality of the harvesting operations that exist in some countries, conditioned by unfavorable soil conditions and difficult accessibility of the logging roads during part of the year. In Nordic countries, harvesting tends to be focused during the winter when the ground is frozen, thus reducing the risk of soil erosion when moving logs out of the forest, while in Chile and in the Mediterranean countries harvesting and transportation are forced to occur mainly during the summer to avoid the rainy decision-making negative impact on the quality of the road network. Moreover, in some countries, sawmills or harvest operations are closed during summer holidays, whereas the pulp and paper mills work throughout the year. This impacts the inventory planning of the assortments. Optimed encompasses a MIP model. Binary variables address where to harvest and whether to upgrade or build a certain road segment in a certain period. Continuous variables are related to the wood flow decisions. The model is solved by a combination of strengthening the LP formulation and heuristic rounding of variables. At least one industry in Chile has been using the DSS since 1994, running every few months and reporting relevant revenue gains.

Beaudoin et al. (2007) address harvest scheduling and wood transportation decisions in a demand-driven multifacility environment. Specifically, the problem consists of maximizing a firm's profit while satisfying demand for end products and wood chips covered under agreements and demand for logs from other companies. The DSS also takes into consideration the movement of machinery from one harvested stand to the next. Equipment transportation is a nonprofit operation that further contributes to the increase of harvesting costs whenever there is a need to hire specific equipment movers for traveling long distances between harvesting units. In some cases of disintegrated forest value chains, the decisions related to the efficient use of the harvesting resources are separated from harvest scheduling as these are the sole responsibility of the subcontractors. The MIP model proposed by Beaudoin et al. (2007) was tailored to the case of productive forestland within the public domain, as in Canada, where the government allocates volumes of timber to mills through timber licenses (TLs) in wood procurement areas. Procurement areas and

TL may be shared among companies and wood exchanges between companies can also occur. The outcome of this model is a 5-year development plan (tactical plan) that identifies blocks to be harvested in each year. It assumes that a strategic plan was produced before and also that an annual plan will follow, including more details on surrounding activities on the harvesting blocks for the first year of the tactical plan. The solution method makes use of Monte Carlo methods to address uncertainty. This approach was successfully applied in a hypothetical case, suggesting an 8.8% increase in profitability when compared with a deterministic model.

FlowOpt addresses the allocation of catchment areas to demand points with the possibility of integrating multimodal transportation planning (truck, train, and vessel) and back-haulage tours for reducing empty driving. The DSS further foresees the possibility of wood bartering among companies. The first version of the system was developed from 2002 to 2004 by the Forestry Research Institute of Sweden (Skogforsk) and was used by Skogforsk in conducting analyses for many Swedish forest companies. The optimization model is based on an LP model with a lot of flexibility provided by many detailed input files. The software has been used to carry out case studies with savings from 5% to 15% (Forsberg et al., 2005, Frisk et al., 2010). In addition, the use of the DSS has led to increased knowledge in the industry about optimization. FlowOpt is also used as an important educational tool in Swedish forest logistics education (Fjeld et al., 2014) and a slightly modified version was used to update the whole transportation and logistics planning of a Swedish forest company after its supply areas were hit by a major storm (Broman et al., 2009).

Carlgren et al. (2006) present a MIP model for harvesting and transportation planning considering alternative strategies for sorting the logs in the forest and the possibility of back-haulage tours. The solution method is based on column generation combined with branch-and-bound techniques. The method was applied in two case studies in Sweden including three pulpwood suppliers working with many pulp mills and sawmills. One case study showed that the introduction of specific demands on pulpwood from thinning by two of the region's pulp mills would lead to a 6% increase in total sorting and haulage costs. By optimizing the use of back-haulage tours, the cost increase could, however, be reduced by 25%.

Similarly, RoadOpt (Karlsson et al., 2006; Flisberg et al., 2014) relies on a MIP formulation for demand-driven annual harvesting and transportation planning with several assortments and road opening decisions, considering variations in road accessibility conditions during the year due to the weather conditions. RoadOpt further addresses harvest team/machinery allocation to each harvesting area, considering skills, home base, and production capacities as well as stand characteristics (e.g., terrain physiography, tree density, height, and stand composition). The model was solved optimally with CPLEX. Alternatively a heuristic approach was proposed for larger problem instances to mimic limited branch-and-bound in CPLEX. This DSS has been applied in case studies for several Swedish companies and has led to promising results. Similarly, Bredström et al. (2010) solves an annual resource planning problem, which includes decisions related to the assignment of the machine systems and teams to the harvesting stands minimizing the harvesting costs over time, taking into account the specific characteristics of the stands as well as home base location for the teams and production capacities, as well as varying weather and road conditions during the year. It also includes variables to decide the sequencing of teams during the seasons. This part is handled by solving the overall problem in two phases. The first phase allocates stands to teams and the second finds a sequencing solution. The system has been further developed to consider also a detailed demand description at mills. Here, variables for transportation flows are also included. The system has been used to support capacity planning in a number of case studies.

Planex combines these machinery assignment decisions with road design. Decisions include the following: which areas to harvest by skidders and which by towers, where to locate the landings for towers, what area should be harvested by each tower, what road to build, and what volume of timber to harvest and transport. The system is highly dependent on geographical data for the stands location and site characteristics. A graphical user interface enables the user to modify and visualize solutions as well as possible location of towers, relevant costs, technical parameters, and maximum slope. The solution approach encompasses a series of heuristics rules for the minimum cost allocation of machinery to harvest sites. Priority is given to areas to be harvested with skidders and towers according to slopes. Then a shortest-path algorithm determines the best new roads to build to link the machinery location to existing roads. A local search routine looks for changes of machine locations to improve the solution. Planex has been in use by Chilean companies since 1996. Savings were US$0.5 to 1.5 per cubic meter and the road network was reduced by as much as 50%.

There are a number of other technical, economical, and ecological aspects affecting harvest scheduling decisions that may be included in the DSS, often as alternative constraints, including budget constraints or producing minimum levels of certain assortments.

It is noteworthy that none of the DSS listed earlier takes into account spatial adjacency constraints. However, when the planning horizon extends up to 5 years, national regulation or silvicultural best practices may impose a maximum allowable size of the clear-cut opening area to minimize the risk of soil erosion. This means that consecutive stands cannot be harvested in the same period if the sum of the areas is higher than the maximum allowed clear-cut opening size (e.g., Clark et al., 2000; Richards and Gunn, 2000; Murray, 1999). Green-up constraints may also be used to assure that there is a minimum number of periods between harvesting two consecutive stands to assure that the vegetation from the first harvested stand covers the bare ground before the neighboring stand can be clear-cut. For additional information about adjacency constraints and spatial harvest scheduling please refer to Baskent and Keles (2005) and Weintraub and Murray (2006).

The level of utilization of the listed DSS is the most diverse. Some of the DSS developed for the Chilean companies (Planex, Optimed, Opticort) have been in use since the 1990s. Some of the DSS developed for the Swedish companies [RoadOpt, Carlgren et al. (2006) and FlowOpt] have also been in use since 2004. FlowOpt has been in use at two of the major Swedish forest companies for monthly transportation planning and in many case studies to support the forest industry with answers to "what if" scenarios (e.g., location of new terminals). The software described in Carlgren et al. (2006) has been used internally in one company for analysis. RoadOpt has been used in several case studies to support the companies with a selection of suitable roads for upgrading. This problem is receiving increasing interest due to deteriorating quality of roads and discussions to increase the truck load limit. In Canada, the FPSuite developed by FPInnovations includes a number of simulation/ planning modules and we discuss two of them. Deployed to over 100 licenses in government, industry, and academics in Canada (Favreau, 2013), the DSS FPInterface is a simulation module allowing the results (e.g., costs, yield, products baskets) to be generated for a given procurement plan entered on the system by a DM. The system's first obvious benefit is the time saved by a DM in assessing the performance of the harvesting plan and Canadian industry has reported gains of over CAD\$0.25/m³ (Favreau, 2013) when using the system. To increase the benefits, the system could be linked to other planning modules supporting the DM such as the transportation module MaxTour (Gingras et al., 2007). This system computes the potential in back-haulage tours within the volume of one or several types of products usually managed by distinct DMs (e.g., round timber/bulk fiber delivered/shipped to/from a sawmill). Its planning method was developed in partnership with researchers at HEC Montréal (Canada) and is based on an adaptation of the well-known savings heuristic of Clarke and Wright (1964). During recent years, a number of analyses have been conducted by FPInnovations on historical transportation data of Canadian forest companies and, in the six most exhaustive cases, potential cost savings (traveling time reduction) between 4% and 7% (5%–9%) have been identified. Also, in a number of the analyses, the proposed back-haulage tours have been used by DMs in Canada to support their manual truck routing (Audy et al., 2012a). When several types of products are jointly planned, multiproduct truck trailers (i.e., logs and bulk fiber trailers) are used in addition to classic (monoproduct) truck trailers. By allowing the transportation of different types of products on the same truck trailer, a multiuse truck trailer increases the number of possibilities for back-haulage tours and thus, additional cost savings can be realized. For example, Gingras et al. (2007) report an additional savings of 1.1% with the addition of multiuse truck trailers in the transportation of timber and bulk fiber in a large network of forests and mills of a Canadian company.

10.4.2 Lumber, Panel, and Engineered Wood Products Value Chains

A typical supply chain in the wood (softwood and hardwood) lumber industry includes sawlog suppliers, sawmills, kilns, added-value products mills, warehouses, retailers, and end customers. The combination of seasonality of supply, log quality variation, customer demand variation, the wood long cycle time (and relatively short transformation cycle time), and the divergent production process with a lack of synchronization and integration between business units makes the planning of the lumber and value-added products value chain a complex task. The planner faces the challenge of defining optimal procurement, sawing, drying, and transportation plans as well as seasonal stock levels for each product, in each location of the value chain, while taking into account all of the procurement, production, transportation, and customer constraints.

This section covers the literature about lumber, panel, engineered wood, and value-added wood products value chains, respectively. There are several articles that describe these value chains, see, for example, Singer and Donosco (2007) and

D'Amours et al. (2008, 2011). The references on planning method/DSS in this value chain that are discussed in this section are listed in Table 10.2 with an indication of the main process(es) covered along the value chain.

Maness and Adams (1993) proposed a model to integrate the processes of bucking and sawing to respond to expected changes in product value or market demand by changing policies with regard to sawing patterns and log consumption. They developed an iterative approach solution based on three models. The first model involves a cutting pattern optimizer that determines the optimal sawing pattern for each log including diameter, taper, and length, according to lumber values. The log bucking model objective includes determining the optimal combination of logs to cut from the stem. The problem can be formulated as a knapsack problem and it can be solved using a dynamic programming approach. The log allocation model acts as the master problem and uses the cutting pattern optimizer and the stem bucking model. Its objective involves distributing logs to different sawmills and selecting optimal bucking and sawing strategies to maximize profit. Maness and Adams (1993) reported that the computational results show between 26% and 36% potential revenue gain due to the integration of the bucking and sawing processes for a large log mill in British Columbia producing export products. Maness and Norton (2002) developed an extension of the model to take into account several planning periods.

TABLE 10.2

Scope along the Lumber, Panel, and Engineered Wood Products Value Chains Addressed by the Reviewed Literature

	Main Processes along the Value Chain			
References	**Procurement**	**Production**	**Transportation/ Distribution**	**Sales**
Carino et al. (1998, 2001a, 2001b)	X	X		
Maness et Adams (1993) ; Maness and Norton (2002)	X	X	X	X
Reinders (1993)	X	X		
CustOpt (Liden and Rönnqvist, 2000)	X	X	X	X
Donald et al. (2001)	X	X	X	X
Farrell et al. (2005)	X	X	X	X
Optitek (Zhang and Tong, 2005; Favreau, 2013)	X	X	X	
FORAC's experimental platform (Frayret et al., 2007; Forget et al., 2008)	X	X	X	X
Ouhimmou et al. (2008, 2009)	X	X	X	X
Singer et al. (2007)	X	X	X	X
Feng et al. (2008, 2010)	X	X	X	X
Marier et al. (2014)	X	X	X	X

Donald et al. (2001) developed two linear programming models for tactical production planning in value-added lumber manufacturing facilities. The first model is designed for nonintegrated value-added facilities (sells its entire lumber production to the market); the second is designed for integrated value-added facilities (resaw and molder) with a sawmill with the ability to produce their own raw materials from their primary operations (sawmill sells only the lumber that is not directed to the value-added facility for further processing). The authors compared the two models to explore the financial benefits for a real sawmill of integrating a value-added lumber manufacturing facility at the back end of the mill. The results showed that net revenue for an integrated value-added sawmill exceeds the net revenue of a nonintegrated one by 10% and also the production decisions in the value-added facility had a significant influence on production decisions in the sawmill. The authors suggested that these results should be validated by practical testing of the model in field use and how easily they can be used and understood by mill personnel with little or no background in mathematical programming.

Liden and Rönnqvist (2003) introduced an integrated optimization system, CustOpt, which allows a wood supply chain to satisfy customer demand at minimum cost. The model considers bucking, sawing, planning, drying, and the classification process. This integrated model aims to maximize the value of various products and secondary products while taking into account harvesting costs, transportation, external buying, production costs (drying, grading, and planning), and internal flow. The system was tested and analyzed in a company using two to five harvesting districts, two sawmills, and two planning mills, and a very detailed log breakdown information with many products. Key decisions at the mill were to decide the production of products for three main customer areas (Japan, Europe, and the United States). From a similar perspective, Singer and Donoso (2007) presented a model for optimizing planning decisions in the sawmill industry. They modeled a supply chain composed of many sawmills and drying facilities, with storage capacities available after each process. In this problem, each sawmill is considered as an independent company, making it imperative to share both the profitable and unprofitable orders as equitably as possible. The model allows transfers, externalizations, production swaps, and other collaborative arrangements. The proposed model was tested on a corporation that consists of 11 sawmilling plants located in southern Chile. Based on the results of the testing, the authors recommend using transfers, despite the explicit transportation costs incurred. They also recommended that some plants focus almost exclusively on the upstream production stages, leaving the final stages to other plants. The authors find an opportunity to increase profits by more than 15% through a higher utilization of the capacity and a better assignment of production orders.

Reinders (1993) developed a prototype for a decision support system called Integral Decision Effect Analysis System (IDEAS) for tactical and operational planning of a centralized conversion site where bucking and sawing operations are performed. The model considers only one sawmill and does not take into account other processes such as planning and drying. IDEAS consists of a database, a model base (bucking process, sawing process, production planning models), and a user interface. The model base is an optimization-based model, based on both dynamic programming and column generation. The author has validated the model in a real case study where a real-world plant in Germany served as a test. The plant uses raw material from company-owned forests and that is purchased on the open market. The author simulated five different policies ranging from service level, profit maximization (production effectiveness), to value recovery (production efficiency) from wood, and so on for conducting. The simulation results show that a trade-off between profitability and value recovery can be made by manipulating stock-out costs.

Farrell et al. (2005) developed a relational database approach to create an integrated linear programming-based DSS that can analyze short- and mid-term production planning issues for a wide variety of secondary wood product manufacturers. The mathematical model takes into account generic constraints related to the secondary wood products industry such as raw material, material balance, recovery, machine capacity, and marketing considerations. They aimed to maximize the profits of the secondary manufacturing operation over a planning horizon. They generated specific reports related to the financial aspect, procurement strategies, machine yield, sales, and so on. The authors did not report any results of the implementation of the DSS on real industrial cases but conclude that due to its generic design, the system can determine product mix, raw material sourcing, production strategies, pricing strategies, and resource evaluation for different configurations of companies in the secondary wood industry.

A DSS called Optitek has been developed by FPInnovations to simulate the whole softwood sawmilling process (bucking, sawing, trimming, and edging) in Canada. The system allows analysis of the impacts on the yield (value or volume) and basket products (including by-products) of modifications to the sawmilling process or in the input log characteristics (Zhang and Tong, 2005). Since the tool required advanced expertise and direct use by industry is often an impediment to gaining the full potential from the system, most sawmills use external resources to conduct such studies. Over 75 Canadian sawmills have been modeled on the system over the last decade and case studies often indicate a potential improvement of more than CAD\$2/m^3 (Favreau, 2013). Optitek has been integrated with FPInterface (FPInnovations) to anticipate the economic value of each harvest area (net value of each bloc) by simulating trees of each harvest bloc in Optitek and allocating them to the right sawmill. On the other hand, Frayret et al. (2007) and Forget et al. (2008) have together proposed an agent-based experimental platform for modeling different lumber supply chain configurations and assessing the impact of different planning approaches. This model represents the sawmilling, drying, and finishing processes as alternative one-to-many processes constrained by bottleneck capacity. The authors used different business case studies to validate the simulation platform and the specific planning models proposed (e.g., linear programming, constraints programming, and heuristics). In addition, simulations were done to evaluate different strategies for the lumber industry, given different business contexts. During the simulation, wood procurement was set as a constraint and demand patterns were stochastically generated according to different spot market and contract-based customer behaviors. The authors did not report any implementations of the simulation platform in a real mill.

Carino and Lenoir (1988) developed a mathematical model to successfully optimize wood procurement for an integrated cabinet-manufacturing company that owns one sawmill and one kiln. The authors used regression equations based on a sample of 25 logs to determine the volume and grade and furniture components yielded from each log diameter and length. They found an optimal wood procurement policy where raw material input should be limited to #2 grade hardwood logs and #2 common green lumber purchased directly from outside suppliers. The model was not used by the company even if the authors estimate the potential savings could reach 32% for raw material purchases.

Carino and Willis (2001a, 2001b) presented an LP model to solve the production-inventory problem inherent in vertically integrated wood products manufacturing operations (hardwood lumber-cabinet). The model aims to maximize mill profitability and provides valuable information for making management decisions related to the desired level of production and end-of-period inventories, desired quantity of products to be sold, level of resource utilization at each stage, and impact of changes in input/output and operating conditions on system profitability. The authors presented the results of a real case study to demonstrate the ability of this model in solving a complex set of production-inventory problems. The objective of the analysis was to determine the optimal sawlog and lumber production-inventory program for the study mill over a specified planning horizon. Their results indicate that mill profit could be maximized by adopting a specific log procurement policy (log volume, sawing patterns, and inventory level). Such a policy could result in profit improvement of up to 156% over the result from the minimum 1-month log inventory policy used by the sawmill. They have also performed a parametric analysis and showed that mill profitability is 1. very sensitive to changes in kiln-dried lumber prices, sawmill conversion efficiency, and lumber drying degradation and 2. moderately sensitive to changes in log supply and, prices and processing costs.

Ouhimmou et al. (2008, 2009) presented a MIP model for planning the wood supply for furniture assembly mills. Their model addresses multisite and multiperiod planning for procurement, sawing, drying, and transportation operations. Assuming a known demand that is dynamic over a certain planning horizon, the model was solved optimally using CPLEX and approximately using time decomposition heuristics. The model was then applied to an industrial case with a high cost-reduction potential (22%), with the objective of obtaining procurement contracts, setting inventory targets for the entire year for all products in all mills, and establishing mill-to-mill relations, outsourcing contracts, and sawing policies. These results have convinced the company to use the tool for the future configuration of its supply chain network. This research project has been extended to develop the DSS called LogiLab (see Section 10.4.6).

Feng et al. (2008) applied the concept of sales and operations planning (S&OP) to the oriented strand board (OSB) supply chain. They used sales decisions to investigate the opportunities of profitably matching and satisfying the demands of a given supply chain, given the chain's production, distribution, and procurement capabilities. They proposed three MIP-based planning approaches of the four processes within the value chain of an OSB company using a make-to-order strategy: fully integrated planning, fully decoupled planning, and integrated sales and production with decoupled distribution and procurement planning. The MIP models were simulated, for a real OSB manufacturing supply chain, with deterministic demand (Feng et al., 2008) and with stochastic demand using rolling horizon planning (Feng et al., 2010). In both cases, the fully integrated planning approach outperformed (e.g., up to 4.5% revenue increase with perfect demand forecasting) the fully decoupled and partially integrated planning approaches. In a similar way, Marier et al. (2014) proposed a linear program for the integrated annual planning of the sales and operations of a network of sawmills. Simulated over the historical data of 12 years, a two-sawmill case study showed that the model would have increased the gross margin by an average of 1.47% of sales revenue. This potential increase is due to adapting production and inventory decisions to market price fluctuations. The authors reported that these results convinced the company to explore ways of implementing sales and operations planning even though they were very skeptical about the benefits of such an approach before the start of the study.

10.4.3   PULP AND PAPER PRODUCTS VALUE CHAIN

The main activities of the pulp and paper value chain are harvesting and transportation, pulp making, papermaking, sales, and distribution. There are several articles that describe this value chain; see, for example, Carlsson et al. (2009) or more recently D'Amours et al. (2014). Harvesting is of course also a part of other value chains. However, in some cases harvesting is driven by one main value chain. For example, in thinning operations a vast majority is focused on pulpwood. In others, the focus is on sawmills, and pulpwood is a secondary coproduct. Moreover, in other situations there is no harvesting. This happens often in Québec (Canada) where virtually all logs flow through sawmills and hence the raw material (wood chips) come directly from sawmills. Pulp making converts pulp logs unless chips are directly transported as mentioned above. Chips of different species are mixed in recipes to get pulp with desired properties. The chips are boiled and washed to separate fibers from lignin in a number of steps. To get the correct brightness level the fibers are blended with different chemicals in a bleaching process. The pulp process is often a continuous process where some parts may be batched, for example, the cooking. Papermaking produces so-called jumbo rolls that are typically 5–8 m wide and many kilometers long. It is also possible to put some coating on the paper depending on the end use of the products. The jumbo rolls are later cut in shorter lengths and smaller widths according to specific customer demand. This cutting is done in order to minimize waste or maximize value when quality is considered. Some of the typical tactical planning decisions made in the pulp and paper (P&P) value chain are wood fiber procurement alternatives (chips vs. pulplogs), defining appropriate pulp recipes with a mix of species, sequence of recipes for pulp production, allocating the right wood fiber grade to processes and end products, and optimal lot sizing in the paper machine. The references on planning method/DSS in this value chain that are discussed in this section are listed in Table 10.3 with an indication of the main process(es) covered along the value chain.

There are many computerized tools in use for operational and process control at the pulp and paper mills. Yet, the number of tactical decision support tools is much lower.

TABLE 10.3

Scope along the Pulp and Paper Products Value Chain Addressed by the Reviewed Literature

References	Main Processes along the Value Chain			
	Procurement	Production	Transportation/ Distribution	Sales
Bredström et al. (2004)	X	X	X	X
Carlsson and Rönnqvist (2005)	X	X	X	X
Bouchriha et al. (2007)		X		
Carlsson and Rönnqvist (2007)	X			
Chauhan et al. (2008)		X	X	X
Rizk et al. (2008)		X	X	
Everett et al. (2010)	X	X	X	X
Dansereau (2013)	X	X	X	X
Carlsson et al. (2014)			X	X

One reason is the uncertainty in the production processes and the fact that there is a limited number of pulp products produced. One system is PIVOT, the developed for Norske Skog to optimize manufacturing, distribution, and sourcing of raw materials in Australia and New Zealand (Everett et al., 2010). It is based on a MIP model and the application was an INFORMS Franz Edelman Award finalist in 2009. Even though the main decisions are on a strategic level, the model considers a tactical decision level. The system has been developed over many years but has been used actively by the company to make both strategic and tactical decisions. The potential savings from the system evaluated at the Franz Edelman competition was estimated at US$100 million each year. This includes operations for all pulp and paper mills at the company.

Södra Cell is a large pulp company that mainly produces pulp for European customers from pulp and paper mills in Sweden and Norway. A number of planning problems are outlined and described in Carlsson and Rönnqvist (2005). This company has tested a number of different tactical planning tools based on OR for their operations. In Bredström et al. (2004), a system for combining procurement, production planning, and sales is tested. It is based on a detailed production planning model where column generation is an important part of the solution process. Large savings are reported by making integrated decisions instead of using a sequential planning process. This paper received the EURO Excellence in Practice Award in 2004. The DSS is at the prototype development stage, but it has been used in some rounds of production planning. Here, it helped the planners to change their behavior even if the DSS was not integrated with the company ERP system. The same company has introduced a vendor-managed inventory (VMI) system. This has put high stress on making sure that the right products are available to customers at all times. A prototype DSS system using robust optimization has been tested to better plan the routing and inventory handling (Carlsson et al., 2014). The VMI system is implemented and in full use but the optimization system has only been used on a case study basis.

Chauhan et al. (2008) describes a DSS to optimize the roll cutting of tambours at the paper mills. It takes customer demand into account to decide how to manage the cutting, including which parent roll should be kept in inventory before the cutting operations once customer orders are known. The model is a MIP model and a column generation approach has been used to solve the problem. The case study provided the company with many insights and the network structure was redesigned. The DSS has been used as a case study but is not implemented for continuous planning. Rizk et al. (2008) expand the model for multiple distribution centers and propose an efficient heuristic sequential solution approach to solve large problem instances. Bouchriha et al. (2007) developed a model for production planning at a single paper machine where the campaigns are fixed in duration.

A tactical planning problem for the wood procurement stage of the supply chain is dealt with in Carlsson and Rönnqvist (2007). The problem was to decide sorting strategies at different catchment areas to best satisfy the demand at paper mills. The model is a MIP model where the alternatives are pregenerated. The system is implemented at one company and used for case studies within the company, in particular when there are larger changes made for production planning and a change in the need or mix of species. Collaboration between a paper mill and its customers has been analyzed by Lehoux et al. (2009). Different contract agreements are simulated and optimized. One result was that depending on the different players, they may prefer different alternatives and this must be considered in the agreements. The study led to some changes in the way business was conducted between the paper company and certain key customers.

Dansereau (2013) proposes a margins-based approach for the profit maximization of a pulp and paper value chain. The framework involves five main components: profit maximization, revenue management, manufacturing flexibility, activity-based cost accounting, and integrated tactical planning optimization. The author has justified the inclusion of each of these components as follows. First, a company should aim to maximize its profitability and not just minimize costs. Second, a company should use revenue management concepts to manage its sales and produce the most profitable product portfolio. Third, manufacturing flexibility should be exploited in order to be able to deal with market volatility and manufacture the most profitable product combination. To analyze the trade-offs between different manufacturing modes, the company should access reliable operating cost estimations for each manufacturing mode. Then the fourth aspect of the proposed planning framework would be about activity-based accounting, which makes it possible to accurately quantify the cost trade-offs between different manufacturing modes. Finally, all these four concepts have been included in an integrated tactical planning model that optimizes the whole supply chain from procurement to production, distribution, and sales. The proposed margins-based planning approach proved to be effective especially in difficult market scenarios; it provides a robust planning approach through exploiting manufacturing flexibility. The model was tested in a real case study of a newsprint manufacturer in North America with overcapacity in its thermomechanical and deinking pulping lines, and which also faces varying wood chips and recycled paper prices. In this case study, the author ran the model under two different process and flexibility configurations. The first configuration represents the current case in the pulp and paper mill. In this configuration, the mill managers select the thermomechanical pulping line and paper machine recipes based on a heuristic which is believed to minimize production costs. In the second configuration, the margins-based approach was used to optimize the recipe selection and throughput of pulping lines and paper machines in order to maximize profitability. These two instances were run in different market scenarios. Utilizing the proposed margins-based planning model showed the mill's earnings before interests, taxes, depreciation and amortization can be increased by up to 35% in some scenarios by adapting pulping production to changing market conditions.

10.4.4 BIOREFINERY VALUE CHAIN

As discussed by Dansereau et al. (2012a), the biorefinery concept appears to be a promising business opportunity for the forest products industry, especially the pulp and paper sector, to diversify its revenue stream and improve its environmental profile. Specifically, the diversification of the traditional product baskets will involve the production of value-added biochemicals and biomaterials as well as biofuels from the renewable forest biomass. This supply will come from traditionally unused biomass such as forest residues (directly from harvest areas or through an intermediate processing site) but also compete for biomass with current customers including bioenergy producers. Because existing pulp and paper mills have been using woody biomass for decades, these facilities represent natural sites to implement biorefineries (as illustrated in Figure 10.1) but selecting the most profitable biorefinery configurations to install in an operating P&P mill is a challenging decision (Dansereau et al., 2012a). The typical tactical planning decisions made in the biorefinery value chain can be summarized as biomass procurement quantities from each supplier, amount of each biomass feedstock used for producing different products through different processes, which recipe to use in each process unit, inventory levels of biomass feedstock and production level in each period, and distribution and transportation mode use and sales to different customers.

We refer to Feng et al. (2012) and Dansereau et al. (2012a) for a description of this value chain. The references on planning method in this value chain that are discussed in this section are listed in Table 10.4 with an indication of the main process(es) covered along the value chain.

These papers have modeled the biorefinery value chain planning problem mostly as a mixed-integer linear programming (MILP)/LP problem. Some papers combined MILP models with simulation modeling while another paper developed a multiobjective optimization model. We have also observed that the sales process has been covered by only two papers due to the lack/nonexistence of data (price, volume, etc.) for new bioproducts. None of these papers reported implementation in the industry, except the one by Dansereau (2013).

Ekşiogğlu et al. (2009) proposed a MIP model addressing both the strategic and tactical decisions about the design and management of a regional network of biorefineries producing biofuels. They test their model over the entire state of Mississippi, USA, using corn stover and woody biomass including pulpwood and sawtimber. They show that transportation cost and biomass availability are the two

TABLE 10.4

Scope along the Biorefinery Value Chain Addressed by the Reviewed Literature

	Main Processes along the Value Chain			
References	Procurement	Production	Transportation/ Distribution	Sales
Ekşioğğlu et al. (2009)	X	X	X	
Ekşioğğlu et al. (2010)	X	X	X	
Santibañez et al. (2011)	X	X		
Faulkner (2012)	X	X	X	X
Dansereau (2013)	X	X	X	X
Meléndez (2015)	X			

main factors affecting value chain design and therefore suggest operating multiple small-size biorefineries instead of one centralized mega-biorefinery. Ekşioğğlu et al. (2010) extended the previous model by considering different modes of transportation including intermodal and exploring how the existence of an intermodal facility affects the biofuel value chain design. Because of the bulky and low-density nature of biomass feedstock, the quantity and volume of a biorefinery's outgoing product (i.e., ethanol) are smaller in comparison to the incoming biomass. This fact justifies the result of testing the MIP model on the same case study, which encourages locating the biorefinery closer to the source of biomass than the market and leads to a 5% reduction in the biofuel delivery cost. Moreover, the case demonstrated that a biorefinery consuming a much larger amount of biomass than is available locally must be located close to a transportation hub (i.e., an intermodal facility) to be economically sustainable. Indeed, this reduces the biofuel delivery cost by as much as by 4.6 times the number of incoming truck shipments when using barges.

Santibañez et al. (2011) proposed a multiobjective optimization approach maximizing the annual profit while minimizing the environmental impact (measured through an indicator based on a life cycle analysis) of the procurement, production, and sales decisions of a biorefinery. A constraint approach is used to find a set of optimal solutions of these two conflictual objectives and thus construct a Pareto curve. Several sources of supply in agricultural biomass and woodchips are available for the production of different biofuels according to specific processing recipes. The proposed methodology was tested to study different scenarios for a biofuel mill located in Mexico.

Dansereau (2013) extended its model presented in Section 10.4.3 (i.e., profit maximization of a pulp and paper value chain) with the addition of a biorefinery within the same industrial complex. Using the same case study, the author studied several configurations of running a P&P mill and biorefinery in parallel and showed that using the proposed margin-based approach can lead to higher revenues and more savings in both P&P and biorefinery product lines. The benefit of feedstock flexibility on the biorefinery operations and of manufacturing flexibility on the integrated

P&P and biorefinery operations is also demonstrated in the case. For instance, a biorefinery line with feedstock flexibility allows increasing the operational profitability by 12%. Also, as a general conclusion, they demonstrated that biorefinery lines have to consider flexibility in their process in order to be able to deal with market volatility and maintain profitability. The proposed model has been used by a newsprint mill in North America that was implementing a parallel biomass fractionation line producing various biochemicals.

Some studies have combined simulation and MILP modeling to solve a biorefinery value chain planning problem. Faulkner (2012) proposed a MILP model that addresses both the strategic and tactical decisions about the value chain design and management of one biorefinery. The author used a simulation model to generate baskets of products using all available biomass in the case study located in Kentucky, USA. The output of the simulation was the input for the MILP model. Despite biomass abundance (including forest residue) and existence of a robust chemical industry (i.e., potential market), testing the model for three different sizes of integrated biorefinery reports no profitable instance. To improve performance of the value chain, two options are proposed: first, using a less expensive mode of transportation (i.e., via pipeline) instead of truck for delivery of the most profitable product, and second, shutting down the mill in the nonprofitable months to negate the truck transportation cost. Meléndez (2015) analyzed the feedstock procurement costs and feasibility of 10 biorefinery scenarios involving two biorefinery technologies and a cogeneration plant. These were deployed at different times and scales of production at an existing P&P mill with the partial or complete shutdown of the paper machines. They also studied the potential savings on procurement costs by changing the forest harvesting technologies. The scenarios focused on fulfilling feedstock demand according to available resources while minimizing procurement costs over the whole scenario lifespan for a financially feasible biorefinery implementation strategy. A MILP optimization model for strategic decision-making along with a forest harvesting techno-economic simulation model for tactical decision-making were proposed and run over a 20-year planning horizon on a case study in Eastern Canada. Each scenario's procurement costs were compared with current practices and amongst themselves to determine which led to the best procurement strategy both for the P&P mill and interacting forest industry during and beyond the transition period.

10.4.5   BIOENERGY VALUE CHAIN

Forest residues are by-products of conventional harvesting operations and production of traditional forest products. In recent years, the conversion of forest residues to bioenergy has gained great interest for two main reasons: (1) it gives communities in forest-based regions access to new sources of revenue, and (2) it provides the opportunity to diversify their energy sources and/or dependency while reducing greenhouse gas emissions, as forest residues are renewable materials with the potential to replace fossil fuels. As discussed by Cambero et al. (2015a), there are several operational and economic challenges that hinder the intensified use of forest residues for energy production such as challenges related to capital investment, feedstock availability, quality, and cost. Since capital costs of energy-producing technologies are high, success of bioenergy projects relies heavily on achieving the economies of scale. This would lead to an increase in the demand for forest residues, which are scattered over vast regions and whose availability varies over time. Also, different quality attributes of different types of biomass influence their procurement, preprocessing, and transportation cost as well as their conversion efficiency. Additionally, due to the low-energy density of forest biomass, collecting, processing, and transporting large amounts of forest biomass over the operational cycle of a bioenergy facility is required. To do so, several types of specialized equipment and logistics strategies are available. Consequently, to install a profitable bioenergy facility, it is necessary to address the optimal design and management of the value chain. Particularly, the main strategic–tactical decisions that affect the overall profitability of the bioenergy value chain are: the sources and types of forest residues, the location of bioenergy plant(s), the type and capacity of technologies, the material flows per period within the value chain and, in the case of uncertain feedstock supply and market conditions, the plant(s) installation period must be determined. We refer to Hughes et al. (2014) for a review on the pellet value chain and Shabani et al. (2013) for a review on the forest biomass energy production value chain. The references on planning method/DSS in this value chain that are discussed in this section are listed in Table 10.5 with an indication of the main process(es) covered along the value chain.

These papers have modeled the bioenergy value chain planning problem mostly as a MILP/LP problem; a few used simulation, multiobjective modeling, and nonlinear

TABLE 10.5

Scope along the Bioenergy Value Chain Addressed by the Reviewed Literature

	Main Processes along the Value Chain			
References	**Procurement**	**Production**	**Transportation/ Distribution**	**Sales**
Eriksson and Björheden (1989)	X		X	
De Mol et al. (1997)			X	
Freppaz et al. (2004)	X	X		
Gunnarsson et al. (2006)	X	X		
Alam et al. (2009)	X			
Kanzian et al. (2009)	X		X	
Mäkelä et al. (2011)		X		
FuelOpt (Flisberg et al., 2012)	X	X	X	
Keirstead et al. (2012)	X	X		
Shabani and Sowlati (2013)	X	X		
Akhtari et al. (2014)			X	
Hughes (2014)	X	X	X	X
Mobini et al. (2014)	X	X	X	X
Shabani et al. (2014)	X	X		
Flisberg et al. (2015)	X		X	

formulation, while only one paper integrated the proposed DSS with a geographical information system (GIS)–based interface. We have also observed that the sales process has not been considered in most of the studies mainly because of the lack/ nonexistence of data (price, volume, etc.) for the bioenergy market. Another reason is that the mills themselves are in fact the final customers. Nevertheless, two papers studied the entire value chain and in order to generate sales (e.g., demand) information they used simulation and forecasting techniques. Furthermore, only Eriksson and Björheden (1989) and Flisberg et al. (2012) reported implementation of the proposed DSS in the industry.

De Mol et al. (1997) developed a simulation model called BIOLOGICS (BIOmass LOGIstics Computer Simulation) and a MIP optimization model to analyze the logistics costs of biomass fuel collection. The optimization model determines the optimal network structure (i.e., inclusion/exclusion of possible nodes and situation of pretreatment) as well as the mixture of biomass types supplied to the energy plant, given the available quantities as a restriction. The simulation model, on the other hand, calculates costs and flows for a given network structure. Testing the proposed models in an energy plant fed with biomass in the Netherlands showed that both models are useful to gain insight into the logistics cost of biomass fuel collection. Indeed, the latter is typically the main cost component when evaluating the feasibility of a biomass conversion energy plant(s) project. That is why many other research projects in different countries are also focused on the logistics cost of the bioenergy value chain; in that respect the next paragraph summarizes three such studies.

Eriksson and Björheden (1989) presented an LP formulation to model the energy value chain of a forest fuel supplier. The model determines optimal annual planning decisions about procurement, processing, and storing of raw material while minimizing the sum of acquisition, processing, and transportation costs of raw material and fuel chips. The proposed DSS was implemented on the energy value chain of Jämtlandsbränslen AB (a subsidiary of the Swedish Cellulose Company), which includes several forest supply regions (consisting of four different types of raw material: chip wood, logging waste, tree sections, and sawmill waste), one central processing site, and one heating plant. The result of this analysis showed that using mobile chippers to produce chips at forest supply regions is more cost efficient than using stationary chipping equipment at the terminals. In fact, when the chips are stored at the terminals an additional transhipment cost would occur, and the results indicate these additional costs would not be paid off by the better quality (better moisture content) of stored biomass at the terminals. Accordingly, the optimal solution of the model recommended chipping 92% of the fuel by mobile chippers and transporting them directly to the heating plant while only 8% of the forest fuel should be chipped and stored at terminals. This problem is also studied by Kanzian et al. (2009) and the authors proposed a model consisting of two submodels (LP and MIP) solved sequentially. The proposed solution method is applied on a case study for a value chain of 16 combined heat and power plants and eight terminal storages in Austria. Results similar to Eriksson and Björheden (1989) were obtained; specifically, direct flow of biomass from forest area to plants proved less expensive than indirect flow via terminals. For instance, supply cost increased by 10% when half of the fuel and by 26%

when all the fuel was sent via terminals. The same problem is studied by Akhtari et al. (2014) in Canada; an LP formulation is proposed and tested on a potential district heating plant in Williams Lake, British Columbia. The results of this case study do not refute those of Eriksson and Björheden (1989) and Kanzian et al. (2009) in general. Particularly, the optimal solution emphasizes that all chipping processes should be done at the forest sites and suggests transporting 90% of annual woodchip demand directly to plants and sending the remaining 10% via storage terminals.

Gunnarsson et al. (2006) developed an integrated MIP model to handle forest fuel for a Swedish forest fuel company. This model includes transportation, comminution (or conversion to wood chips) at terminals, and inventory. The aforementioned DSS FlowOpt has recently been extended to address the procurement logistics of forest biomass, in particular comminution and selection of areas for production of forest fuel (Flisberg et al., 2012). Named FuelOpt, the DSS relies on a MIP model because there is a need to select harvest areas as well as a machine system. The FuelOpt system is implemented at the Forestry Research Institute of Sweden (Skogforsk) in Sweden and has been used in several large case studies at Swedish forest companies. The savings are about 5%–15% compared with existing manual planning. One of the case studies for Stora Enso Bioenergi included 86 heating plants, six assortments, six truck types and five chipping systems, 12 periods (months), 72 terminals of which 8 have train transport possibilities, and 1,256 supply areas. The energy consumption was 3.6 TWh corresponding to 1.5 million metric tons of wood chips. The initial model had 16.4 million variables and 4.6 million constraints. Some aggregation of supply areas reduced the size to 5.9 million variables and 0.5 million constraints. The total cost of using the executed system was SEK 508.8 million (US$ 62.5 million) and with optimization it was reduced to SEK 477 million (US$ 58.7 million).

To make an optimal biomass exploitation plan for thermal and electrical energy conversion plants, Freppaz et al. (2004) developed a mathematical model accompanied with a GIS-based interface and tested the proposed tool in a consortium of municipalities in an Italian mountain region. The objective was to optimize costs and benefits of the energy value chain including collection, transportation, harvesting, and plant installation and maintenance costs together with benefits from the sale of thermal and electrical energy. The local authority of the region under study set a target of satisfying at least 10% of the overall energy needs of the area with biomass exploitation and in that regard, the optimal result made use of only 1.9% of the total biomass available in the region, which provided about 14% of the whole energy demand. More importantly, the optimum cost was 63% higher than the cost for receiving the same amount of energy from combustibles other than forest biomass. The authors analyzed this extra contribution of cost according to the environmental impact of the proposed solution. The same problem of optimization of an urban energy supply system was addressed in Keirstead et al. (2012); specifically, it assessed various biomass conversion technologies. A MIP model is developed based on a resource-technology network where resources are materials involved in provision of energy for a city and technologies represent processes converting a set of input resources to a set of output resources. The model was tested on a case study in an eco-town in UK, evaluating five scenarios of different types of conversion technologies [i.e., grid fuels, biomass boilers, biomass combined heat and power (CHP) plants with internal combustion engine (ICE), or organic Rankine cycle (ORC) and all technologies]. Results showed that, since finished wood chips have higher energy density than forest residues, importing them is economically more beneficial than importing forest residues to be converted into chips within the eco-town. The results also confirmed that using biomass domestic boilers alone is more expensive than the traditional gas-fired systems, whereas biomass CHP systems offer up to 15% cost savings over the gas-fired boiler scenario. Moreover, since the CHP systems make full use of the biomass fuel, these technologies are recognized as the most energy-efficient scenarios; for instance, compared to the gas boiler scenario, the CHP technologies consume 15%–19% less energy per capita. Also, from the environmental point of view, CHP scenarios had 80%–87% fewer emissions compared with the gas boiler scenario, meeting the regulation of the eco-town for 80% reduction in CO_2.

Shabani and Sowlati (2013) modeled the value chain optimization problem of a forest biomass power plant as a mixed-integer nonlinear programming problem. The proposed model calculates a monthly amount of biomass to buy from each supplier, burn, and store, and it determines whether or not to produce extra electricity to maximize the total profit. The model is solved by the AIMMS Outer Approximation algorithm. Testing the proposed tool on a real case study in Canada reduced the biomass procurement cost by 15%, when compared with the current situation where the company managers conduct tactical planning based solely on their own experience. Biomass procurement cost and transportation cost contributed to 63% and 33% of the total cost of the power plant, respectively. Additionally, evaluating various scenarios of biomass supply availability and investment in a new ash recovery system showed investing in a new ash recovery system is beneficial from both the environmental and economic aspects. Shabani et al. (2014) reformulated the mixed-integer nonlinear programming model developed by Shabani and Sowlati (2013) into a MIP model which determines the monthly consumption and storage variables of biomass as well as monthly generated electricity in a one-year planning horizon. The authors integrated procurement, storage, production, and ash management decisions in a single framework, maximizing profitability while considering uncertainty in the amount of available biomass. First, the proposed model was solved by means of a two-stage stochastic programming approach; then the authors developed a weighted bi-objective model to balance risk and profit within the value chain. Profit variability index and downside risk (the probability that the real profit is less than a certain threshold) are the two risk measures considered. Testing the model in the case of a Canadian power plant resulted in an annual profit of CAD\$16.2 million, calculated based on perfect information about suppliers' monthly available biomass. However, in reality, the amount of available biomass varies and implementing the average scenario, while other scenarios occur, led to a CAD\$0.4 million reduction in the expected profit. This amount could be improved by CAD\$0.2 million if uncertainty in biomass availability was taken into account in the model and the stochastic programing approach was used to solve it. Moreover, when downside risk was reduced, the probability of having high profit in the range of CAD\$17–18 million or low profit between CAD\$12–12.9 million became zero and the total expected profit of the power plant decreased.

Procuring wood biomass for bioenergy production in a sustainable and economical way is by itself a complex task. Alam et al. (2009) specifically focused on procurement activities involved in bioenergy production, modeled this problem as a multiobjective optimization problem, and solved it with a pre-emptive goal programming technique. The three objectives considered were minimizing the total biomass procurement cost, minimizing the total distance for biomass procurement, and maximizing biomass quality in terms of its moisture content. The authors demonstrated the application of the model in a biomass power plant consuming harvesting residues and poplar trees collected from three forest management zones (FMU) in northwestern Ontario, Canada. The problem is solved sequentially based on the DM's prioritization of the three objectives and the solution includes optimal weekly quantities of wood biomass to be collected from each FMU.

Alternatively, forest industry profitability can be improved by producing value-added products, that is, by more efficient utilization of by-products in energy application such as wood pellets. Mäkelä et al. (2011) addressed the problem of maximizing profit for Finnish sawmills with a fixed production capacity aiming at pellet production. The authors developed a static partial equilibrium model as a mixed complementarity problem. The proposed model optimizes the use of wood and by-products, which determines the optimal output mix (i.e., sawnwood, heat and power, and pellet) as well as decisions about investments in increasing the production capacity of sawnwood, heat, CHP, and pellet. Testing the model on 30 large-scale Finnish sawmills revealed the fact that with the pellet price at the time of study in the Finland sawmill industry, pellet production would not be profitable. It suggests slightly increasing pellet price or applying modest political support can make pellet production in sawmills a financially feasible business. In that respect the authors studied the application of input, investment, and production subsidies where the last two proved to be the most efficient policy instruments in promoting pellet production. Recently, in Canada, Hughes (2014) studied the pellet value chain planning problem under uncertain demand conditions over a 1-year planning horizon with the objective of gross margin maximization. The author generated stochastic demand information by means of the exponential smoothing forecasting method and proposed three optimization models based on different operating conditions (i.e., with/without an inventory management system and with variable/fixed production rate). The models have been tested on a case study of a wood pellet producer in northern Ontario, Canada. Results show the model with an inventory management system and variable production rate outperforms the other models and this is because it enables the pellet producer to account for deviation in demand according to its operational environment. In addition, the result of a sensitivity analysis indicates fluctuations in supply and demand have the highest influence on the gross margin.

In another recent work by Mobini et al. (2014), the integration of torrefaction into wood pellet production is evaluated; the authors used a simulation model called the pellet supply chain (PSC) proposed by Mobini et al. (2013). The outputs of PSC are the amount of energy consumed in each process, its related CO_2 emissions, and the cost components of delivered wood pellets to customers. The underlying model combines discrete event and discrete rate simulation approaches and has taken into account uncertainties, interdependencies, and resource constraints along the value chain. More precisely, uncertainty in parameters such as quality and availability of raw materials, processing rates and equipment failure, and electricity/fuel consumption is taken into account. The model was tested in an existing wood pellet value chain, located in British Columbia, Canada, to assess the cost of delivered torrefied pellets to different markets. Also, energy consumption and carbon dioxide emissions along the supply chain were compared with those of regular pellets. The result of this case study shows, due to increased energy density and reduced distribution costs compared with regular pellets, the delivered cost of torrefied pellets ($/GJ) to Northwest Europe decreases by about 9%. Moreover, in terms of energy consumption and CO_2 emissions along the value chain, the result of this study indicates that torrefied pellets are superior to regular pellets. Hence, the success of integration of torrefaction into wood pellet production depends on trade-offs between the increased capital and operating costs and the decreased transportation cost. For example, when long transportation distance is involved, torrefied wood pellets are more economical in terms of lower cost of delivered energy content.

Flisberg et al. (2015) analyzed all transport of forest biomass in Sweden for a year. There are 200,000 transports of eight assortments from 58,000 harvest areas to 647 heating plants included in the case study. The authors use the FlowOpt system for the analysis, which also includes 61 companies. Of these companies, 28 have volumes exceeding 10,000 tons and are treated as single companies whereas the others are aggregated. The largest model includes 100 million variables and 1.2 million constraints. Some cost allocation methods are proposed and analyzed. One of the problems with cost allocation is that the number of coalitions is 536 million, which means that many standard game theoretical models based on core stability are not practical. The actual transports are registered and by changing delivery time, changing assortments, and collaborating, different levels of savings can be obtained. Collaboration in itself can save 12% and together with the other options up to 22%.

10.4.6   INTEGRATED VALUE CHAINS

Some planning methods/DSS are designed to combine two or more value chains in an attempt to avoid suboptimization. The references discussed in this section are listed in Table 10.6 with an indication of the main process(es) covered along the value chain, as well as which value chains they address.

The DSS LogiLab has been under development by researchers at the FORAC Research Consortium, Université Laval, since 2009 (Lemieux, 2014). The system enables the tactical modeling and optimization of a FVCN from the supply areas up to the final customers. The user-friendly modeling is done through either the fulfillment of an Excel spreadsheet (that will be imported on the system by the user) or a schematic/geographical representation where the user adds the different locations of its network one by one, and defines for each a set of mandatory/optional parameters (e.g., geographical location, inputs and outputs according to the transformation process involved, processing capacity, demand, etc.). The current material flow between the locations and the traveling distances are also defined. Then the DSS optimizes the value creation of the network by maximizing the profit of the whole

TABLE 10.6
Scope and Value Chains of the FVCN Addressed by the Reviewed Literature

	Main Processes along the Value Chain						Value Chain			
References	Procurement	Production	Transportation/ Distribution	Sales	Forest	Lumber, Panel, and Engineered Wood Products	Pulp and Paper Products	Biorefinery	Bioenergy
Kong et al. (2012)	X	X	X	X	X		X		X
Kong et al. (2015)	X	X	X	X	X		X		X
FPInterface—Optitek-LogiLab (Morneau-Pereira et al., 2013, 2014)	X	X	X		X	X			
FPInterface—Optitek-ForestPlan (Kryzanowski, 2014)	X	X	X		X	X			X
Kong and Rönnqvist (2014)	X	X	X	X	X	X	X		X
LogiLab-SilviLab (Simard, 2014)	X	X	X		X	X	X	X	X
Troncoso et al. (2015)	X	X	X	X	X	X	X		X

network while reducing transportation, inventory, and production costs. Therefore, the DSS allows answering two main questions: (1) what is the most profitable wood fiber allocation among the FVCN entities? (2) can we increase profitability of as-is VCN with a given what-if scenario? A number of case studies have been conducted with the DSS LogiLab; we discuss one of them and also report its combinations with other DSS.

Elleuch et al. (2012) used the system to compute the potential profitability of implementing three interfirm collaboration approaches (i.e., regular replenishment, VMI, and collaborative planning, forecasting, and replenishment) in a FVCN of five sawmills and one pulp and paper mill in Eastern Canada. Each approach was computed according to four what-if scenarios (e.g., opening of two shutdown mills, consideration of chip freshness and sorting rules, external chip supplier) and for a base case scenario. Through a column generation method, the optimization model of the DSS LogiLab (master problem) has been combined with the optimization model of SilviLab (subproblem), a strategic forest management DSS also developed by the FORAC Research Consortium. Through an iterative process, this tactical–strategic combination allows the tactical planning to ask for modifications to the forest management plan (strategic planning) to increase FVCN profitability. A case study of an FVCN (i.e., six sawmills and one pulp and paper mill in Eastern Canada) demonstrated the potential gains of such an integrated approach from forest management to production and sales decisions. For instance, an increase from 23% to 92% of a sawmill production capacity utilization rate (while still respecting the annual allowable cut) leads to a lumber demand satisfaction increase of 13% and whole network profit increase (Simard, 2014). A case study involving an FVCN of three sawmills is presented by Morneau-Pereira et al. (2013) to demonstrate the combination of the aforementioned simulation tools FPInterface and Optitek with the DSS LogiLab. The two simulation tools allow generating the required data on different harvesting and sawing scenarios (e.g., costs, yield, product baskets) that is the input for optimization. Assuming no limit on the assortment sorting at the forestland, the potential profitability of the annual optimized plan is on average 55.6% better than the ones generated by a heuristic rule that mimics a typical DM planning behavior. This impressive gain comes from a better selection of the harvesting blocks and a better allocation of the wood to the sawmills but again, supposes no restriction on the assortment sorting rule in the forest. The simulation tools FPInterface and Optitek were also combined with the ForestPlan, which uses LP to maximize the annual plan profitability of a company-wide forest value chain. Developed in 2013 by FPInnovations and Dalhousie University, the DSS was tested on two industrial cases in Western Canada (Kryzanowski, 2014). The application case involved eight sawmills with a wide range of domestic and international customers (lumber, logs, chips, hog fuel, shavings, sawdust). Results show a potential to increase profit by 13% by selecting a different mix of harvesting blocks to meet the demand in comparison to the 691 harvesting blocks (spanning over 16,000 hectares) in the current annual harvest plan (Ristea, 2015).

Troncoso et al. (2015) studied how sequential planning tools for harvesting, transportation, production, and sales can be integrated to find better solutions in comparison with using a sequential planning process. They report savings of between 5% and 8.5% with integrated planning. This is due to the fact that better log types are connected to appropriate sawmills and final prices are implicitly integrated already in the harvesting planning. Kong and Rönnqvist (2014) took the same models and proposed strategies to establish coordination prizes between the sequential planning steps so that the DSS can be operated in a sequential approach but achieve an overall integrated solution. The strategies to find efficient coordination prizes are based on various dual and Lagrangian dual schemes.

Kong et al. (2012) combined the forest, pulp and paper, and bioenergy value chains. In Sweden, the roundwood (sawlogs and pulpwood) chains are integrated but the forest fuel for energy production is planned independently. However, as there is more and more pulpwood used directly for energy production, it is interesting to study how they impact each other depending on, for example, the supply situation and relative prices for lumber, paper, and energy. The problem becomes nonlinear as the demand from the customer follows a demand based on the purchasing cost. In the paper, the authors study an industrial case from a major Swedish forest company and conudct an analysis based on a number of scenarios. Substantial benefits and savings from integration are reported. Kong et al. (2015) expands the previous work where the selection of harvest areas also is included as decision variables. In addition, different settings of market prices are tested.

10.5 DISCUSSION

10.5.1 Gaps and Trends in DSS Development

The scientific community worldwide has been developing DSS for the forest value chain for many years. The wiki page of the Forest DSS Community of Practice (www.forestDSS.org) reports 62 DSS for forest management developed in over 23 countries, covering a wide range of forest systems, management goals, and organizational frameworks. Yet, only 18 of them addressed medium- and/or short-term decisions; some of them originated from internal development of forest companies. In fact, we observe that on the one hand, DSS for tactical/operational planning are more recent developments and still more rare than DSS for strategic planning. On the other hand, DSS for tactical planning are often tailored to the needs of a specific industry and country, which makes them unique, flexible, and scalable and also more likely to be utilized outside the scientific publications. We can argue that DSS are usually research-driven proofs-of-concept, developed by researchers and gradually introduced to the end user in practice. This may explain the way they are developed as prototypes rather than real commercial software where the focus is on the modeling/optimization rather than DSS features such as a friendly graphical user interface, support, maintenance, and upgrades. This jeopardizes the implementation and is most of the time the main reason behind the failure and also why forest companies do not adopt such DSS in practice. The lack of scalability and flexibility of such DSS to meet new needs of the end user can be another issue. This mismatch between DSS features and the needs of the end users leads them to cease using such DSS. This mismatch is also due to the long cycle time of developing a DSS where a large gap arises between the original user's needs at the development phase and his current needs at the implementation phase. Also, end users use the DSS for other purposes completely different from the initial ones for which the DSS has been designed, which leads to another mismatch. We should also note that we limit our comments to the DSS that are published in the scientific literature. There are software programs used by many companies, but their solution methodologies are not known.

Despite the large number of DSS developed in forest planning, some studies (e.g., Reynolds et al. 2007; Menzel et al., 2012) emphasized the need for a clear focus on the target users, therefore acknowledging the human dimension in information systems. Stakeholders' participation may be instrumental in developing a DSS that might effectively address the business specificities (Sousa and Pereira 2005). This is a critical success factor for DSS (Arnott and Dodson 2008).

Most of the research addresses the forest-to-mill part of the FVCN or from the mill to the market in each respective value chain (decoupled). There is a need to better integrate the forest value chain with the following value chains of the FVCN and in this way, to better use the information flow from the different markets in the earlier stages of the FVCN. Also, there is a lack of integration between the tactical planning with upper and lower levels (strategic and operational) that leads to misalignment between the three planning levels. We state that current DSS that cover the full FVCN are still rare, with the exception of biomass where recent DSS have been developed. No forest value chain planning methods/DSS discuss the sales process. Other issues typically included in logistics such as stockyard management and inventory management are also poorly addressed. We refer to Rönnqvist et al. (2015) for a review of research challenges (open problems) related to the application of OR in the FVCN, mainly on the forest-to-mill part.

10.5.2 Issues and Challenges in Implementation

Different issues related to DSS adoption are discussed by Audy et al. (2012a) and Rönnqvist (2012). To implement a DSS there are many practical questions that arise. In the article, a number of seemingly easy questions become difficult in implementing full DSS.

DSS are data intensive and are not always integrated with GIS and ERP systems; they also require a lot of data and connections with other systems to be fully utilized. These missing connections and gateways are expensive and complex due to lack of expertise, time, or funding to perform them in an appropriate way. Sometimes, end users do not see the value to justify such investments and efforts to replace their current practices with the new alternatives. Also, end users view DSS as black boxes and cannot follow the reasoning behind them; consequently, they are hesitant to accept and trust the results/outcomes of such DSS. Requiring high competencies (e.g., in OR, analytics, databases) to be used at their full potential (and thus provide the highest benefits), several DMs give confidential mandates to specialized resources for conducting advanced analysis using the DSS to help them in their tactical decisions. The DM will then be free to decide whether or not to use the recommendations derived from these studies. Such time-consuming support for the DMs would not be conceivable with DSS designed for operational level decisions.

The individual competencies and training of the end user are often neglected during the implementation process of DSS where he is expected to be capable, ready to use, and understand the reasoning behind the DSS, and finally interpret the results and outcomes of the DSS. The lack of support and continuous improvement of DSS after implementation is another factor that leads to failure due to the disconnection between the development and implementation teams that belong to university and industry, respectively.

Expectations are very high regarding what DSS can deliver. Most people expect that DSS can solve problems for them which a DSS is not aimed to do: DSS by itself does not solve the problem. One reason could be that DSS are presented as game changers and very sophisticated tools based on advanced optimization techniques combined with technology, which may lead end users to think that they can really solve problems and are more than just systems aiming to help them. There is a need to draw business models built on collaboration between companies (or departments within the same company) which may be supported in the DSS (Audy et al., 2012b).

10.5.3 FUTURE RESEARCH PATHS FORWARD

Stakeholders including the public are paying ever more attention to how forest resources are managed and utilized, which poses new challenges for the new generation of DSS in respect to its comprehensiveness but also simplicity. Economic performance is no longer the ultimate goal as other environmental and social aspects gain greater importance. Among the key drivers that will influence the research in DSS in tactical planning in forestry are big data and Internet, sustainability, group decision-making by stakeholders, uncertainty, interfirm collaboration, integrated planning, and multidisciplinary research approaches.

The rapid development of the Internet and the use of advanced technologies have led to the explosive growth of data in the forest industry. Currently, data sources include large spatial data sets, GIS information, ERP systems, ecological information, social and environment-related data sets, government regulations, GPS-based solutions and sensors to track products/machines in real time, and so on. These sources generate a huge amount of data across the value chain ready to be used by DSS. An illustrative example for such a platform is being developed in the EU project FOCUS—Advances in Forestry Control and Automation Systems in Europe (www.focusnet.eu). The next generation of DSS must be able to handle and process these raw data and turn them into valuable information and pertinent decisions. The Internet of Things (IoT), where all devices will be connected to the Web, will enable DSS to be web-based applications and available on new mobile platforms such as smartphones, tablets, and so on. Big data and IoT will be key drivers in the development of the next generation of DSS and this requires research in new methodologies to fill the gap between existing DSS and these new technologies (Bettinger et al., 2011; Vacik and Lexer, 2014).

The social acceptability and environmental impact of the forest industry should be integrated in tactical planning in the next generation of DSS for a truly sustainable forest value chain. For instance, the development of new bioenergy and biorefinery products in the last decade, in conjunction with new regulations and policies, requires the combination of existing and new assessment methods such as life-cycle assessment and multiobjective optimization that must be integrated in DSS (Boukherroub et al., 2015; Cambero et al., 2015b).

Forest planning affects and involves many stakeholders (industry, governments, landowners, communities, etc.) with different goals and objectives. The Internet has contributed and facilitated interactions between groups, including the public, making them more active in forest planning and problem solving. This shows the limitations of current DSS to support this interactive planning approach and raises the need to propose new frameworks to design a new decision theater to support coordination and interactions among stakeholders and integrate them into new group DSS (Kangas, 1992; Donaldson et al., 1995; Azouzi and D'Amours, 2011).

Uncertainty is an inherent phenomenon in forestry due to many social, economic, biological, and technological factors. New technologies and big data show promise in reducing these uncertainties but need to be economically sound. Depending on planning level, different approaches are more appropriate to deal with uncertainty (e.g., pooling, hedging, stochastic programming, robust optimization). In some cases deterministic methods where uncertainty is considered through, for example, safety stock levels are most appropriate due to the model size and solution times. In others where it is possible to generate a number of scenarios and where the best expected result is wanted, stochastic programming is an interesting path. For others where feasibility is critical, it is better to use robust optimization approaches. For each of these alternatives it is important to evaluate them through agent-oriented simulation approaches (Palma and Nelson, 2009; Ouhimmou et al., 2010; Feng et al., 2012; Shabani et al., 2014; Abasian et al., 2015).

Collaboration across value chains has been proven to reduce overall cost considerably. However, there are many questions regarding how confidential data is used, and how cost allocation schemes are agreed on and put into contracts (Marques et al., 2016). There are also open questions about how the coalitions should be formed and managed (Audy et al., 2012c; Guajardo and Rönnqvist, 2015). The collaboration has traditionally looked at vertical integration and lately at horizontal collaboration. What is next is to study cross-chain integrations.

Most DSS have been developed by researchers through case studies and gradually introduced to the end user. The researcher's background has a big impact on the DSS structure where forestry, management science, industrial engineering, and operations research are the most dominant disciplines. Recently, more researchers from computer science, graphics, software, and social sciences have been involved in developing such DSS. Because of the complexity and multidisciplinarity of forest-integrated planning, new DSS must be designed by multidisciplinary research teams in a collaborative approach to be more successful in the future.

10.6 CONCLUSION

This chapter provides a broad overview of a number of planning methods and DSS for tactical decisions in the FVCN. A generic mathematical model is introduced to illustrate the typical tactical decisions to be made in a value chain. About 60 methods/DSS were discussed regarding what decisions (planning problems) were made, their applications (e.g., results reported, level of implementation), and the solution approach used. We note that they almost always rely on OR-based solution approaches and they focus on one of the value chains within the FVCN. However, in recent years, a growing number of methods/DSS have been integrating two or more value chains. Also, despite the promising results reported (e.g., case studies), it appears that a relatively low number of planning methods/DSS has been adopted/ used in practice by the DMs. This raises the need to better understand the adoption impediments and success factors in such a way to enhance in that regard the development-to-implementation innovation process followed by the researchers and practitioners. Other trends and future research directions are also presented. Social and environmental impacts have recently been added in DSS and will be fully integrated in the next generation of DSS. Integration with GIS and development of graphical user interfaces have always been a big challenge to DSS but many recent experiments have been attempted to overcome such difficulties. Big data and IoT, where all devices will be connected to the Web, is a challenge and tremendous opportunity for the next generation of DSS to have access to more accurate data in real time and to be used by more stakeholders in collaborative and group decision approaches for a truly sustainable forest value chain. A new era for research will involve developing and implementing new innovative, fast methods and algorithms to deal with a huge amount of uncertain data for multiobjective and multiple stakeholders' decision-making in forest planning.

ACKNOWLEDGMENTS

The authors would like to acknowledge the partial financial support of the Natural Sciences and Engineering Research Council of Canada Strategic Research Network on Value Chain Optimization as well as the valuable cooperation of the persons cited as personal communication.

REFERENCES

Abasian, F., Rönnqvist, M., Ouhimmou, M., 2015. Designing forest biomass value chain under uncertainty. *11th International Industrial Engineering Conference*, 26–28 October, Canada.

Akhtari, S., Sowlati, T., Day, K., 2014. Optimal flow of regional forest biomass to a district heating system. *International Journal of Energy Research*, 38(7): 954–964.

Alam, Md. B., Shahi, C., Pulkki, R., 2009. Wood biomass supply model for bioenergy production in Northwestern Ontario. *1st International Conference on the Developments in Renewable Energy Technology (ICDRET)*, 17–19 December, Bangladesh, pp. 1–3.

Arnott, D., Dodson, G., 2008. Decision support systems failure. In Burstein, F., Holsapple, C.W. (Eds.), *Handbook on Decision Support Systems 1: Basic Themes*, 763–790, Chapter 34, Berlin: Springer Berlin Heidelberg.

Audy, J.-F., D'Amours, S., Rönnqvist, M., 2012a. Planning methods and decision support systems in vehicle routing problems for timber transportation: A review. CIRRELT research paper 2012-38, CIRRELT, Montreal, Canada.

Audy, J.-F., D'Amours, S., Rönnqvist, M., 2012c. An empirical study on coalition formation and cost/savings allocation. *International Journal of Production Economics*, 136(1): 13–27.

Audy, J.-F., Lehoux, N., D'Amours, S., Rönnqvist, M., 2012b. A framework for an efficient implementation of the logistics collaborations. *International Transactions in
Operational Research*, 19(5): 633–657.

Azouzi, R., D'Amours, S., 2011. Information and knowledge sharing in the collaborative
design of planning systems within the forest products industry: Survey, framework, and
roadmap. *Journal of Science and Technology for Forest Products and Processes*, 1(2):
6–14.

Bare, B.B., Briggs, D.G., Roise, J.P., Schreuder, G.F., 1984. A survey of systems analysis
models in forestry and the forest products industries. *European Journal of Operational
Research*, 18(1): 1–18.

Baskent, E.Z., Keles, S., 2005. Spatial forest planning: A review. *Ecological Modelling*,
188(2–4): 145–173.

Beaudoin, D., LeBel, L., Frayret, J.-M., 2007. Tactical supply chain planning in the forest
products industry through optimization and scenario-based analysis. *Canadian Journal
of Forest Research*, 37(1): 128–140.

Bettinger, P., Cieszewski, C.J., Falcão, A., 2011. Perspectives on new developments of decision
support systems for sustainable forest management. *Mathematical and Computational
Forestry and Natural Resource Sciences (MCFNS)*, 3(1): 15–17.

Bouchriha, H., Ouhimmou, M., D'Amours, S., 2007. Lot sizing problem on a paper machine
under cyclic production approach. *International Journal of Production Economics*,
105(2): 318–328.

Boukherroub, T., LeBel, L., Ruiz, A., 2015. A framework for sustainable forest resource allocation: A Canadian case study. *Omega*. http://dx.doi.org/10.1016/j.omega.2015.10.011
Available online 27 October 2015.

Bredström, D., Jönsson, P., Rönnqvist, M., 2010. Annual planning of harvesting resources in
the forest industry. *International Transactions in Operational Research*, 17(2): 155–177.

Bredstrom, D., Lundgren, J.T., Ronnqvist, M., Carlsson, D., Mason, A., 2004. Supply chain
optimization in the pulp mill industry—IP models, column generation and novel constraint branches, *European Journal of Operational Research*, 156: 2–22.

Broman, H., Frisk, M., Rönnqvist, M., 2009. Supply chain planning of harvest operations and
transportation after the storm Gudrun. *Information Systems and Operational Research*,
47(3): 235–245.

Cambero, C., Sowlati, T., Marinescu, M., Roser, D., 2015a. Strategic optimization of forest residues to bioenergy and biofuel supply chain. *International Journal of Energy
Research*, 39(4): 439–452.

Cambero, S., Sowlati, T., Pavel, M., 2015b. Economic and life cycle environmental optimization of forest-based biorefinery supply chains for bioenergy and biofuel production. *Chemical Engineering Research and Design*. http://dx.doi.org/10.1016/j.
cherd.2015.10.040 Available online 9 November 2015.

Carino, H.F., LeNoir, C.H. Jr., 1988. Optimizing wood procurement in cabinet manufacturing.
Interfaces, 18(2): 10–19.

Carino, H.F., Willis, D.B. III, 2001a. Enhancing the profitability of a vertically integrated
wood products production system: Part 1. A multistage modeling approach. *Forest
Products Journal*, 51(4): 37–44.

Carino, H.F., Willis, D.B. III, 2001b. Enhancing the profitability of a vertically integrated
wood products production system: Part 2. A case study. *Forest Products Journal*, 51(4):
45–53.

Carlgren, C.-G., Carlsson, D., Rönnqvist, M., 2006. Log sorting in forest harvest areas integrated with transportation planning using backhauling. *Scandinavian Journal of Forest
Research*, 21(3): 260–271.

Carlsson, D., Rönnqvist, M., 2005. Supply chain management in forestry—case studies at
Södra Cell AB. *European Journal of Operational Research*, 163(3): 589–616.

Carlsson D., Ronnqvist, M., 2007. Backhauling in forest transportation—models, methods and practical usage. *Canadian Journal of Forest Research*, 37(12): 2612–2623.

Carlsson, D., D'Amours, S., Martel, A., Rönnqvist, M., 2009. Supply chain planning models in the pulp and paper industry. *Information Systems and Operational Research*, 47(3): 167–183.

Carlsson, D., Flisberg, P., Rönnqvist, M., 2014. Using robust optimization for distribution and inventory planning for a large pulp producer. *Computers and Operations Research*, 44: 214–225.

Chinese, D., Meneghetti, A., 2009. Design of forest biofuel supply chains. *International Journal of Logistics Systems and Management*, 5(5): 525–550.

Chauhan, S.S., Martel, A., D'Amours, S., 2008. Roll assortment optimization in a paper mill: An integer programming approach. *Computer and Operations Research*, 35(2): 614–627.

Clark, M.M., Meller, R.D., McDonald, T.P., 2000. A three-stage heuristic for harvest scheduling with access road network development. *Forest Science*, 46(2): 204–218.

Clarke, G., Wright, J.W., 1964. Scheduling of vehicles from a central depot to a number of delivery points. *Operations Research*, 12(4): 568–581.

Dansereau, L.-P., 2013. Cadre de planification intégrée de la chaîne logistique pour la gestion et l'évaluation de stratégies de bioraffinage forestier. [Integrated planning framework for supply chain management and evaluation of forest biorefinery strategies]. Ph.D. Thesis, École Polytechnique de Montréal, Canada.

Dansereau, L.-P., El-Halwagi, M.M., Stuart, P.R., 2012a. Value-chain management considerations for the biorefinery. In Stuart, P.R., El-Halwagi, M.M. (Eds.), *Integrated Biorefineries: Design, Analysis and Optimization*, 195–250, Chapter 7, Boca Raton, FL: CRC Press/Taylor & Francis.

Dansereau, L.-P., El-Halwagi, M.M., Stuart, P.R., 2012b. Value-chain planning in the forest biorefinery: Case study analyzing manufacturing flexibility. *Journal of Science and Technology for Forest Products and Processes*, 2(4): 60–69.

D'Amours, S., Carle, M.-A., Rönnqvist, M., 2014. Pulp and paper supply chain management. In Borges, J.G., Diaz-Balteiro, L., McDill, M.E., Rodriguez, L.C.E. (Eds.), *The Management of Industrial Forest Plantations: Theoretical Foundations and Applications, Managing Forest Ecosystems*, 33: 489–514, Chapter 17, Netherlands: Springer.

D'Amours, S., Epstein, R., Weintraub, A., Rönnqvist, M., 2011. *Operations Research in Forestry and Forest Products Industry*. New Jersey: Wiley Encyclopedia of Operations Research and Management Science.

D'Amours, S., Rönnqvist, M., Weintraub, A., 2008. Using operational research for supply chain planning in the forest products industry. *Information Systems and Operational Research*, 46(4): 265–281.

De Meyer, A., Cattrysse, D., Rasinmäki, J., Van Orshoven, J., 2014. Methods to optimise the design and management of biomass-for-bioenergy supply chains: A review. *Renewable and Sustainable Energy Reviews*, 31: 657–670

De Mol, R.M., Jogems, M.A.H., Beek, P.V., Gigler J.K., 1997. Simulation and optimization of the logistics of biomass fuel collection. *Netherlands Journal of Agricultural Science* 45(1): 217–228.

Donald, W.S., Maness, T.C., Marinescu, M.V., 2001. Production planning for integrated primary and secondary lumber manufacturing. *Wood and Fiber Science*, 33(3): 334–344.

Donaldson, T., Preston, L.E., 1995. The stakeholder theory of the corporation: Concepts, evidence, and implications. *Academy of Management Review*, 20(1): 65–91.

Ekşioğğlu, S.D., Acharya, A., Leightley, L.E., Arora, S., 2009. Analyzing the design and management of biomass-to-biorefinery supply chain. *Computers and Industrial Engineering*, 57(4): 1342–1352.

Ekşioğlu, S.D., Li, S., Zhang, S., Sokhansanj, S., Petrolia, D., 2010. Analyzing impact of intermodal facilities on design and management of biofuel supply chain. *Transportation Research Record: Journal of the Transportation Research Board*, 2191: 144–151.

Elleuch, M., Lehoux, N., LeBel, L., 2012. Collaboration entre les acteurs pour accroître la profitabilité: étude de cas dans l'industrie forestière [Collaboration among stakeholders to increase profitability: A case study in the forest industry]. *9th International Conference on Modeling, Optimization & SIMulation—MOSIM'12*, 6–8 June, France.

Epstein, R., Morales, R., Seron, J., Weintraub, A., 1999a. Use of OR systems in the Chilean forest industries. *Interfaces*, 29(1): 7–29.

Epstein, R., Nieto, E., Weintraub, A., Chevalier, P., Gabarro, J., 1999b. A system for the design of short term harvesting strategy. *European Journal of Operational Research*, 119(2): 427–439.

Eriksson, L.O., Björheden, R., 1989. Optimal storing, transport and processing for a forest-fuel supplier. *European Journal of Operational Research*, 43(1): 26–33.

Everett, G., Philpott, A., Vatn, K., Gjessing, R., 2010. Norske Skog improves global profitability using operations research. *Interfaces*, 40(1): 58–70.

FAO. Food and Agriculture Organization, 2014a. FAO Yearbook of Forest Products 2008–2012. United Nations.

FAO. Food and Agriculture Organization, 2014b. State of the World's Forests: Enhancing the socioeconomic benefits from forests. United Nations.

Farrell, R.R., Maness, T.C., 2005. A relational database approach to a linear programming-based decision support system for production planning in secondary wood product manufacturing. *Decision Support Systems*, 40(2): 183–196.

Faulkner, W.H., 2012. Economic modeling and optimization of a region specific multi-feedstock bio-refinery supply chain. Master Thesis, University of Kentucky, United States.

Favreau, J., FPInnovations, personal communication, September–November 2013.

Feng, Y., D'Amours, S., Beauregard, R., 2008. The value of sales and operations planning in oriented strand board industry with make-to-order manufacturing system: Cross-functional integration under deterministic demand and spot market recourse. *International Journal of Production Economics*, 115(1): 189–209.

Feng, Y., D'Amours, S., Beauregard, R., 2010. Simulation and performance evaluation of partially and fully integrated sales and operations planning. *International Journal of Production Research*, 48(19): 5859–5883.

Feng, Y., D'Amours, S., LeBel, L., Nourelfath, M., 2012. Integrated forest biorefinery supply chain network design using mathematical programming approach. In Stuart, P. R., El-Halwagi, M. M. (Eds.), *Integrated Biorefineries Design, Analysis and Optimization*, 251–282, Chapter 8, Boca Raton, FL: CRC Press/Taylor & Francis.

Fjeld, D., D´Amours, S., Eriksson, L.O., Frisk, M., Lemieux, S., Marier, P., Rönnqvist, M., 2014. Developing training for industrial wood supply management. *International Journal of Forest Engineering*, 25(2): 101–112.

Fleischmann, B., Meyr, H., Wagner, M., 2008. Advanced planning. In Stadler, H., Kilger, C. (Eds.), *Supply Chain Management and Advanced Planning: Concepts, Models, Software and Case Studies*, 4th ed., 81–106, Chapter 4, Berlin: Springer-Verlag.

Flisberg, P., Frisk, M., Rönnqvist, M., 2012, FuelOpt: A decision support system for forest fuel logistics. *Journal of the Operational Research Society*, 63(11): 1600–1612.

Flisberg, P., Frisk, M., Rönnqvist, M., 2014. Integrated harvest and logistic planning including road upgrading. *Scandinavian Journal of Forest Research*, 29(1): 195–209.

Flisberg, P., Frisk, M., Rönnqvist, M., Guajardo, M., 2015. Potential savings and cost allocations for forest fuel transportation in Sweden: A country-wide study. *Energy*, 85(1): 353–365.

Forget, P., D'Amours, S., Frayret, J.-M., 2008. Multi-behavior agent model for planning in supply chains: An application to the lumber industry. *Robotics and Computer-Integrated Manufacturing Journal*, 24(5): 664–679.

Forsberg, M., Frisk, M., Rönnqvist, M., 2005. FlowOpt: A decision support tool for strategic and tactical transportation planning in forestry. *International Journal of Forest Engineering*, 16(2): 101–114.

Frayret, J.-M., D'Amours, S., Rousseau, A., Harvey, S., Gaudreault, J., 2007. Agent-based supply chain planning in the forest products industry. *International Journal of Flexible Manufacturing Systems*, 19(4): 358–391.

Freppaz, D., Minciardi, R., Robba, M., Rovatti, M., Sacile, R., Taramasso, A., 2004. Optimizing forest biomass exploitation for energy supply at a regional level. *Biomass and Bioenergy*, 26(1): 15–25.

Frisk, M., Göthe-Lundgren, M., Jörnsten, K., Rönnqvist, M., 2010. Cost allocation in collaborative forest transportation. *European Journal of Operational Research*, 205(2): 448–458.

Gingras, C., Cordeau, J.-F., Laporte, G., 2007. Un algorithme de minimisation du transport à vide appliqué à l'industrie forestière [An algorithm for minimizing unloaded transportation applied to the forest industry]. *Information Systems and Operational Research*, 45(1): 41–47.

Gold, S., Seuring, S., 2011. Supply chain and logistics issues of bio-energy production. *Journal of Cleaner Production*, 19(1): 32–42.

Guajardo, M., Rönnqvist, M., 2015. Operations research models for coalition structure in collaborative logistics. *European Journal of Operational Research*, 240(1): 147–159.

Gunnarsson, H., Rönnqvist, M., Carlsson, D., 2006. A combined terminal location and ship routing problem. *Journal of the Operational Research Society*, 57: 928–938.

Hughes, N.M., 2014. Modeling uncertain demand in wood pellet supply chains: A case study from Northern Ontario. Master Thesis, Lakehead University, Canada.

Hughes, N.M., Shahi, C., Pulkki, R., 2014. A review of the wood pellet value chain, modern value/supply chain management approaches, and value/supply chain models. *Journal of Renewable Energy*, 2014: 1–14

Kangas, J., 1992. Multiple-use planning of forest resources by using the analytic hierarchy process. *Scandinavian Journal of Forest Research*, 7(1–4): 259–268.

Kanzian, C., Holzleitner, F., Stampfer, K., Ashton, S., 2009. Regional energy wood logistics—optimizing local fuel supply. *Silva Fennica*, 43(1): 113–128.

Karlsson, J., Rönnqvist, M., Frisk, M., 2006. RoadOpt—A decision support system for road upgrading in forestry. *Scandinavian Journal of Forest Research*, 21(7): 5–15.

Keirstead, J., Samsatli, N., Pantaleo, A.M., Shah, N., 2012. Evaluating biomass energy strategies for a UK eco-town with an MILP optimization model. *Biomass and Bioenergy*, 39: 306–316.

Kong, J., Rönnqvist, M., 2014. Coordination between strategic forest management and tactical logistic and production planning in the forestry supply chain. *International Transactions in Operational Research*, 21(5): 703–735.

Kong, J., Rönnqvist, M., Frisk, M., 2012. Modeling an integrated market for sawlog, pulpwood and forest bioenergy. *Canadian Journal of Forest Research*, 42(2): 315–332.

Kong, J., Rönnqvist, M., Frisk, M., 2015. Using mixed integer programming models to synchronously determine production levels and market prices in an integrated market for roundwood and forest biomass. *Annals of Operations Research*, 232: 179–199.

Kryzanowski, T., 2014. ForestPlan software helps increase profit margins for forest products industry. *Logging and Sawmilling Journal*, 2014: 32.

Lehoux, N., D'Amours, S., Langevin, A., 2009. Collaboration and decision models for a two-echelon supply chain: A case study in the pulp and paper industry. *Journal of Operations and Logistics*, 2(4): 1–17.

Lemieux, S., 2014. Université Laval, personal communication, September 2014.

Liden, B., Ronnqvist, M., 2000. *CustOpT—a model for customer optimized timber in the wood chain*. Proceedings of the 12th Annual Conference for Nordic Researchers in Logistics, NOFOMA 2000, Aarhus, Denmark, 421–441.

Liden, B., Ronnqvist, M., 2003. *Customer optimized timber in the wood chain- a model.* Proceedings of the International Union of Forest Research Organizations, New Zealand, 1–25.

Mäkelä, M., Lintunen, J., Kangas, H.-L., Uusivuori, J., 2011. Pellet promotion in the Finnish sawmilling industry: The cost effectiveness of different policy instruments. *Journal of Forest Economics*, 17(2): 185–196.

Maness, T. C, Adams, D. M., 1993. The combined optimization of log bucking and sawing strategies. *Wood Fiber Sciences*, 23: 296–314.

Maness, T. C, Norton, S. E., 2002. A multiple period combined optimization approach to forest production planning. *Scandinavian Journal of Forest Research*, 17: 460–471.

Marier, P., Bolduc, S., Ben Ali, M., Gaudreault, J., 2014. S&OP network model for commodity lumber products. CIRRELT research paper 2014–25, CIRRELT, Montreal, Canada.

Marques, A.F., de Sousa, J.P., Rönnqvist, M., Jafe, R., 2014a. Combining optimization and simulation tools for short-term planning of forest operations. *Scandinavian Journal of Forest Research*, 29(1): 166–177.

Marques, A.F., Olmo, B., Audy, J.F., Rocha, P., 2016. A comprehensive framework for developing inter-firm collaboration—A study in the forest-based supply chain. *Journal of Science and Technology for Forest Products and Processes*. Submitted for publication.

Marques, A.S., Audy, J.-F., D'Amours, S., Rönnqvist, M., 2014b. Tactical and operational harvest planning. In: Borges, J.G., Diaz-Balteiro, L., McDill, M.E., Rodriguez, L.C.E. (Eds.), *The Management of Industrial Forest Plantations: Theoretical Foundations and Applications, Managing Forest Ecosystems*, 33: 239–267, Chapter 7, the Netherlands: Springer.

Meléndez, J., 2015. Biomass procurement cost minimization for implementation of a retrofit biorefinery in a pulp and paper mill. Ph.D. Thesis, École Polytechnique de Montréal, Canada.

Menzel, S., Nordström, E.-M., Buchecker, M., Marques, A., Saarikoski, H., Kangas, A., 2012. Decision support systems in forest management—requirements for a participatory planning perspective. *European Journal of Forest Research*, 131(5): 1367–1379.

Mobini, M., Meyers, J., Trippe, F., Sowlati, T., Froehling, M., Schultmann, F., 2014. Assessing the integration of torrefaction into the wood pellet production. *Journal of Cleaner Production*, 78(1): 216–225.

Mobini, M., Sowlati, T., Sokhansanj, S., 2013. A simulation model for the design and analysis of wood pellet supply chain. *Applied Energy*, 111: 1239–1249.

Morneau-Pereira, M., Arabi, M., Gaudreault, J., Nourelfath, M., Ouhimmou, M., 2013. An optimization and simulation framework for integrated tactical planning of wood harvesting operations, wood allocation and lumber production. CIRRELT research paper 2013-48, CIRRELT, Montreal, Canada.

Murray, A.T., 1999. Spatial restrictions in harvest scheduling. *Forest Science*, 45(1): 45–52.

Murray, A.T., Church, R.L., 1995. Heuristic solution approaches to operational forest planning problems. *Operations Research Spektrum*, 17(2–3): 193–203.

NRC. Natural Resources Canada, 2014. Ten key facts on Canada's natural resources.

Ouhimmou, M., D'Amours, S., Ait-Kadi, D., Beauregard, R., Chauhan S.S., 2008. Furniture supply chain tactical planning optimization using a time decomposition approach. *European Journal of Operational Research*, 189(3): 952–970.

Ouhimmou, M., D'Amours, S., Beauregard, R., Ait-Kadi, D., Chauhan, S.S., 2009. Optimization helps Shermag gain competitive edge. *Interfaces*, 39(4): 329–345.

Ouhimmou, M., Raulier, F., Fortin, M., D'Amours, S., 2010. Robust optimization approach to consider tree growth model uncertainty in forest strategic planning. *3rd International Conference on Information Systems, Logistics and Supply Chain (ILS 2010)*, 13–16 April, Casablanca, Morocco.

Palma, C.D., Nelson, J.D., 2009. A robust optimization approach protected harvest scheduling decisions against uncertainty. *Canadian Journal of Forest Research*, 39(2): 342–355.

Reinders, M.P., 1993. Tactical planning for a cutting stock system. *Journal of the Operational Research Society*, 44(7): 645–657.

Reynolds, K.M., Twery, M., Lexer, M.J., Vacik, H., Ray, D., Shaom G., Borges, J.G. 2007. Decision support systems in natural resource management. In Burstein F., Holsapple C. (Eds.), *Handbook on Decision Support Systems*. Berlin: Springer Verlag International.

Richards, E.W., Gunn, E.A., 2000. A model and tabu search method to optimize stand harvest and road construction schedules. *Forest Science*, 46(2): 188–203.

Ristea, C., FPInnovations, personal communication, March 2015.

Rizk, N., Martel, A., D'Amours, S., 2008. Synchronized production-distribution planning in a single-plant multi-destination network. *Journal of the Operational Research Society*, 59: 90–104.

Rönnqvist, M., 2003. Optimization in forestry. *Mathematical Programming*, 97(1–2): 267–284.

Rönnqvist, M., 2012. OR challenges and experiences from solving industrial applications. *International Transactions in Operational Research*, 19(1–2): 227–251.

Rönnqvist, M., D'Amours, S., Weintraub, A., Jofre, A., Gunn, E., Haight, R.G., Martell, D., Murray, A.T., Romero, C., 2015. Operations research challenges in forestry: 33 open problems. *Annals of Operations Research*, 232(1): 11 40.

Shabani, N., Akhtari, S., Sowlati, T., 2013. Value chain optimization of forest biomass for bio-energy production: A review. *Renewable and Sustainable Energy Reviews*, 23: 299–311.

Shabani, N., Sowlati, T., 2013. A mixed integer non-linear programming model for tactical value chain optimization of a wood biomass power plant. *Applied Energy*, 104: 353–361.

Shabani, N., Sowlati, T., Ouhimmou, M., Rönnqvist, M., 2014. Tactical supply chain planning for a forest biomass power plant under supply uncertainty. *Energy the International Journal*, 78: 346–355.

Simard, M., 2014. Université Laval, personal communication, February and September 2014.

Singer, M., Donoso, P., 2007. Internal supply chain management in the Chilean sawmill industry. *International Journal of Operations and Production Management*, 27(5): 524–541.

Sousa, P., Pereira, C.M., 2005. Enterprise architecture: Business and IT alignment. *2005 ACM Symposium on Applied Computing*, 13–17 March, Santa Fe, NM, 1344–1345.

Troncoso, J., D'Amours, S., Flisberg, P., Rönnqvist, M., Weintraub, A., 2015. A mixed integer programming model to evaluate integrating strategies in the forest value chain—A case study in the Chilean forest industry. *Canadian Journal of Forest Research*, 45(7): 937–949.

Vacik, H., Lexer, M.J., 2014. Past, current and future drivers for the development of decision support systems in forest management. *Scandinavian Journal of Forest Research*, 29(suppl 1): 2–19.

Weintraub, A., Murray, A.T., 2006. Review of combinatorial problems induced by spatial forest harvesting planning. *Discrete Applied Mathematics*, 154(5): 867–879.

Zak, J., 2010. Decision support systems in transportation. In Jain, L.C., Lim, C.P. (Eds.), *Handbook on Decision Making: Techniques and Applications*, Vol. 1. 4: 249–294, Chapter 11, Berlin: Springer Berlin Heidelberg.

Zhang, S.-Y., Tong, Q.-J., 2005. Modeling lumber recovery in relation to selected tree characteristics in jack pine using sawing simulator Optitek. *Annals of Forest Science*, 62(3): 219–228.

11 Key Aspects of Forest Woody Biomass Flows within the Canadian Forest Value Chain

Luc LeBel, Reino Pulkki, Riadh Azouzi, and Denis Cormier

CONTENTS

11.1 INTRODUCTION

Woody biomass usually refers to wood fiber available as residues. It is usually divided into three groups: primary, secondary, and tertiary residues. Primary residues are by-products of conventional forest operations. Secondary residues are by-products of industrial processes, including bark, sawdust, shavings, and chips. Tertiary residues are by-products of demolition, construction, and packaging processes. Materials that can be obtained from early thinning and from stands killed by disturbances such as fire, disease, or insects can also be considered as forest biomass (see Figure 11.1). In this chapter, the focus is on the primary residues. They are interchangeably referred to as "forest woody biomass" (FWB). The main sources of FWB are the logging residues. These materials historically had a low value and could not be sold for traditional forest products (namely pulp, lumber, wood-based panels, and engineered wood products). They are organic materials that if used in a sustainable manner have the potential to improve the economic viability of forest products firms and contractors, as well as the economic sustainability/health of rural communities, while improving air quality, substituting for nonrenewable resource extraction, and reducing greenhouse gas emissions. Note that Figure 11.1 positions forest biomass with respect to other wood-based raw materials used for fuel production. Basically, these are fast growing species grown in short rotations. Forest biomass remains the prevalent source of wood-based energy (Röser et al., 2008).

Canadian forests have a large potential for biomass production. Paré et al. (2011) compared estimates made by various authors of available FWB as well as the volume of harvested roundwood in 2007 (Figure 11.2). These authors emphasized the fact that, in general, estimates of available residues are reported without taking into consideration the operational and economic feasibility of recovering these residues or the amount that needs to be left on site to maintain ecosystem functions. These factors are likely to significantly lower the volume of FWB available for further processing in the value chain. In an effort to figure out a highly aggregated estimate of FWB potential at the world scale and over the long term, Smeets and Faaij (2007) distinguished several key variables. These variables included the demand for industrial roundwood and wood fuel, plantation establishment rates, and various technical, economic, and ecological limitations related to the supply of wood from forests. In brief, the limitations were applied by excluding certain forest areas and applying certain recognized increments to wood production potential of forests.

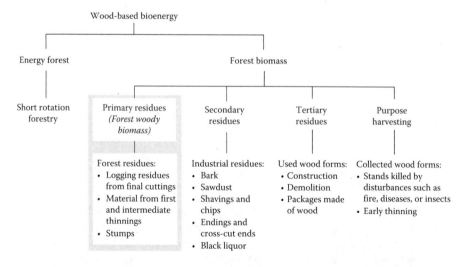

FIGURE 11.1   Positioning forest woody biomass with respect to other wood-based fuels.

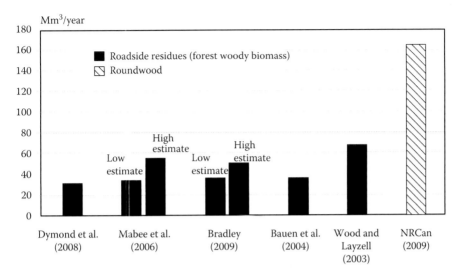

FIGURE 11.2 Comparison of estimates made by various authors of available volumes of forest woody biomass (Mm³/year) with total harvested roundwood volume in 2007.

At the end, one rapidly understands that a limited fraction of total forest area might be expected to yield FWB available for industrial processes. Logistical aspects such as the handling, storage (in one or more intermediate locations), forest transportation, and road transportation were not considered by Smeets and Faaj (2007). These logistical aspects have a higher contribution to FWB-based product costs than any other aspect (Rentizelas et al., 2009).

This chapter is not aimed at providing a precise estimate of the available forest biomass in Canada, nor at detailing the various bioproducts or processes related to FWB. Rather, it provides a supply chain perspective of the key aspects allowing for the utilization of this renewable resource in the Canadian context. More specifically, this chapter gives a broad characterization of woody biomass flows in terms of the aspects that are crucial to design and manage sustainable FWB supply chains. The chapter is organized as follows: Section 11.2 introduces the Canadian forest products value chain at a general scale. Following this, Section 11.3 illustrates both the promises and challenges of integrating FWB into regular forest product flows. A review and discussion of the key aspects for the design and sustainable management of FWB supply chains are presented in Section 11.4. Finally, in Section 11.5, conclusions are drawn and suggestions made for future research.

11.2 FOREST PRODUCTS VALUE CHAIN

In Canada, approximately 93% of forest land is owned by governments: 77% of this is under the jurisdiction of the provinces and 16% is under federal purview (Natural Resources Canada, 2011b). The provinces organize harvesting under Sustainable Forest Licenses. License holders are allocated an "annual allowable cut" (AAC) and pay the province a stumpage charge based on roundwood harvested.

Forest harvesting in Canada is largely fully mechanized. The predominant harvest systems involve full-tree harvesting, usually using a feller buncher and skidders, with delimbing taking place roadside with a stroke delimber. In western Canada, processing of full trees into logs also occurs roadside with dangle-head processors. In all these systems, the roadside residues can be chipped or hogged with a chipper or horizontal grinder and discharged into a van. Cut-to-length (CTL) harvest systems are also used widely and have become the most common tree harvesting method in several regions. With CTL, trees are cut and then processed at the stump (delimbed and cut into assortments).

Harvest residues are probably the largest potential source of forest biomass. They belong to the provinces and are currently seldom used. In fact, they often are burnt roadside. Wood residues resulting from industrial processes are the property of the companies and they are already fully utilized in forest products (pulp, chipboard) or as a source of energy in mills (Paré et al., 2011). In an attempt to generate value from residues left in the forest, several provinces have developed biomass harvesting and valorization frameworks (referred to as expressions of interest, policies, guidelines, or calls for proposals depending on the province). In general, these frameworks are focused on developing the bioenergy sector, which is capital intensive. The Canadian federal government has put in place the Investments in Forest Industry Transformation program (IFIT) that provides funding for projects that implement innovative technologies in the forest sector, leading to a diversified, higher-value mix of products, including bioenergy, and renewable power, as well as biomaterials, biochemicals, and other bioproducts (Natural Resources Canada, 2011a). The success of new biomass-based products is closely related to the success of the biomass supply chain.

A general description of the forest products value chain is presented in Figure 11.3. Basically, the woody residues include slash (top and branches) left after stand harvesting, slash, and small trees from thinning and cleaning, and nonmerchantable wood (including standing residuals). These materials can be collected and processed in order to be made available to transformation sectors such as the energy and biorefining sectors. The type of biorefinery may range from biomass to heat and power or to production of extractives and solid or liquid fuels. The earlier stages of the conversion process might involve combustion, gasification, or a form of either heat or chemical decomposition technologies. The technologies that could be used at later stages of the conversion process include boilers, turbines, generators, and combustion engines.

Figure 11.3 also presents the logistics networks or critical links that connect FWB to the refineries. These networks are broadly composed of four stages: collection, transport, preprocessing, and storage. The harvest residues may be located at the stump, roadside, or at a landing. During its collection and flow, biomass might go several times through transportation and preprocessing stages. Storage might take place in one or more intermediate locations. Preprocessing may include a single or multiple stages of size reduction, fractionation, sorting, and densification. The storage of wet biomass may also impart biochemical and physical modifications (referred to as in-store preprocessing). The preprocessed woody biomass is then transported to a biorefinery where it can be stored or converted immediately. The arrows on

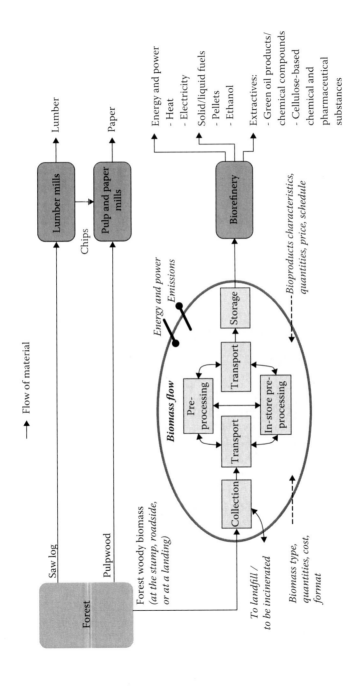

FIGURE 11.3 Forest products value chain in Canada.

the diagram show the flow of material and information. The information flow from biorefinery to woody biomass supply enterprise includes quality specifications for biomass, that is, moisture content (MC) and particle size. Important information for logistics includes quantities, delivery schedules, and price. In response to demands, the production side provides biomass to the supply system. The supply system uses energy and power to collect, preprocess, and transport biomass. The system will give off emissions that need to be minimized.

In addition to new, standalone, business opportunities offered by untapped volumes of woody biomass available in Canadian forests, there are also other drivers for the transformation of this resource. Indeed, Cormier (2010) asserts that the collection of FWB can mitigate for the relatively high fixed costs associated with forest harvesting. In addition, lowering the quantities of debris left after harvest can significantly reduce the costs of reforestation and stand tending activities (such as planting, cleaning, and precommercial thinning) (Dubeau et al., 2012) and reduce fire risk. It could also produce a new revenue stream for forest companies in terms of carbon credits. However, it is recognized that these opportunities do not come without constraints. Supply is tightly linked with the sequence and dispersion of conventional harvesting, highly exposed to seasonal variability affecting quality and volumes, and subjected to stringent regulatory requirements. These constraints make collection very complex and its variable cost—the cost of delivered biomass—very high. Studying bioenergy made from forest biomass, Roberts (2008) stated that high variable cost is a distinct economic hurdle that other forms of renewable energy do not have (e.g., solar and wind power). Especially in eastern Canada where average stand volume is relatively low, woody biomass supply sites are of various sizes and spread over large areas, and the number of suppliers can be large. The challenges of forest biomass harvesting in a Canadian context can be summarized by stating three of its characteristics as a feedstock: (1) high MC, (2) dispersed, and (3) remote. These greatly influence procurement costs.

The variable cost can be broken into the following four components: prepiling, communition, handling, transportation, and other (Kellogg et al., 2006). It is difficult to establish an estimate of the cost of delivered FWB from the literature since not all harvesting and cost studies delineate costs for each component. Figure 11.4 shows an estimation of the proportions of the cost components associated with forest biomass supply operations in Quebec's public forests with full-tree harvesting and 120 km transport, one way, using live-floor chip vans. These estimates were calculated using the BiOS model (Cormier and Ryans, 2006). This model is an operational planning tool that includes a module that can be used to estimate the costs of various operations, taking into account the harvest, recovery, and transport of FWB. It can be seen that the transportation and the prestacking costs are more than two times that of the communition cost. Although the transportation and prestacking operations are critical to the conversion of biomass to a useful form, they both do not add value to biomass. In this context, effective ways of sharing costs for guaranteeing the long-term biomass supply capacity of the forest need to be developed. Several reasons point toward the integration of biomass supply chains with the supply chains of the traditional forest products industries as a key to successful biomass valuation.

But first, it is important to point out that cost component estimates for delivered FWB (such as in Figure 11.4) are often reported by the literature without clear

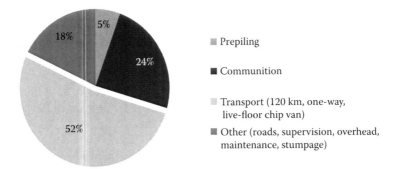

FIGURE 11.4 Estimation of the proportions of the cost components associated with forest biomass supply operations in Quebec's public forests with full-tree harvesting and 120 km transport, one-way, using live-floor chip vans. (From Cormier, D., Ryans, M., The BiOS model for estimating forest biomass supply and costs. Internal Report IR-2006-03-22. 32 p. Quebec, Canada: FPInnovations-FERIC Division, 2006.)

explanation about the factors that influence each cost component. In fact, cost components could be influenced by a variety of factors: factors associated with the characteristics of the machinery (such as material handling, hauling capacity, speed, or maneuverability), worksite conditions (such as the nature of the land, the method used for prior or simultaneous logging, distance to landing, and the amount of biomass left at the logging site), operators (wage system, productivity, and competencies), and the organization of the supply system (such as the size of the companies of the organization and the level of integration) (Asikainen et al., 2002). Proper organization requires that independent successive operations be integrated, whenever possible, to yield more efficient system output.

11.3 IMPORTANCE OF BIOMASS AND PRIMARY FOREST PRODUCTS INDUSTRIES SUPPLY CHAINS INTEGRATION

Several studies reported in the literature have already culminated in recommendations for the integration of the biomass supply chain with the existing primary forest products industries. For instance, the USGuild (Evans, 2008) conducted 45 case studies of how woody biomass is removed from forests and used across the United States. The case studies combined several considerations including economics, collaboration, ecology, implementation, and regional differences. The woody biomass removal projects were found to rarely be a source of income for the landowner and a profitable venture for entrepreneurs. However, the case studies demonstrate that some project managers generated a profit by combining multiple forest products in the removal, taking advantage of fluctuations in the biomass market, and selling to established outlets. Some project managers hired more than one contractor to take advantage of each contractor's expertise, and some others tested new technologies. More recently, the BioBusiness Alliance of Minnesota (BioBusiness Alliance of Minnesota, 2010) published the results of a one-year analysis engaging more than 100 experts from industry, academia, and government who studied the issues required for moving the forest-based biomass industry forward in a way that would benefit multiple stakeholders. It is noted that "there is a clear understanding that the forest biomass supply chain must be integrated with the existing primary forest products industries." However, the authors emphasized the need for an "unprecedented level of cooperation, where communities ... pursue opportunities for biomass processing as part of an integrated statewide strategy that marries old with new, and high value with lower value." In Canada, the Biopathways project reached a similar conclusion [Forest Products Association of Canada (FPAC), 2010]. In reality, the preestablished primary forest industries themselves have a great need to integrate with the emerging biomass industry and create more opportunities for value creation from their supply sources.

Indeed, the forest industry in Canada has been traditionally dominated by large companies. These companies have the management capabilities required to coordinate complex relationships with suppliers and customers. These firms have invested significantly in equipment and infrastructure, and their governance involves considerable costs related to monitoring and enforcement. They are generally centered on the manufacture of traditional wood products such as lumber or pulp. The Canadian forest sector is still plagued by a long lasting and profound crisis, probably its worst in history. This crisis is largely caused by the decreased demand for commodity products made of Canadian wood (Government of Canada, 2009). In times of crisis, firms need to be transformed and to adopt new business models built on collaboration, coinvestment, and knowledge sharing with producers, suppliers, distributors, and retailers across several supply chains of different forest and forest product sectors. The firms producing traditional forest products still have a significant influence on the Canadian forestry sector. However, it is in their best interest to seek integration of their supply chains with the supply chains of other emerging industries in the forestry system. The biomass sector is emerging within or in parallel with the forestry system. It offers a unique opportunity to create increased value out of the forest without reducing current product flows. However, integrating a new product flow within an existing system presents challenges. Furthermore, some mills within the industry may not see favorably the emergence of a new production unit competing for the same resource. This raises the question of how to optimize the design of an integrated bioenergy and forest products supply chain, which has proved to be a complex problem (Feng et al., 2011).

Because integration may seem like a theoretical concept, four case studies are presented to illustrate both the promises and challenges of integrating biomass into regular forest product flows. The first three cases reflect documented situations from three forest territories in the province of Quebec, while the fourth case reflects a series of tests conducted in northwestern Ontario in order to study the conversion of existing power plants from fossil fuels to FWB.

11.3.1 CASE 1: MAURICIE REGION AT THE SOUTHERN LIMIT OF THE BOREAL FOREST ZONE

This region has an extensive forest cover consisting of hardwood (deciduous), mixed, and softwood (coniferous) forests (17%, 44%, and 32%, respectively). Proportionally, few eastern regions have such diverse forest (Ruel, 2010). In addition, its relatively high forest density (m³/ha), relatively short transportation distances, and good quality hardwood make it interesting for forest products companies. Hence, in 2003, the region was the top paper-producing region of Quebec, with nearly 15% of the province's paper shipments and six pulp, paper, and paperboard mills (Ruel, 2010). In addition, there were nearly 55 sawmills, two plywood/veneer mills, one wood-based panel mill, one factory of turned and shaped wood, and one cogeneration plant (combined heat and power) (Ministère des Ressources naturelles et de la faune, 2004). Some of these mills have since shut down due to poor market conditions and high wood procurement costs.

Despite the unique characteristics and the apparently flourishing and creative forest industrial sector, the factories operating in the territory have difficulty controlling their supply chains. First of all, it is relevant for the reader to keep in mind that, in Quebec, as can be found in other provinces, the public forest is divided into procurement area (also referred to as management units [MU]), and it is the government which allocates volumes of timber from crown forests to mills through timber licenses (TL). A TL specifies, on a yearly basis, the general areas from which wood for the mill can be procured, with predefined volumes of one or more tree species. A mill may hold a TL on more than one MU, and several TLs may be awarded to different mills on the same MU, even for the same tree species. The mills must pay the corresponding fees (stumpage forest fund and forest protection agencies) and commit themselves to regeneration of harvested forests. In Beaudoin et al. (2008), it is explained how the wood procurement planners (usually one per company) typically agree among themselves to solve the different planning and coordination issues in a successive manner. They summarize the current approach in four successive steps: "(1) identification of the companies that will conduct forest operations on a given MU (often referred to as the 'mandatary' or the 'operators'), (2) allocation of the blocks to be harvested by each of the companies identified for conducting the forest operations, (3) wood allocation from the blocks on a given MU among the companies having rights according to their TL, and (4) wood procurement activities coordination among the companies." Interested parties first meet to address steps 1 through 3. Interfirm coordination arises afterward when planners express their preferences in order to organize their respective wood procurement activities. The main source of conflict involves the question of transaction prices to provide procurement services among the various companies. With this approach, the appointed company, typically the largest in the region, is at the same time a beneficiary and an intermediary. Thus, consensus may be reached in a simple and inexpensive way; however, it is most likely far from being optimal for all the beneficiaries. Considering the large number of actors in the case of the Mauricie region, including not only mills and other factories but also the forest entrepreneurs such as logging and silviculture contractors, and transportation and equipment operators, the issue of optimization becomes more complex to solve. It gets particularly complex if we consider the need for valuing the hardwood "trapped" within the softwood in mixed stands and the huge quantities of biomass that can be recovered from the entire territory in order to improve the economy of the lumber, pulp, and paper operations. In that context, Beaudoin et al. (2010) demonstrate the benefits (available volume, lower procurement cost, and higher profits) made possible by collaborative forest planning and, by extension, integration. Unless the Mauricie region can introduce new wood-consuming industries, the very survival of its remaining mills is in jeopardy.

11.3.2 Case 2: Quebec's Abitibi-Temiscamingue Region, South of James Bay

The territory is within the continuous boreal forest zone and subzone in its northern part and within the Nordic mixed temperate forest zone in its southern part. According to Collini (2006), the forest cover has changed subtly but irrevocably. Overall, across the three decadal forest stand inventories, the proportion of the area of softwood (coniferous) stands has been reduced and conversely, the area of mixed hardwood–softwood stands has been steadily increasing. The transition from softwood and mixed stands dominance toward mixed and hardwood (deciduous) stands dominance is the most important change that has occurred during the last 20 years in the territory. Between the first and the third stand inventories, the softwood coverage has decreased by 7% (from 47% to 40%), and the mixed coverage has increased by 4% (from 32% to 36%). During the same period, the number of highly dense softwood stands has also decreased by 15% (from 49% to 34%). In short, it is now the medium-density stands that dominate the coniferous cover (42% of the area). These significant changes are attributed to natural disturbances and to a lesser extent, human interventions that affect large areas of territory. As a consequence of this stand cover change, forest operations in the region now generate a larger amount of low-quality hardwood. At the same time, the harvest cost has increased as operations take place increasingly at remote areas. This has contributed to a decrease of 10.3% in allowable cut of the primary softwood species for the territory. In general, hardwood of these forests is of poor quality and therefore has a low commercial value in the region. The industrial structure needs to adapt.

Currently, a manufacturer of plywood panels is the single large user of hardwood fiber in the region. Low-quality hardwood is of little interest for such industry. Moreover, knowing that most sites in the territory are harvested using logging methods in which stems are delimbed and topped on site to avoid debris accumulation roadside, the amount of biomass left at the stump is quite significant. In order to reflect the shifting resources in the territory in a business shift, the Quebec government issued in 2011 a request for proposals to allocate available forest biomass in the region for bioproducts and bioenergy projects. But to this day, it has not received serious responses from entrepreneurs. This is probably due to high costs of collection and transportation of the FWB. In such forests, significant gains can be achieved if the FWB supply chain is fully integrated into the traditional logging industry in this territory; the biomass will be transformed and the traditional loggers will share the cost for the extraction of larger quantities of fiber and will share the value generated by these quantities. One would ask why this is not happening now. The situation observed in the Abitibi-Temiscamingue region may be explained by the facts that the traditional forest products industry is still battling poor economic conditions, and that some companies have shut down their operations, reducing opportunities for joint operations that would allow sharing infrastructure (roads) costs and employing current contractors to full capacity.

11.3.3 CASE 3: QUEBEC'S NORTH SHORE (CÔTE-NORD) REGION

This territory is the largest forested area in the province; most of it is covered by boreal forest, almost exclusively composed of softwood (balsam fir, black spruce). The primary processing sector is limited to newsprint and lumber production, with five independent sawmills and one integrated complex (a sawmill and a paper mill owned by the same company). Sawmills sell the chips they generate to the only paper mill still in operation. Chip revenues are critical for their survival. As stated by Moore et al. (2011), referring to a consultant's report, the cost of log hauling in this territory is the highest in Quebec, considering the distance between the mills and the harvest areas for this region (generally located to the north of the mills), which is often over 200 km. This explains why the region has experienced several permanent and temporary mill closings. A direct consequence of this is that the paper mill is itself threatened by the high cost it pays for its chip supplies. The paper mill is also threatened by the poor quality of chips it receives from the sawmills. Indeed, many trees that come out of this territory are of poor quality (a mixture of small and old stems with lots of decay, butt rot, sweep, crook, etc.). These trees are processed at the mills, and the resulting chips show high variations in size distribution, density, MC, brightness, bark content, decay rates, and so on. These feedstock variations significantly affect the performance of the papermaking process in terms of energy consumption, quantities of bleaching chemicals needed, as well as in terms of the grade of the produced paper. This situation is unsustainable in a highly competitive commodity industry such as newsprint. The challenge is, therefore, to efficiently harvest and process these stands so that chip quality can be monitored and optimized to fit the paper mill's production needs. This is all the more important given that the fiber quality that could be obtained from the slow growing black spruce stands available in the region can be considered among the best in the world for papermaking and lumber production. However, increased screening to produce prime chips would generate a high volume of poor quality chips (e.g., pins and fines, low density, etc.) produced from the mixture of small and old stems with lots of decay, butt rot, sweep, crook, and so on, coming out from the territory. It is, therefore, necessary to find a market for these low quality chips. Such a market does not exist at the moment in this region.

Considering the territory, there are few opportunities for reducing costs such as transport or harvesting (Moore et al., 2011). Then, the most promising way to improve the region's economic situation consists in developing a collaborative approach between the stakeholders to manage the supply chains of the chips utilized by the paper mill. The idea here is to supply the paper mill only with acceptable chips—chips of known characteristics (species, MC, freshness, whiteness, bark content, size distribution, etc.) and managed in a way that reduces their variability and optimizes the performance of the papermaking processes (Azouzi et al., 2009). Balsam fir has been singled out as a species that lowers the current value chain competitiveness. That species may account for 20% or more of the AAC. New regional users or marketing channels need to be developed in order to recover all the remaining residues (chips that are not consumed by the paper mill anymore). Here the energy market clearly stands out as a potential user.

11.3.4   CASE 4: ATIKOKAN POWER GENERATING STATION CONVERSION TO FWB, NORTHWESTERN ONTARIO

Atikokan is located in Northwestern Ontario, 209 km west of Thunder Bay. In 2011, Atikokan had a population of 2730 while in 2006 it was 3230, a drop of 15.5% (Statistics Canada, 2012). The economy of Atikokan has been based on forestry, the Atikokan Power Generating Station (APGS), government services, retail services, tourism, mining, and a mixture of light manufacturing businesses. During 2007, the leading employers in Atikokan were Atikokan Forest Products (random length/width lumber and woodchip mill), Fibratech Manufacturing Ltd. (particle board plant), and the APGS (The Township of Atikokan, 2008). At present only the APGS is running. With the closures of Atikokan Forest Products and Fibratech there was a loss of approximately 350 direct jobs. The APGS is a peak demand generating station, has a capacity of 211 MW, and currently burns low sulfur lignite coal brought in by rail from Saskatchewan. It provides 90 good-paying jobs and significant tax revenue for the township. In 2007, the Ontario Provincial Government passed legislation to phase out the generation of electricity from coal at Ontario Power Generation's (OPG) generating stations by December 31, 2014. If the APGS were to close, without its revenue and jobs, the Township of Atikokan would face a huge financial problem.

Since 2005, OPG has been investigating the use of woody biomass as a coal offset option. In a series of tests at the APGS during January to July 2008, a total of 1,622 tons of commercial grade pellets were used at various levels of cofiring and 100% pellet-based feedstock (Marshall et al., 2010). The first pellet-based test at the APGS was during January 2008. This test consisted of 26 tons of wood pellets that were cofired with coal. In March 2008, a second cofiring test was conducted with 100% pellets on a single burner row. In this test, 181 tons of pellets were used and accounted for 20% of the furnace energy input level (Marshall et al., 2010). In May 2008, a third cofiring test with 177 tons of pellets was run. During July 2008, a series of tests were conducted over the month to assess the plant's potential to operate on 100% wood pellet fuel. During early to mid-July, 796 tons of pellets were used in various tests with one of the tests being conducted using 100% pellets in mid-July. On July 31, a 100% run of pellets was made and 442 tons were used. The tests showed that the conversion of the APGS to fire 100% wood pellets as feedstock is technically feasible, and resulted in significantly reduced SO_2, NO_x, and heavy metals emissions (Marshall et al., 2010). In March 2010, OPG issued a "request for indicative prices" for 90,000 tons/year of wood fuel pellets for the APGS (OPG, 2011). In August 2010, the Ontario Government gave permission for the conversion of the APGS from coal to woody biomass with a generating capacity of 21 MW using 90,000 tons/year of pellets. This capacity corresponds to 10% of the peak demand for which the plant is designed. It may be increased if power demand rises in Northwestern Ontario. Atikokan Renewable Fuels (ARFuels) is now in full control of the former Fibratech Mill. ARFuels is investing an initial $15 million to renovate the plant to produce 140,000 tons/year of superpremium (DIN+) quality wood fuel pellets (ARFuels, 2012) that could potentially be used in the APGS. In January 2011, ARFuels received a wood allocation of 179,000 m³/year of white birch and trembling aspen under the Ministry of Northern Development, Mines and Forestry's wood supply competition. In addition to the new allocation ARFuels retains 100,000 m³ from Fibratech, for a total guaranteed wood supply 279,000 m³/year. When the wood pellet operation is up and running it is anticipated to generate 95 jobs in the Atikokan area (Atikokan Progress, 2012).

This case study has shown that the use of woody biomass to generate bioenergy in small rural communities in Northwestern Ontario can benefit the communities and industries through (1) creating long-term employment, (2) improving the financial stability of all forest operations, (3) generating income for the community and industry, (4) improving community and industry stability and sustainability, and (5) reducing negative environmental impacts.

Each of these cases demonstrate the strategic importance of finding new users for lower value wood or untapped biomass. While the Quebec cases mostly highlight the promises of forest biomass utilization, the western Ontario case (case 4) demonstrated that biomass utilization at a fairly large scale is technically and economically possible in Canadian boreal forest. Also, the Ontario experience shows the importance of wood fuel quality standards. In that case, woody biomass was generated from mature hardwood trees. The integration of woody biomass supply chains with the existing primary forest products industries can be successful if new business models are built on collaboration, coinvestment, and knowledge sharing, and adopted by the suppliers of wood fiber. It should also be extended to the processors of this fiber and the distributors of the resulting products. Azouzi et al. (2012) analyzed several models for integrating interdependent (legally and financially independent but in the same chain) businesses. Some models are driven by producers (e.g., cooperatives), some by buyers (e.g., processing and retail companies), and some supported by intermediaries (e.g., forest contractors). While integration seems an obvious objective, it remains easier said than done and the factors making it possible should be investigated.

At this point, it should be mentioned that in Nordic countries, the integration of the supply chains for FWB and for traditional forest products (i.e., roundwood) has a longer history. This is particularly true when the FWB is used for bioenergy (Röser et al., 2008). Indeed, in countries like Finland and Sweden, bioenergy is becoming competitive because of the high environmental taxes applied to fossil fuels. Asikainen et al. (2002) discussed the organization of the FWB supply chain in these countries. They characterized systems in regard to low and higher levels of integration. In low level integrated systems, FWB recovery is carried out as a separate, subordinate operation. Existing technologies are used, possibly with simple modifications. Organizational relationships with traditional harvesting are limited. Such two-pass systems could not maximize the utilization of the available economic potential. In systems with higher levels of integration, harvesting operations of FWB and traditional forest products are conducted in parallel using coordinated one-pass harvesting, and new methods and technologies are used. For instance, feller-skidders/forwarders and whole-tree trucks might be preferred for short-haul operations to terminals or industrial conglomerates. However, in remote areas where wood-consuming industries are far and dispersed, it is often feasible to separate some or all the assortments at the logging site. Finally, at the organizational level, it is believed that if the FWB supply chain is integrated into larger roundwood supply chains (and thus able to use existing managerial tools, procurement practices, and information systems), any extra marginal administrative costs could approach zero. For more details on these systems and for examples of what is considered a "highly integrated system," the reader is referred to Asikainen et al. (2002).

11.4 KEY ASPECTS OF AN INTEGRATED SUPPLY SYSTEM FOR FOREST BIOMASS

In FPAC's Biopathways report (Forest Products Association of Canada, 2010), it is asserted that a FWB supply chain's best chance of success presupposes an integration into the traditional forest industry. It is stated that "… the research shows that bioenergy and bioproduct opportunities are stronger economically and socially when integrated within the traditional industry's operations rather than on a stand-alone basis. Both the traditional and emerging bioenergy and bioproducts operations enjoy higher economic returns and perform better when integrated." The three case studies presented in the previous section support these conclusions. Yet, integration of complex systems remains a challenging task for which models are few, no matter the product. From the literature, we can broadly define system integration as bringing together subsystems so that they efficiently contribute to the performance of the global system (Barker et al., 2001). Thus, a variety of aspects should be considered and analyzed. There is a consensus that all three aspects of sustainable development should be accounted for when considering biomass utilization: economic, social, and environment.

Building on the three dimensions of sustainable development, Johnson (2011) proposes a framework focusing on the different segments of the forest biomass supply chain (Figure 11.5). This framework details technical, behavioral, economic, and environment aspects, the coordination of which, according to this author, is "paramount to success." Johnson (2011) briefly described a few of these aspects in a case study about torrefied and pelletized biomass for the electric power industry. She concluded that there is a need for systems modeling across the different aspects (or dimensions) in order to provide a comprehensive analysis and a powerful management decision support system. This is rather complex and the first step is to focus on understanding the key aspects for building sustainable biomass supply chains. Based on the above, we propose to describe in further detail six elements: (1) residues (feedstock) value, (2) residues harvesting, (3) residues preprocessing, (4) transportation, (5) impact on conventional industry, and (6) environmental and social impact.

11.4.1 Residues Value

FWB has long been considered unmerchantable material. Until recently, and except for a short period in the late 1970s and early 1980s following the second oil crisis, no coordinated effort has been made to extract value from residues at a large scale (one can consult the rich Canadian research literature generated by the federal energy from the forest [ENFOR] program). Increasing and fluctuating prices of fossil fuels,

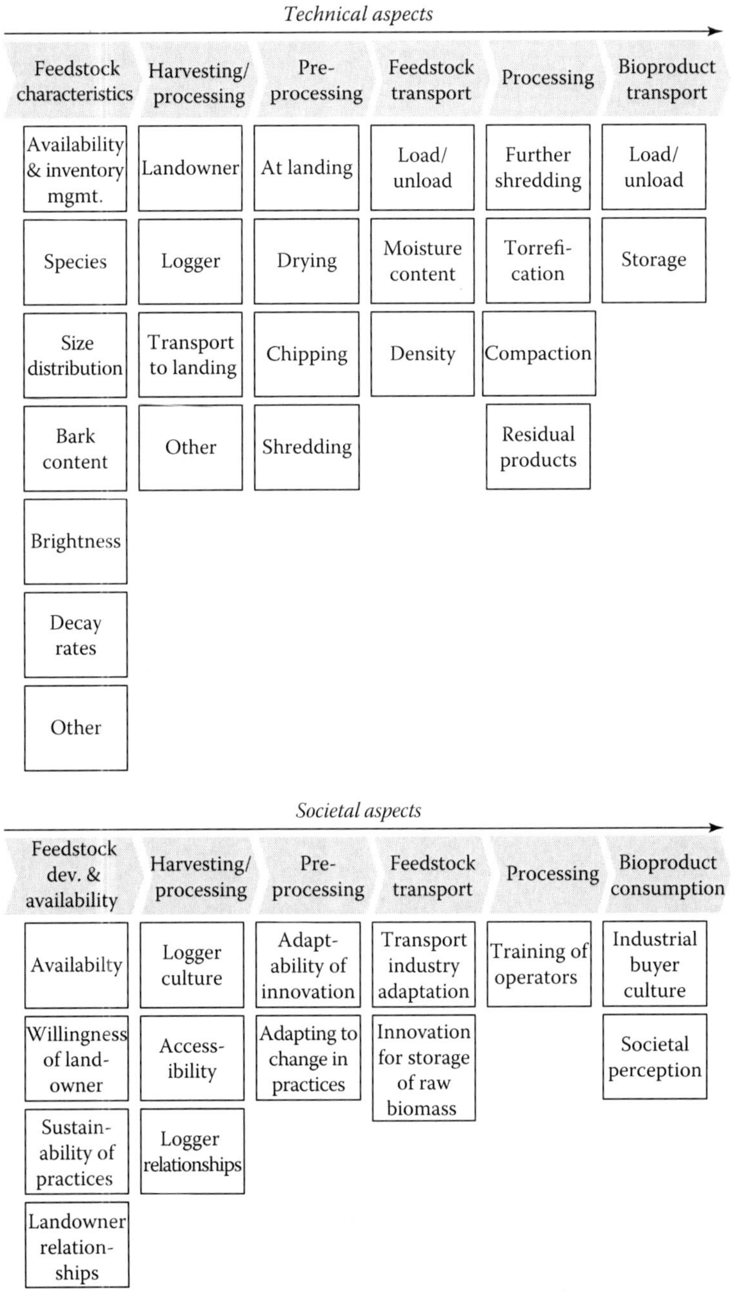

Technical aspects →

Feedstock characteristics	Harvesting/ processing	Pre-processing	Feedstock transport	Processing	Bioproduct transport
Availability & inventory mgmt.	Landowner	At landing	Load/ unload	Further shredding	Load/ unload
Species	Logger	Drying	Moisture content	Torrefi-cation	Storage
Size distribution	Transport to landing	Chipping	Density	Compaction	
Bark content	Other	Shredding		Residual products	
Brightness					
Decay rates					
Other					

Societal aspects →

Feedstock dev. & availability	Harvesting/ processing	Pre-processing	Feedstock transport	Processing	Bioproduct consumption
Availabilty	Logger culture	Adapt-ability of innovation	Transport industry adaptation	Training of operators	Industrial buyer culture
Willingness of land-owner	Access-ibility	Adapting to change in practices	Innovation for storage of raw biomass		Societal perception
Sustain-ability of practices	Logger relationships				
Landowner relation-ships					

FIGURE 11.5　Biomass supply chain aspects: technical, societal, economic, and environmental aspects. (*Continued*)

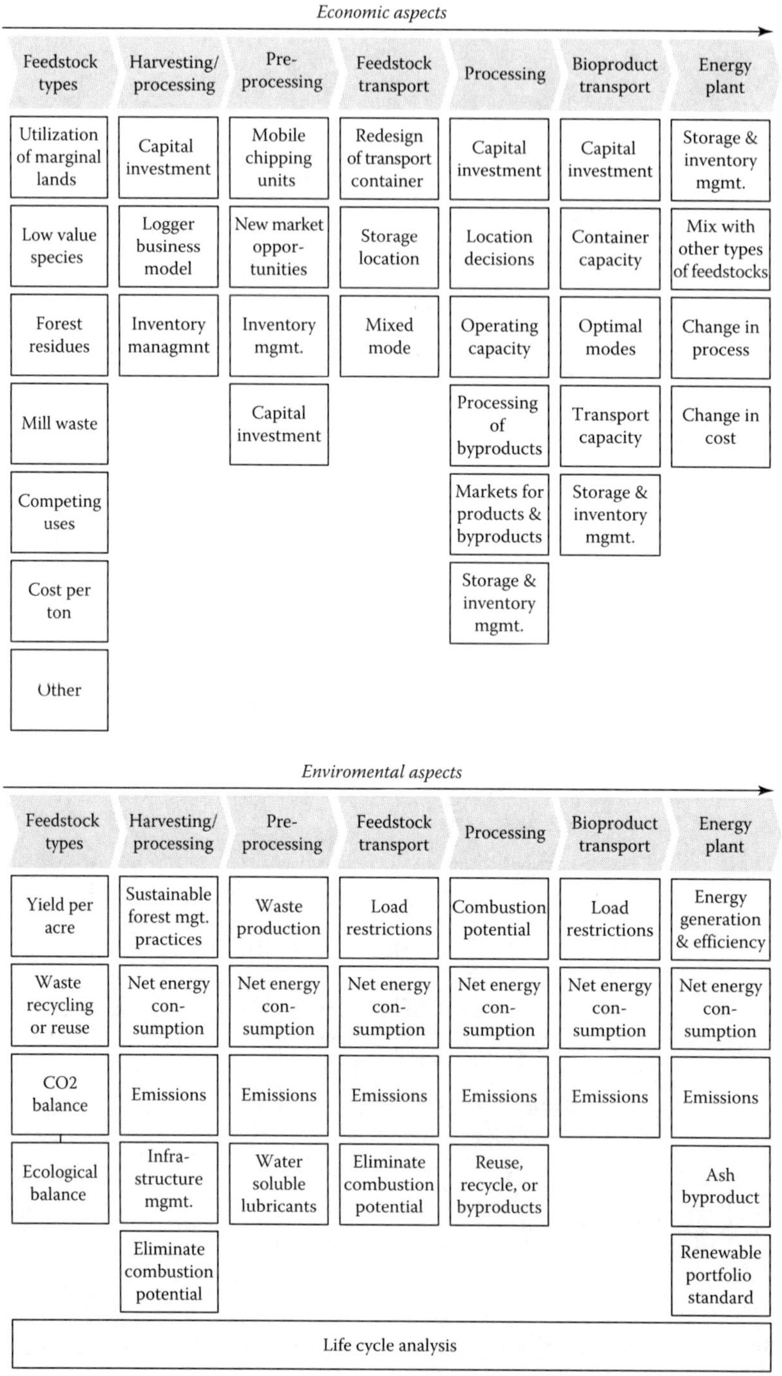

Economic aspects

Feedstock types	Harvesting/ processing	Pre-processing	Feedstock transport	Processing	Bioproduct transport	Energy plant
Utilization of marginal lands	Capital investment	Mobile chipping units	Redesign of transport container	Capital investment	Capital investment	Storage & inventory mgmt.
Low value species	Logger business model	New market opportunities	Storage location	Location decisions	Container capacity	Mix with other types of feedstocks
Forest residues	Inventory managmnt	Inventory mgmt.	Mixed mode	Operating capacity	Optimal modes	Change in process
Mill waste		Capital investment		Processing of byproducts	Transport capacity	Change in cost
Competing uses				Markets for products & byproducts	Storage & inventory mgmt.	
Cost per ton				Storage & inventory mgmt.		
Other						

Enviromental aspects

Feedstock types	Harvesting/ processing	Pre-processing	Feedstock transport	Processing	Bioproduct transport	Energy plant
Yield per acre	Sustainable forest mgt. practices	Waste production	Load restrictions	Combustion potential	Load restrictions	Energy generation & efficiency
Waste recycling or reuse	Net energy consumption	Net energy consumption	Net energy consumption	Net energy consumption	Net energy consumption	Net energy consumption
CO_2 balance	Emissions	Emissions	Emissions	Emissions	Emissions	Emissions
Ecological balance	Infrastructure mgmt.	Water soluble lubricants	Eliminate combustion potential	Reuse, recycle, or byproducts		Ash byproduct
	Eliminate combustion potential					Renewable portfolio standard

Life cycle analysis

FIGURE 11.5 (Continued)   Biomass supply chain aspects: technical, societal, economic, and environmental aspects.

as well as its environmental costs, have transformed forest biomass to an alternative source of energy. We have shown earlier that it is necessary to develop cost-effective operations to harvest and recover woody feedstock so that it becomes a viable alternative. Before comparative studies can be made, the real value of FWB as an energy source must take into account its available heat content, and the investment and operating costs of the plants needed to handle and convert it to usable energy. Thus, FWB cannot be assessed in the same way that wood is traditionally assessed, that is on a volume (m^3) or weight (odt = oven-dry tons, gmt = green tons) basis. When FWB is used for energy purposes, it should be assessed on the basis of the amount of energy that can be obtained when burning (kWh = kilowatt-hour, MWh = megawatt-hour, GJ = gigajoules).

It is a well-known fact that the amount of water present in wood can have a direct effect on the weight, strength, physical characteristics (dimensional stabilities), and process efficiencies (Röser et al., 2011b). The MC of woody biomass also affects the energy content of the biomass and the efficiencies of the technologies to extract the energy. The technologies can be a combination of wood combustion, wood gasification, cogeneration, and cofiring depending on the energy application. The effect of MC on biomass combined heat and power (CHP) technologies is well explained in U. S. E. P. Agency (2007). Suppliers must find ways to deliver biomass with a low and uniform MC. With dryer biomass, not only do they deliver more energy using the same truck load but also they do it with lower transportation cost due to a lower weight. Thus, it becomes more interesting for suppliers to get paid a price based on energy content rather than weight. This would improve the overall efficiency of the FWB supply chain since the transactions between all actors of the value chain become easier, transparent, and more reliable. These factors are key to collaborative actions within a complex supply chain (Asikainen et al., 2002; Visser et al., 2010). Several scientific papers report on methods to dry woody biomass (Gautam et al., 2012; Pettersson and Nordfjell, 2007; Röser et al., 2011b).

To determine a price based on energy content one needs to first determine the MC of the material and then calculate the amount of energy it contains. However, the MC is subject to a host of factors that can and do cause it to fluctuate widely, both within a species and among species, and within a tree section and among tree sections. Moreover, the conditions of transportation, storage, and even systems of measurement can introduce additional variations. As a consequence, the MC might vary significantly from one supplier to another and even from one truckload to another. Accordingly, there is a need for decision support systems or tools that can be used to measure very rapidly the MC of a given quantity of biomass (even when frozen) quickly and accurately, and for tools that can use these measures to predict the energy content of this same quantity of biomass. Researchers at FPInnovations are developing efforts in this direction (Volpé, 2012). Recently, an electronic platform labeled FPJoule was developed to facilitate the conversion of conventional sales units of biomass into energy units. This tool calculates the amount of energy available in the forest biomass according to three parameters: the species group (softwood, hardwood, or mixed species); the part of the tree that is used (stem, bark, full tree), and the MC. Then, using the traditional method of payment for biomass, either dollars per ton or cubic meters, or the monetary equivalent in dollars per gigajoules or megawatt-hours is calculated. FPJoule is expected to stimulate regional forestry development by allowing suppliers to obtain a fair price for their forest biomass based on an energy basis. It is user-friendly and is accessible free-of-charge over the Internet (www.fpinnovations.ca/FPJoule). FPJoule developers recognize that instruments that can provide a quick and precise reading of the MC in FWB, especially when frozen, are still lacking (Hanson, 2012).

Tools like FPJoule determine the market price of a quantity of biomass of a given quality. They do not explicitly consider the investment and operating costs of the plants needed to handle and convert it to usable energy. These are fixed costs that are known to be very high and should be minimized. Special attention should be paid to the movement of the biomass along the different segments of the supply chain (from its source to final user of the end products). In the next two sections, Sections 11.4.2 and 11.4.3, two of the segments that are believed to be very critical are explored.

11.4.2 Residues Harvesting

Cut-to-length operations, which are the norm in Nordic harvesting systems, also dominate in eastern Canada. In cut-to-length systems, as mentioned earlier, the trees are delimbed and bucked at the stump, leaving the harvest residues at the harvesting site. The residues must then be forwarded roadside before being comminuted (Ralevic et al., 2010). Residue forwarding generally occurs after all merchantable assortments have been extracted, and may occur weeks or even several months later. Notice that in Nordic countries (such as Sweden or Finland), residues forwarding could be deferred to the end of summer or to early autumn, after residues have dried and needles have fallen, but before reabsorption of moisture during colder months.

However, recent studies by FPInnovations (Volpé and Desrochers, 2010) have shown that if the forwarding of fresh residues is integrated with the forwarding of traditional products, then significant economies can be made. It was observed that during wood harvesting, in a work team composed of a harvester and one forwarder, the latter is in general more productive than the former. Then, one way to maximize the utilization of the forwarder and reduce the total cost would be to forward fresh residues using this excess capacity. By doing so, the forwarding operations can be 29–63% less costly when the extraction of FWB is integrated with traditional forest products. At the same time, forwarder utilization is improved by up to 46%. However, the studies recommend that the use of the harvester (feller-processor) for the handling of the residues should be strictly minimized. Also, it is specified that any excess residues should be left to dry on site and recovered later with a dedicated residues forwarder, and the recovery should be delayed if site type could lead to potential nutrient depletion. Note that the use of a dedicated forwarder in the deferred recovery mode could potentially lead to 15–20% in cost savings.

11.4.3 Residues Preprocessing

Before they are transported to the conversion facility, the residues are typically preprocessed at the landing. Two important preprocessing steps occur: communition and drying. As explained previously, drying increases the value of the biomass. For its part, communition consists basically of size reduction to improve transport efficiency. It is an energy intensive process. The characteristics of the comminuted material determine the equipment required in all subsequent steps of the supply chain (Wilkerson and Perlack, 2008). In Canada, communition is usually conducted using grinders, which come in assorted configurations. The grinder typically breaks down the residues with blunt impact hammers or teeth mounted on a rotor in sizes that fit desired products (according to specifications of the biorefinery). Many models of grinders are available, with some belt fed and others top fed.

Compared to other means of material size reduction (such as chippers or top-fed grinders), horizontal grinders are more versatile machines (they can handle various feedstocks) that are commonly used in recycling yards and that are known to be less sensitive to contaminants and appropriate for older piles of delimbing residues. In a study by FPInnovations, it is shown that up to 25 oven-dried tons per productive machine hour (odt/PMH) and 72% utilization at an efficiency rate of 68% can be expected if a horizontal grinder is used in the recovery of roadside residues. However, the use of such a grinder has some logistics limitations: the equipment requires a lot of space to operate; it is discharged using a conveyor that is loading directly in a chip van positioned at a right angle to the road; and it is essential to prepile the residues around the equipment using an excavator with a proper grapple (bulldozers are not recommended). Also, the grinder will offer better off-road mobility if it is on tracks instead of wheels. Finally, it is noted that a better utilization of the grinder can be achieved if the chips are dropped directly on the ground (cold deck). However, this comes with 8–14% loss on the ground, a higher risk of contamination, and an additional cost due to the machinery needed to load the truck.

11.4.4 TRANSPORTATION

Biomass transportation from its source to the biorefinery represents technical and economic challenges. To summarize, biomass is a low value material with a low bulk density. If, at the preprocessing operation, the grinder discharges directly in a van, the biomass can be hauled directly to the end user. Figure 11.6 shows that, at a high MC, the comminuted biomass reaches the legal load weight before filling the van because

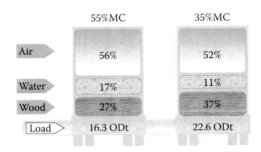

FIGURE 11.6 Impact of moisture content on truck load of white birch chips. (From Ryans, M., Pay for Energy, Canadian Biomass Magazine, March–April 2010.)

of the added weight of water. Then, distance from the plant is a major cost determinant. Tools for estimating the cost of delivering the biomass to the biorefinery can be helpful for determining the maximum travel time for a given biomass (e.g., Alam et al., 2012; Cormier and Ryans, 2006; Röser et al., 2011b). In order to make the long distance transportation of biomass economically viable, more efficient preprocessing methods for drying the biomass should be considered. This additional preprocessing step will be even more interesting considering the fact that it increases the energy content of the biomass (and thus the value of the biomass as explained previously).

11.4.5   Impact on Conventional Industry

In Section 11.3, we have stated that, for maximum benefits, the supply systems for traditional forest products and for FWB-based products should be integrated. Integration means harmonized forest operations and collaborative procurement activities. For this to happen and to reduce cost, integration will require new business models, advanced decision support systems/tools, and improvements in harvesting and processing technology.

Figure 11.7 summarizes these requirements. The literature testifies to the intensity of research designed to improve processes, material, and decision-making. A reduction in the total cost of supplied biomass will certainly increase interest in biomass-based products, especially energy. One avenue considered to reduce the cost of FWB production is to be more selective with raw material. As long as the demand for traditional forest products is higher than the demand for products related to forest residues, the biomass sector will compete directly with traditional products. To the contrary, integration plays in favor of a complementary role of one industry with respect to the other. Figure 11.8 depicts the possible trade-off scenarios explained in the following.

In general, product value is tightly linked to supply and demand. Thus, as demand and price for traditional products decrease, its industry will first abandon lower value stands. Biomass-based industries may then gain access to new volumes found in these stands. These volumes are expected to improve the quality of the FWB as the proportion of roundwood material increases, and as a result, the content in terms of moisture, dirt, ash, bark, foliage, and other contaminants decreases. This improvement in

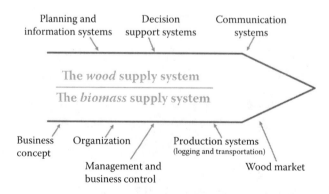

FIGURE 11.7   Integrated wood/biomass supply system.

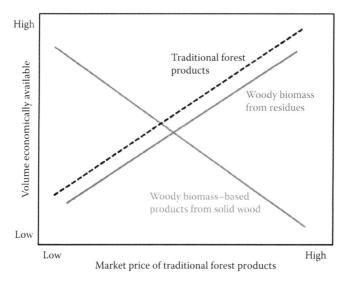

FIGURE 11.8 Lower demand for traditional forest products is expected to result in improvement in the quality of woody biomass and wider opportunity for the biomass industry.

FWB quality will help reduce production costs for that type of biomass. As a result, with higher quality biomass, the FWB industry may pursue a wider range of market opportunities. In contrast, when traditional products are in high demand, the FWB industry will be limited to lower quality supplies. It might, however, expect residues to be available in larger quantities and, potentially, at a lower cost. These trade-offs are only possible if volumes can be reassigned without constraint from one industry to another, which is not always the case when supplying from public forest. However, this situation is changing. Recent studies show that the trend is toward less distinct uses of roundwood and FWB, and that there will be no more clear demarcation between the two markets (Jackson et al., 2010). Then one could expect the cost curves for traditional forest products and biomass-based products to converge in the longer term.

11.4.6 ENVIRONMENTAL AND SOCIAL IMPACT

Environmental and social issues associated with woody biomass extraction are complex and diverse. This chapter does not aim at providing detailed analysis of the environmental and social impacts of FWB harvest. Yet, it is impossible not to account for the most obvious benefits and consequences as they may have to be accounted for when analyzing the supply chain related to its utilization.

The environmental and social considerations need to be assessed along with economic aspects of sustainability and within a wider analysis of the entire FWB supply chain. In regard to the environmental aspect, one important issue is related to the need for the conservation of nutrients on the site and the protection of soil quality. These are critical parts of the physiological requirement of trees that determine the rate of FWB production. On one hand, large quantities of nutrients are removed in FWB harvest, both in the stems and in the branches and foliage. On the other hand, soil quality can be significantly degraded with an increase of the activities related to FWB harvest-ing. Thus, sustainable nutrient retention level and soil protection measures need to be established and observed by implementing regulations and/or efficient management practices to prevent excessive loss of nutrient stocks and limit the extent of soil dis-turbance (Damery et al., 2009). FWB harvesting guidelines on public forest suggest that a given proportion of the total logging residues produced should be left on site. According to Titus et al. (2009), these proportions range from 0% to 100%. As a conse-quence of such guidelines, the inventories of available biomass are usually reduced. In Quebec, for instance, the calculated estimates of the total available FWB are generally reduced by 20–25% to guarantee a residue retention rate on each site. An additional volume of FWB is not allocated for harvesting to consider the technical efficiency of the wood harvesting systems in use. Efficiency rates of 65% and 85% are applied for cut-to-length harvesting and for full-tree harvesting, respectively. As a consequence, the FWB recovery rate could vary between 52% and 68% (Government of Quebec, 2011). Moreover, different sites are likely to have different constraints regarding FWB removal. Accordingly, the regulations and management practices need to be consistent with the resources to be managed because they are tightly linked with the conditions of the harvesting site (some soils are more sensitive than others, the species and their distribution might differ, etc.), the FWB quantity removed, and the type of machinery used (machine movement affects biodiversity, etc.). Ensuring this consistency requires an adaptive approach (Stupak et al., 2008). In Sweden, for instance, guidelines for careful FWB harvesting, taking the negative effects into account, were first published in 1986 and reviewed periodically, and then established as forest protection rules (transposed into national legislation) around 2008 (Asikainen et al., 2002). Excellent explanations of the complexity of the problem and the challenges in creating adaptive approaches can be found in Lattimore et al. (2010). These authors identified a number of mechanisms based on an adaptive forest management framework and related tools and processes that can be used to support the sustainable management of the complex FWB energy production systems.

Regarding the social aspects, it is critical that the public accepts FWB harvest and utilization. The establishment of clear and efficient adaptive forest manage-ment approaches, discussed above, could facilitate this acceptability. But beyond that, the role of a well-developed FWB-based energy sector in the lives of people in Canada should be clearly demonstrated. Particularly, people living in remote north-ern regions in Canada will likely to be able to reduce their cost of heating, access more stable employment, and improve the economic viability of existing forestry operations because the human resources as well as the equipment could be better utilized. Meanwhile, the availability of skilled labor (labor with expertise in FWB harvest) in such remote regions will be a challenge.

11.5 CONCLUSIONS

Canada is a country that is rich in forest resources. These resources have long been transformed to create economic value. Today, maximizing the value of Canadian forests implies new challenges. First, traditional markets have declined, especially those related to paper and newsprint. Forests are part of a world heritage and their management is subject to important public scrutiny. The decline for traditional forest products arrives at a moment when new opportunities are emerging in regards to renewable energy and other bioproducts. On the other hand, contrarily to many European countries, Canada has an ample supply of hydropower, a source generally considered renewable. In addition, Canada also possesses important oil and gas reserves and thus, it does not have the same economic pressure to use woody biomass to reduce its dependence on fossil fuel as other world regions do. These economic advantages provide many opportunities for the Canadian forest economy, but also add many technical difficulties that make forest value extraction and value chain optimization challenging.

This chapter reviewed the value chain of Canadian forest woody biomass-based products, and discussed the key aspects for their designs and sustainable management. The discussed aspects point toward the integration of biomass supply chains with the supply chains of the traditional forest product's industries. With this integration comes the opportunity to optimize the use of each product extracted from the forest. New business models, advanced decision support systems/tools, and improvements in harvesting and processing technology will be required so that cost does not fall on a single product and is shared among different stakeholders. But first, from an environmental perspective, a clear demonstration must be made that harvesting levels are sustainable. Important research efforts have been conducted and a clear picture of harvest intensity that various sites can tolerate has been drawn. Yet, this is a sensitive issue and still more in-depth knowledge is required. From a social perspective, clear indications are required from the governments regarding the need for a diversified energy portfolio. If this is the case, biomass from woody residues has several interesting characteristics. This requires forest and energy policies to be aligned. For many communities affected by the slow but continuous decline of traditional forest industries, waiting for the promise of new and innovative products is not a sustainable option. Shorter-term solutions seem to converge to more basic bioproducts such as pellets and power plant feedstock. Still key questions arise regarding the actions to setup competitive supply chains for those markets. One of these concerns the mechanism and decision support systems to favor a more efficient flow of all forest products, including woody biomass.

Optimization and simulation decision-making tools are needed at the operational, tactical, and strategic levels to conceive collaborative networks that build on each other's strength and complementarity, and that translate into an industrial complex producing a wide array of products, supplied by integrated procurement systems. These tools should allow integrating spatial data and performing analyses over large and small areas. They should also allow consideration of possible variations in the logistics system. In particular, more studies need to be devoted to the question of MC variation and the question of sharing forest resources between different competing users. Also, studies are needed to establish better estimates of the potential reduction in the total harvesting costs when the extraction of FWB is integrated with traditional forest products and to optimize the redistribution of cost among the different products. The latter studies are challenging because cost distribution is not necessarily aligned with value distribution. In addition, logistical issues such as the location of concentration yards, the choice of trucking paradigms that maximize the transportation of energy rather than weight, and improved trucking scheduling are all key questions that should be studied and optimized thoroughly. Another key question that has not been addressed explicitly in this chapter is FWB storage. Storage can be an effective method for the reduction and stabilization of the MC of the FWB. Several conditions, such as the selection of the appropriate sites (at the stump, roadside, at a landing, etc.), form of the FWB, and storage procedure, need to be optimized and controlled for chip quality, cost, and availability (Röser et al., 2011a).

Considering that integrated operations, from a larger perspective, are known to be hardly competitive with other types of energy and that combustion is the only real option now available for FWB, subsidies, tax incentives, energy quotas, and other forms of government policies are likely needed to create a favorable economic setting in the short term.

REFERENCES

Alam, B., Pulkki, R., Shahi, C. 2012. Woody biomass availability for bioenergy production using forest depletion spatial data in northwestern Ontario. *Canadian Journal of Forest Research*, 516:506–516.

ARFuels. 2012. Atikokan Renewable Fuels awarded fibre supply through Ministry of Northern Development, Mines and Forestry. Released February 1, 2011. Available at: http://www.arfuels.ca/(Accessed: November 28, 2011).

Asikainen, A., Björheden, R., Nousiainen, I. 2002. Cost of wood energy. In J. Richardson, R. Björheden, P. Hakkila, A. T. Lowe, and C. T. Smith (Eds.), *Bioenergy from Sustainable Forestry: Guiding Principles and Practice*, 125–157. The Netherlands: Kluwer Academic Publishers.

Atikokan Progress. 2012. Full wood allocation, plus $1.25 million, for Atikokan Renewable Fuels. Released February 9, 2011. Available at: http://www.atikokanprogress.ca/2011/02/09/full-wood-allocation-plus-1-25-million-for-atikokan-renewable-fuels/(Accessed: April 26, 2016).

Azouzi, R., D'Amours, S., Dessureault, Y. 2009. A vision of a future industry model for the valorization of forest fiber resources in Quebec. Communication at the 13th Symposium on Systems Analysis in Forest Resources, Francis Marion Hotel, Charleston, South Carolina, May 26–29.

Azouzi, R., Lebel, L., D'Amours, S. 2012. *Restructuring the Forest Value Chain Using Intermediaries: A Methodology with Application to Community-Managed Forests.* CIRRELT, CIRRELT-2012-02. Available at: http://www.cirrelt.ca/DocumentsTravail/CIRRELT-2012-02.pdf (Accessed April 26, 2016).

Barker, R. M., Dos Santos, B. L., Holsapple, C. W., Wagner, W. P., Wright, A. L. 2001. Tools for building information systems. In G. Salvendy (Ed.), *Handbook of Industrial Engineering: Technology and Operations Management*, 3rd edn. New York, NY: John Wiley & Sons.

Bauen, A., Woods, J., Hailes, R. 2004. Biopowerswitch! A biomass blueprint to meet 15% of OECD electricity demand by 2020. WWF International and Aebiom. Available at: http://assets.panda.org/downloads/biomassreportfinal.pdf (Accessed April 26, 2016).

Beaudoin, D., Frayret, J.-M., Lebel, L. 2008. Hierarchical forest management with anticipation: An application to tactical-operation planning integration. *Canadian Journal of Forest Research*, 38, 2198–2211.

Beaudoin, D., Frayret, J.-M., Lebel, L. 2010. Negotiation-based distributed wood procurement planning within a multi-firm environment. *Forest Policy and Economics*, 12(2):79–93.

BioBusiness Alliance of Minnesota. 2010. Minnesota's forest biomass value chain: A system dynamics analysis. Available at: https://bioenergykdf.net/system/files/Minnesotas_Forest_Biomass_Value_Chain_ASystem_Dynamics_Analysis.pdf (Accessed April 26, 2016).

Bradley, D. 2007. Canada-sustainable forest biomass supply chains. Available at: http://www.bioenergytrade.org/downloads/sustainableforestsupplychainsoct192007.pdf (Accessed April 26, 2016).

Bradley, D. 2009. Canada report on bioenergy 2008. Environment Canada and IEA Bioenergy Task 40. Available at: http://www.bioenergytrade.org/downloads/canadacountryreport jun2008.pdf.

Collini, M. 2006. Portrait des Ressources Forestières. Québec, Canada: Observatoire de l'Abitibi-Témiscamingue. Available at: http://www.observat.qc.ca (Accessed November 28, 2011).

Cormiers, D. 2010. Forest feedstock supply chain challenges for a new bioeconomy. FPInnovations Presentation, Québec, QC, Canada.

Cormier, D., Ryans, M. 2006. The BiOS model for estimating forest biomass supply and costs. Internal Report IR-2006-03-22. 32 p. Quebec, Canada: FPInnovations-FERIC Division.

Damery, D., Benjamin, J., Kelty, M., Lilieholm, R. J. 2009. Developing a sustainable forest biomass industry: Case of the US northeast. *Ecosystems and Sustainable Development VII*, 122:141–151.

Dubeau, D., LeBel, L., Imbeau, D. 2012. Statistical models to predict brush cutters time consumption in regeneration release operations. *Northern Journal of Applied Forestry*, 29(4):173–181.

Dymond, C. C., Titus, B. D., Stinson, G., Kurz, W. A. 2010. Future quantities and spatial distribution of harvesting residue and dead wood from natural disturbances in Canada. *Forest Ecology and Management*, 260:181–192.

Evans, A. M. 2008. Synthesis of knowledge from woody biomass removal case studies. Forest Guild. Available at: http://www.forestguild.org/publications/research/2008/Biomass_Case_Studies_Report.pdf (Accessed April 26, 2016).

Feng, Y., D'Amours, S., LeBel, L., Nourelfath, M. 2011. Bio-refinery supply chain network design using mathematical programming approach. In M. El-Halwagi and P. Stuart (Eds.), *Integrated Bio-refineries: Design, Analysis, and Optimization*. Boca Raton, FL: CRC Press/Taylor & Francis, 2012.

Forest Products Association of Canada. 2010. *Transforming Canada's Forest Products Industry: Summary of Findings from the Future Bio-Pathways Project*. Ontario, Canada: FPAC.

Gautam, S., Pulkki, R., Shahi, C., Leitch, M. 2012. Fuel quality changes in full tree logging residue during storage in roadside slash piles in Northwestern Ontario. *Biomass and Bioenergy*, 42:43–50

Government of Canada. 2009. *The Canadian Forest Sector: Past, Present, Future*. Elsevier. Available at: http://publications.gc.ca/pub?id=395451&sl=0 (Accessed April 26, 2016).

Government of Quebec. 2011. Analyse sur l'accès aux ressources forestières pour la production d'énergie par les communautés rurales, p. 82. Available at: http://www.mamrot.gouv.qc.ca.

Hanson, B. 2012. Dry biomass makes a difference. The Working forestry newspaper. February 10. Available at: http://www.workingforest.com/dry-biomass-makes-difference/ (Accessed November 28, 2011).

Jackson, S. W., Rials, T. G., Taylor, A. M., Bozell, J. G. 2010. Wood 2 energy: A state of the science and technology, 56 p. Available at: http://www.wood2energy.org/StateofScience.htm (Accessed April 26, 2016).

Johnson, D. M. 2011. Woody biomass supply chain and infrastructure for the biofuels industries. Industry Studies Conference, Pittsburg, CA. Available at: http://www.industrystudies.pitt.edu/pittsburgh11/documents/Papers/PDF%20Papers/1-3%20Johnson.pdf (Accessed November 28, 2011).

Kellogg, L., Davis, C., Vanderberg, M. 2006. Identifying and developing innovation in harvesting and transporting forest biomass. Proceedings of the Forest Products Society 60th International Convention, June 25–28, 2006, Newport Beach, CA.

Lattimore, B., Smith, T., Richardson, J. 2010. Coping with complexity: Designing low-impact forest bioenergy systems using an adaptive forest management framework and other sustainable forest management tools. *The Forestry Chronicle*, 86(1):18–21.

Mabee, W. E., Fraser, E. D. G., McFarlane, P. N., Saddler, J. N. 2006. Canadian biomass reserves for biorefining. *Applied Biochemistry and Biotechnology*, 129–132:22–40.

Marshall, L., Fralick, C., Gaudry, D. 2010. OPG Charts Move from Coal to Biomass. POWER: Business and Technology for the Global Generation Industry. April 1, 2010. 6p.

Ministère des Ressources naturelles et de la faune. 2004. *Portrait forestier de la région de la Mauricie*. Available at: http://www.mrn.gouv.qc.ca/publications/forets/portraits-forestiers/PortraitForestier04.pdf (Accessed November 28, 2011).

Moore, T. Y., Ruel, J.-C., Lapointe, M.-A., Lussier, J.-M. 2011. Evaluating the profitability of selection cuts in irregular boreal forests: An approach based on Monte Carlo simulations. *Forestry* (November 1):1–15. doi:10.1093/forestry/cpr057.

Natural Resources Canada. 2006. Canada's energy outlook: The reference case 2006. Available at: http://www.nrcan-rncan.gc.ca/com/resoress/publications/peo/peo-eng.php (Accessed April 26, 2016).

Natural Resources Canada (NRCan). 2009. L'État des forêts au Canada. Annual. Canadian Forest Service, Ottawa, Canada.

Natural Resources Canada. 2011a. Investments in forest industry transformation. Available at: http://forest-transformation.nrcan.gc.ca (Accessed April 26, 2016).

Natural Resources Canada. 2011b. The state of Canada's forests. Annual Report, Canadian Forest Service, Headquarters, Ottawa, Canada. Available at: http://publications.gc.ca/collections/collection_2011/rncan-nrcan/Fo1-6-2011-eng.pdf (Accessed April 26, 2016).

OPG. 2011. Atikokan Generating Station Biomass Repowering Project: Fact Sheet. Toronto, Canada: Ontario Power Generation. p. 2.

Paré, D., Bernier, P., Thiffault, E., Titus, B. D. 2011. The potential of forest biomass as energy supply for Canada. *The Forestry Chronicle*, 87(1):71–76.

Pettersson, M., Nordfjell, T. 2007. Fuel quality changes during seasonal storage of compacted logging residues and young trees. *Biomass Bioenergy*, 31:782–792.

Ralevic, P., Ryans, M., Cormier, D. 2010. Assessing forest biomass for bioenergy: Operational challenges and cost considerations. *The Forestry Chronicle*, 86(1):43–50.

Rentizelas, A. A., Tolis, A. J., Tatsiopoulos, I. P. 2009. Logistics issues of biomass: The storage problem and the multi-biomass supply chain. *Renewable and Sustainable Energy Reviews*, 13:887–894.

Roberts, D. G. 2008. Convergence of the fuel, food and fibre markets: A forest sector perspective. *International Forestry Review*, 10(1):81–94.

Röser, D., Asikainen, A., Stupak, I., Pasanen, K. 2008. Forest energy resources and potentials. In D. Röser, A. Asikainen, K. Raulund-Rasmussen and I. Stupak (Eds.), *Sustainable Use of Forest Biomass for Energy: A Synthesis with Focus on the Baltic and Nordic Region*, 2–28. New York, NY: Springer, p. 255.

Röser, D., Mola-Yudego, B., Sikanen, L., Prinz, R., Gritten, D., Emer, B., Väätäinen, K., Erkkilä, A. 2011a. Natural drying treatments during seasonal storage of wood for bioenergy in different European locations. *Biomass and Bioenergy*, 35(10):4238–4247.

Röser, D., Sikanen, L., Asikainen, A., Parikka, H., Väätäinen, K. 2011b. Productivity and cost of mechanized energy wood harvesting in Northern Scotland. *Biomass and Bioenergy*, 35(11):4570–4580.

Ruel, J. 2010. Sectoral outlook 2010–2012—Mauricie. Service Canada Administration. Available at: http://publications.gc.ca/collections/collection_2011/servicecanada/SG2-1-12-2010-eng.pdf

Ryans, M. 2010. Pay for Energy. *Canadian Biomass Magazine*. March–April 2010.

Smeets, E. M. W., Faaij, A. P. C. 2007. Bioenergy potentials from forestry in 2050. An assessment of the drivers that determine the potentials. *Climate Change*, 81:353–390.

Statistics Canada. 2012. Atikokan, Ontario (Code 0028) and Ontario (Code 35) (Table). Census Profile. 2011 Census. Statistics Canada Catalogue no. 98-316-XWE. Ottawa, Canada. Released February 8, 2012. Available at: http://www12.statcan.ca/census-recensement/2011/dp-pd/prof/index.cfm?Lang=E (accessed May 25, 2012).

Stupak, I., Asikainen, A., Röser, D., Pasanen, K. 2008. Review of recommendations for forest energy harvesting and wood ash recycling. In D. Röser, A. Asikainen, K. Raulund-Rasmussen and I. Stupak (Eds.), *Sustainable Use of Forest Biomass for Energy: A Synthesis with Focus on the Baltic and Nordic Region*, 155–196. New York: Springer.

Titus, B. D., Maynard, D. G., Stinson, G., Kurz, W. A. 2009. Wood energy: Protect local ecosystems. *Science*, 324:1389–1390.

The Township of Atikokan. 2008. Atikokan strategic community and economic development plan: Final report. Atikokan, Ontario: The Township of Atikokan.

U. S. E. P. Agency 2007. Biomass combined heat and power catalog of technologies. Report prepared for U.S. Environmental Protection Agency combined heat and power partnership. September 7, 113 p. Available at: http://www.epa.gov/chp/documents/biomass_chp_catalog.pdf

Visser, R., Hall, P., Raymond, K. 2010. Good practice guide: Production of wood fuel from forest landings. EECA. 44 p. Available at: http://www.eeca.govt.nz/sites/all/files/production-wood-fuels-from-forest-landings-4-10.pdf (Accessed November 28, 2011).

Volpé, S. 2012. *FPJoule*. FPInnovation Advantage Report, 13(4). Available at: http://www.feric.ca (Accessed November 28, 2011).

Volpé, S., Desrochers, S. 2010. Intégration du débardage de la biomasse aux opérations de récolte. FPInnovation Advantage Report, 12(12). Available at: http://www.feric.ca (Accessed November 28, 2011).

Wilkerson, E. G., Perlack, R. D. 2008. Resources assessment, economics and technology for collection and harvesting. In B. D. Solom and V. A. Luzadis (Eds.), *Renewable Energy from Forest Resources in the Unites States*, 69–91. New York: Routelage.

Wood, S. M., Layzell, D. B. 2003. A Canadian biomass inventory: Feedstocks for a bio-based economy. Final report. Kingston, Ontario: BIOCAP Canada Foundation, 42 p. Available at: http://www.agwest.sk.ca/bioproducts/documents/BIOCAP_Biomass_Inventory_000.pdf (Accessed November 28, 2011).

12 Overview of Wood Transportation and Operations Research Methods in This Area

Bernard Gendron, Reino Pulkki,
Marius Posta, and Jean Favreau

CONTENTS

12.1   INTRODUCTION

This chapter presents a survey of the literature dealing with applications of operations research (OR) methods for planning wood transportation, one of the most important aspects of the forest value chain. By transportation, we mean *long-distance* transportation, which refers to transportation from the raw material source to the point of utilization. Long-distance transportation applies to both transportation of the raw material to the mill and transportation of the finished product to the consumer. In the case of roundwood, it refers to the transportation from roadside to the point of utilization (e.g., the mill). The transportation from the stump area to roadside is referred to as *primary transportation*, also known as terrain transportation, off-road transportation, skidding, forwarding, yarding, and so on. *Secondary transportation* is concerned with the transportation of logs to primary plants and of wood products to secondary plants, as well as moving residues, chips, and so on. For the purpose of this chapter, maritime international transportation of wood and wood products will not be included in our definition of long-distance transportation. Therefore, when the term *wood transportation* is used, it will refer to the transportation of wood (round-wood, chips, and residues) from the raw material source to the mill.

The three main transportation modes are road, railway, and water transportation. If the distance from roadside to the mill or the upper landing where the wood is transferred to water or rail is short, the vehicle used in primary transportation can be used (the forwarder being the best suited), which is referred to as secondary intermediate transportation (Ackerman and Pulkki, 2004). The above modes can be used singly or in combination, which is referred to as *multimodal or intermodal transportation.*

Wood transportation is a prominent part of *wood procurement*, a wide area of activities that begin with wood purchasing, including raw wood and wood residues, and end with delivery of the wood to the mill (LeBel et al., 2009; Uusitalo, 2010). Wood transportation is only a part of this process and to link all the various pieces together is a complex problem. Wood procurement is a logistic problem with the stages interacting with each other and with the external business parties. A simplified view of wood procurement flow for a company is shown in Figure 12.1, where the different transportation modes are displayed: road (primary, secondary intermediate, secondary), water, and railway transportation. It is estimated that transportation accounts for up to 50% of the delivered wood cost, a fact that fully justifies any effort in improving wood transportation planning through the development of OR methods.

This chapter is organized as follows. Section 12.2 provides an overview of the main issues that arise when planning multimodal wood transportation. Section 12.3 is dedicated to a survey of the literature on OR methods for wood transportation planning. We conclude in Section 12.4 by identifying the main research challenges that arise when applying OR methods for wood transportation planning.

12.2   OVERVIEW OF MULTIMODAL WOOD TRANSPORTATION

This section gives an overview of wood transportation, with an emphasis on the selection of transportation modes. Sections 12.2.1 through 12.2.3 are general expositions on water, railway, and road transportation, respectively. Section 12.2.4 discusses the

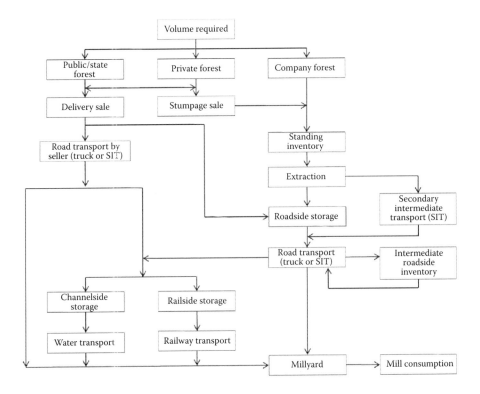

FIGURE 12.1 Wood procurement flow options for a company.

issues to consider when comparing the different transportation modes. As discussed above, wood transportation is an important activity along the value chain, but it has to be considered in relation with the other elements of the value chain. Section 12.2.5 discusses the relationships between wood transportation, inventory, and scheduling. Throughout Section 12.2, whenever relevant to the discussion, we highlight the situation of multimodal wood transportation systems in Canada.

12.2.1 WATER TRANSPORTATION

Water transportation can be thought of as the oldest form of transportation by humans, aside from using muscle power to physically carry or drag objects, and is still used widely in various regions of the world (e.g., Southeast Asia). In the past, water transportation was the only means of wood transportation. With the development of the "wood industry" in Europe and North America from basically a fuel wood supplier, to manual sawmilling, to water-power sawmilling, to steam-powered sawmilling, and finally, to the modern forest industry as we know it, water transportation became more important, and more wood assortments were floated. The forest industries in Europe and North America were based on the use of waterways in wood transportation and for subsequent transportation of finished products.

With the advent of railway transportation in the 1830s, the development of land transportation modes has continuously increased. However, it was not until after 1910 and especially after 1920 that road construction really began. Due to the development of extensive overland transportation networks in both Europe and North America, the importance of water transportation is no longer what it used to be. Since World War II, floating has been abandoned in Europe. Finland and Russia are the last countries in Europe using floating to any degree. Similarly, in Canada, river driving has been abandoned. The last river drives were phased out in Ontario and Quebec during the early 1990s. However, water transportation with barges, as well as with bundled wood in rafts, is quite important along the British Columbia coast.

In the mid-1970s, with the advent of the energy crisis, water transportation experienced a brief renaissance. This is because water transportation is the most energy efficient means for moving wood. For example, road transportation uses 17 times and railway transportation four times more energy than water transportation to move the same mass over the same distance.

In water transportation of wood, we are concerned with bundle and free-floating, barging (pushed and towed), and motor-powered dry goods vessels. Special self-propelled barges are included under the heading of motor-powered dry goods vessels. All of the above can be used in wood transportation. Bundle and free-floating are not applicable to transport of finished products, while motor-powered dry goods vessels are the most applicable for maritime transportation.

In water transportation, we have waterways that require no improvement, partial improvement, and complete improvement (Toivonen, 1979). Large lakes and rivers, coastal waters, seas, and oceans require little improvement work, except for some navigational aids. Therefore, they can be seen as free capital and bundle floating, rafting, barging, and shipping can occur with little or no investment. Other waterways require some improvement to allow water transportation. In free-floating (i.e., river driving), some partial improvements are usually required. These could include the clearing of a few bad rapids, placing of glance and holding booms, the construction of some dams on smaller streams, and so on. In bundle floating, rafting, barging, and shipping, the improvements could include some dredging and channel construction work. A waterway, which is completely improved, includes canals, locks, dredged channels, and so on. The amount of channel improvement work required is directly related to the volume of wood transported and the volume of water available. Clearly, water transportation is most favorable when little or no improvement work is required and large volumes can be transported.

12.2.2 Railway Transportation

With the rapid development of the forest industry during the early 1900s, especially in North America, problems were experienced in wood transportation in areas where there were no waterways suitable for water transportation. This was due to the general lack of waterways, rough terrain resulting in fast flowing water, large size of timber that could not be floated in shallow water, and water flowing in the wrong direction. In these circumstances, special *logging railways* running directly from the forest to the mills were built. Also, often, they joined up with common-carrier railways. The first steel rail logging was attempted in Michigan in the early 1870s, but it did not become popular until the turn of the century (MacKay, 1978). Logging railways were used in eastern North America, but not to any great extent because river driving was much cheaper. It was on the West Coast of North America that logging railways were used to any extent. For example, in 1924, there were 1120 km of logging railways in British Columbia and 148 locomotives. As road transportation became much more flexible and developed, logging railways were gradually abandoned, mainly because of the expensive costs of construction and maintenance (Pulkki, 1982). In 1969, there were only two logging railways left in British Columbia compared to 80 in 1924 (MacKay, 1978). Today, there is only one left, the 122-km-long Englewood Railway operated by Western Forest Products.

Common-carrier railways are now commonly used throughout the world for wood transportation, but more specifically for the transportation of finished forest products. These railways do not go directly to the source of wood and are used for other transport. For railways to be economically competitive in wood transportation, a large volume of continuous and guaranteed wood has to be transported over long distances. In addition, preliminary transportation to the railway must be economical and cars must be loaded and unloaded quickly, thus resulting in fast turnaround times, which in turn reduces the amount of rolling stock required. Finally, to ensure fast delivery of wood, it is advantageous to have trains that carry only wood to a small number of destinations (Pulkki, 1982).

A *unit train* is used for the bulk carriage of a single commodity, generally on a continuous basis between two points, where the entire train, from locomotives to the final car, stays together. A *special train* refers to a train similar to the unit train, but that can shrink and grow (i.e., add or drop cars). Special trains are used mainly for the carriage of mixed commodities, but must have a minimum of 90 cars and are pulled by at least three locomotives. Special trains are about 1.8 km in length and have a payload of about 6100 tons.

Compared to other modes of transportation, railways have several advantages (Pulkki, 1982). First, it should be noted that without railways, it would be practically impossible to supply large, centrally located mill complexes with the wood volume required from many directions, especially if the mill is located in a built-up area. Compared to road transportation, railway transportation is more economical when distances are long; it requires much less energy to transport the same mass over the same distance and it is not influenced to such a great extent by weather or season (road transportation must be limited on lower class highways and side roads during the spring breakup period in Canada). Compared to water transportation, railway transportation is more flexible and can handle high quality logs over long distances with less risk of damage.

In Canada, there are 63 freight railways with tracks. Statistics Canada (2012) reports 68,092 km of railway tracks in 2009, of which just over 57,000 km are under the control of the two largest railway companies, Canadian National (CN) and Canadian Pacific (CP). In 2012, CN and CP combined have approximately 223,000 rail cars online, of which about 152,000 were active. Generally, the freight rate is negotiated between the railway and the shipper; it is generally based on freight type, weight, volume, distance, condition of sidings, and in some cases on the number of pieces. Discounts or better freight rates can be negotiated if a large volume is shipped from concentrated points to a single destination.

In addition to raw wood, the railways transport finished products. The forest industry plays an important role in supplying freight to the railways. Therefore, from a national economic point of view, if the forest industry were to suspend use of railway transportation completely, there would be serious consequences to the entire railway infrastructure, and thus, to the national economy.

12.2.3 ROAD TRANSPORTATION

Road transportation refers to all transportation on roads, whether it is by truck, tractor, or horse-drawn carriage. In almost all cases, some road transportation is required (e.g., to get the wood from the forest to the rail or waterway terminals). Only when wood is extracted directly from a location that is beside a railway terminal or waterway is preliminary transportation by road not required. There is also a recent development in increased delivery of clean chips from satellite and centralized wood processing facilities.

The use of trucks in wood transportation did not begin until after about 1910. The first trucks were not suited to the conditions experienced in logging operations, since large loads could not be hauled and the truck technology was primitive (e.g., open cabs, inadequate brakes, small engines). It was not until after World War II that truck transportation began to be of any significance, as much denser road networks (both forest and public) were developed. In Canada, river driving has been gradually abandoned in favor of truck transportation due to the need for greater flexibility, year-round operations, and faster delivery of wood to the mills. From the end of the 1950s, when single and tandem axle trucks capable of payloads from 8 to 12 tons were used, truck technology has quickly advanced to multiaxle vehicles capable of carrying payloads of over 100 tons.

In Canada, each province has its own rules in place for determining the maximum gross vehicle weight (GVW) for trucks or tractor-trailer units, as well as maximum vehicle dimensions (height, width, and length). In general, maximum GVW is determined based on number of axles, space between axles, design of the front axle, base length of the vehicle, tire width, and whether dual or single tires are used.

Trucks and trailers of various designs can be used to haul any form of wood, whether it is full trees, tree lengths, saw logs, pulpwood, stumps, logging residues, or chips. Choosing the right truck type is based on several constraints: transported products, scale of the operation, physical limitations, and type of road. The main configurations we find are: five-axle trucks, tractor-trailers, semitruck trailers, and B-train tractor trailers. The type, size, and configuration of cargo space have to be adapted to the transported product (logs, chips, etc.). Most transportation activities use public roads, but there are also specific trucks (off-road trucks) that can only be used on forest roads and usually exceed payload limits of provincial legislation.

Most forest companies hire transportation contractors to handle hauling activities. These contractors usually serve multiple companies at the same time, but can dedicate a part of their fleet to a specific customer. Some transportation companies offer their services at a regional level and are capable of transporting different products (roundwood, chips, sawdust, and finished products). These companies usually have their own fleet of loaders that they can dispatch to pickup and delivery sites. Transportation rates are determined based on the hauled product. For roundwood, green ton is the commonly used unit to calculate trucking rates based on the delivered quantities. For wood chips and pulpwood, dry ton is used instead. To ensure stable rates, correction factors are applied and adjusted according to season. These factors take into account the content of the residues included in the truckload, the wood moisture, the transported species, and the bark proportion. In particular, seasons affect the moisture content of wood. In winter, trees carry more water than in summer and wood is then heavier to haul. Specific agreements can be made between forest companies and transportation contractors, which lead to different ways to calculate trucking rates. These rates are based also on total cycle time per trip including loading, unloading, and waiting time.

The Canadian forest road network is composed of different road categories (Desautels et al., 2009). Primary roads are usually linked to the public road network or to a mill. Primary roads access the forest area and they have the highest level of durability (10 years or more). Commonly, speed is limited to 70 km/h in this category of roads. Secondary roads are branched to the primary network and can usually be used during the whole year. Their life span is typically 3–10 years. Tertiary roads constitute the end of the road network branches. They are used during shorter periods and they are usually located in the annual harvesting sectors. In Canada, hauling distances between forests and mills frequently exceed 150 km. Forest characteristics, weather, and land size make forest road networks expensive to build. In particular, weather conditions affect the design and usage of forest road networks. For instance, there is no transportation during the spring period as the soil is thawing and many roads are impassable. Falling rains also can bring several perturbations to transportation activities. Winter roads, which are specific to Nordic countries such as Canada, take advantage of the freezing soil to minimize the impact of trucking activities on the land. Different materials including stems and snow are used to ensure good and secure driving conditions on these roads.

12.2.4 Comparing Transportation Modes

When comparing transportation modes, several important factors must be accounted for. Throughout the world where there is mountainous or hilly terrain, the main transportation arteries are located in the valleys. Therefore, we often have a railway, a highway, or a waterway running parallel to each other. Although the presence of waterways on flat terrain is less pronounced, the same relationship holds. According to an international rule of thumb, highways only make up 10% of the entire road network. In addition, 80% of highway traffic is on 20% of the highways (Wibstad, 1979). The forest roads and other transportation network infrastructure leading to the main transportation arteries are required for access and wood transportation, no matter which major transportation artery is used. The transportation network can be compared to a watershed with the small streams flowing into larger ones, which in turn flow into the main river. The main river increases in size as it flows toward its mouth.

Forest operations are spread over very large geographic areas and the road infrastructure to access these is almost always required to get to processing plants or to transfer points for further transportation by either water or railway. One of the few exceptions occurs on the West Coast where helicopter logging can deposit logs directly from a logging site to water. Transportation of workers and equipment can also be made by barge. In general though, when making strategic decisions with respect to transportation modes only the main arteries should be compared (Wibstad, 1979). In addition, wood transportation traffic is generally sparse on most of the road network, with traffic congestion mostly occurring near cities, processing plants, concentration yards, and multimodal transfer points.

Two major economic points of view can be used concerning the comparison of transportation modes: the business economic and the national economic disciplines. *Business economic factors* in wood transportation planning are often related to operational planning. They are concerned with issues such as the availability of wood for purchasing and cutting; the harvesting schedule (e.g., the availability of wood at roadside); the time required to transport wood from the forest to the mill; the presence of suitable transportation routes; wood holding costs (including deterioration and interest costs); and wood demands at the mill. *National economic factors* in wood transportation planning are often related to strategic/tactical planning. They are concerned with issues such as the investment required to establish, maintain, and operate the different transportation systems; the productivity of the different resources (e.g., labor, energy, capital), the effect of the different transportation modes on the environment; rural and regional employment considerations; accident risk; and traffic congestion near built-up areas.

The business economic competitiveness of a transportation mode refers to the actual cost of the mode to a firm. There are both direct and indirect costs arising from transportation. For example, when water transportation is examined and only direct costs are used, it appears competitive with the other modes. However, due to its nature, water transportation imparts additional costs (e.g., interest on wood tied up in inventory, environmental cost) to the other stages in the wood procurement chain, and thus reduces its competitiveness to some degree. The same holds true for other transportation modes.

When examining the transportation situation from the national economic point of view, the significance of the individual firm is forgotten and the economy of the country becomes the focal point. Thus, the need for foreign imports, such as equipment, parts, energy, and so on, come into question, as well as the cost of building and maintaining roads, railways, and waterways. Rural and regional employment, other transport artery users, pollution/effect on the environment, safety, and other values are also examined. Therefore, depending on the cost calculation used, the competitiveness of the transportation modes can vary considerably. Great care must be taken to make sure of what is actually being compared.

Although it is best to view the available transportation modes as forming an integrated transportation system, it may also be beneficial to view the modes as competitors, from the point of view of both forest companies and public authorities. In this way, competition is stimulated between the modes and costs are kept better in line. If one mode becomes more competitive, the others must follow suit or lose their share of the transportation volume.

12.2.5 Multimodal Wood Transportation, Inventory, and Scheduling

Wood transportation is considered in strategic, tactical, and operational planning of forest operations. Strategic planning is mostly used to decide which harvesting areas to cut. In this context, transportation planning is necessary to match the harvesting capacity. Wood transportation at the tactical planning level is used to determine capacity requirements and hire contractors on a seasonal basis. At the operational level, supervisors or planners who are in charge of harvesting areas define their needs for the short term and manage fleet dispatching in collaboration with transportation contractors.

The decision to build a forest road at a certain location has major impacts, since forest management, including inventory, harvesting, and transportation activities, all depend on this decision. All stakeholders must share their information to ensure the sustainability of the road and the efficiency of forest management. Therefore, the planning of the different activities—inventory, harvesting, and transportation—must be integrated.

When choosing the most appropriate transportation policy, a firm should strive to achieve the best overall economic result over the long term. The inventory policy has a direct effect on the applicable transportation modes. For example, if a firm wants to take a large risk of running out of wood, but keep holding costs at a minimum, it would choose minimum inventories and thus hot-logging techniques, for which road transportation is most appropriate. Wood storage duration also has a direct impact on wood quality and processing costs.

There are many factors affecting the choice of the most appropriate inventory policy. Finding the best balance between the cost of storing wood and the benefits gained through easier planning and no wood shortage is a complex problem. Variations in wood supply and demand throughout the year and between years are an important source of extra costs in wood procurement. One major way to reduce uncertainty in wood demand and supply and thus wood inventory levels is through quick exchange of information between all value chain actors. An efficient system for monitoring standing and felled inventory levels is a crucial factor in the above.

The volumes and schedule of wood available for purchase by other firms or individuals is a major factor. For example, in many regions private forests and many independent contractors supply the majority of the wood used by the forest industry. The more uniform the wood deliveries are throughout the year and matched to the mill's wood use requirements, the better the situation is for the firm.

In a firm's own wood harvesting, the timing when wood is cut plays an important role. For example, although in past years, there has been an attempt to schedule harvesting operations throughout the year, the majority still occurs during the winter, as otherwise the environmental impact, as well as harvesting costs, are excessive (the ground is too wet during the rainy season in fall, while forest workers are required for planting during spring and in late summer).

The harvest methods and systems in use also have an effect. All solutions in wood procurement should fit into the entire value chain (from the forest to the finished product) and not cause unreasonable costs elsewhere. In most cases, transportation forms a service function and must meet the demands imposed upon it by the wood assortments required at the mill and the conditions prevalent at the forest end. Seldom is wood harvested in a form to minimize only transportation costs.

Due to the many factors mentioned above, the transportation mode giving the lowest direct transportation cost is not always the one chosen. However, direct cost forms one of the basic criteria in transportation mode choice. The direct cost of truck transportation to the mill is straightforward; we have costs of loading, transportation to mill, and mill receiving. When dealing with railway or water transportation, we have the additional cost of transportation from the forest to the terminal, as well as terminal operational and maintenance costs. In bundle floating, there are also the additional costs of bundle bindings, bundling, and raft formation and dismantling.

Road transportation is the most flexible of the three modes. This is because we are dealing with small truckload units. In railway transportation, cost savings can be obtained by dispatching larger volumes of wood at a time and throughout the year. However, time is needed to gather the volume required at a railway terminal. Water transportation has the largest delay time of the three modes. One reason is due to seasonal factors; in northern inland areas, the waterways freeze over during winter and floating is only possible for 5–6 months of the year. In some areas, the navigational period can be extended through the use of barges.

The additional cost of interest on the wood investment is one factor affecting the transportation policy. For small, nonintegrated operations, road transportation is solely relied on. For small firms, only a small buffer inventory is required and they are flexible to variations in market demands and wood supply. Since they have a low capital investment, as opposed to a pulp and paper mill, occasional wood shortages do not pose a serious consequence. Also, the area from which wood is procured is quite small and thus suited to road transport. The cost of interest thus forms a much larger criterion for smaller firms.

When dealing with a large mill, and complex and thus large wood requirements, the importance of solely interest costs on money invested in wood decreases. The cost of capital invested in the mill (i.e., cost of an idle mill), shutdown and start-up costs, and the cost of idle labor require the presence of a buffer storage to ensure against wood shortages. The size of the buffer storage depends on the degree of uncertainty present in wood procurement and wood demands at the mill (i.e., market demand). Storage is also important to even out discontinuity between wood demanded and wood available on the market (i.e., lead time) between seasons and years. For example, trying to juggle the wood sales from many different wood suppliers throughout the year and between years is a very complex operation. As mentioned earlier, the seasonal nature of logging (e.g., winter logging, spring breakup) also results in wood inventories. Another point to remember when dealing with large wood volumes is that even though road transportation is the quickest method to get a load of wood to the mill, the movement of an entire inventory by other modes can be competitive.

The choice of a transportation mode also assumes that the mode services both the point of dispatch and the destination. Once wood is transferred from road transportation to actual railway or water transportation, it should be delivered by that mode to the destination, as much as possible. This is due to the high interface cost when transferring from one mode to another. This high interface cost is one of the reasons for the widespread use of containers and intermodal transportation today.

The use of special wood concentration or centralized processing yards is also possible. These could be located at railway or water transportation terminals, or when merchandizing or sorting is done. The use of centralized processing is a viable alternative in areas where truckload restrictions to the terminal are not critical, manpower is scarce and expensive (e.g., there is a high level of mechanization), many wood assortments are required, and there are many destinations. The use of special concentration yards to service only road transportation is questionable due to the high costs involved when compared to the potential savings.

Certain poorly floating wood assortments, for example, chips, fresh hardwood pulpwood, hardwood logs, and small diameter fresh softwood pulpwood, limit the use of bundle floating if additional flotation or barges are not used. The extent of wood deterioration over time differs between transportation modes also. Wood deterioration has a direct effect on the final product value and processing costs. Generally, if wood must be stored, it should be stored in water to minimize deterioration due to fungi and insects, and wood drying.

Wood stored on land (i.e., at a railway terminal, mill yard, or roadside) could be continuously sprinkled with water or treated in some other way (e.g., ends painted or waxed, chemically treated, plastic encapsulation). In general, direct delivery of wood to the mill would give the best results. However, if storage is required, water storage most often gives the best results. In regard to storage, the availability of space at the mill must also be accounted for.

In this section, we have given a high-level overview of wood transportation planning and its relationships to the planning of the other activities along the forest value chain. Section 12.3 will look at wood transportation planning from an OR perspective. It will identify the optimization problems involved at different planning levels—strategic, tactical, and operational—along with a survey of the literature on the methods used to solve these problems.

12.3   SURVEY ON OPERATIONS RESEARCH METHODS

This survey of the literature on OR methods for wood transportation planning covers strategic, tactical, and operational problems in the forest industry that involve transportation in a significant manner. Section 12.3.1 covers the modeling and analysis of the forest products value chain, while Section 12.3.2 covers the construction and upkeep of the forest road network, for which there is a dedicated body of literature. The latter is often closely coupled with harvest planning and stand access problems, which we cover in Section 12.3.3. Tactical transportation planning problems are covered in Section 12.3.4; these problems involve mostly route planning and cost analyses investigating the possible savings to be derived from backhauling. Finally, Section 12.3.5 covers the operational planning horizon, which in this context corresponds to log-truck scheduling problems. An overview of many of these problems can also be found in Epstein et al. (2007).

12.3.1   Value Chain Modeling

Value chain modeling is a subfield of OR that has been generating a strong interest in the last decade in many industries, and forestry is no exception. A notable particularity of forest value chain models is the importance that transportation plays in them, which is what brings them to our interest here. Indeed, the relatively high costs involved in transportation of forest products unsurprisingly have a large impact on optimal long-term planning decisions. For good general introductions to the forest value chain literature, we refer the reader to Carlsson et al. (2008), D'Amours et al. (2003), and D'Amours et al. (2008). The following case studies are also broad enough to serve as good examples as to what is involved: Cochran et al. (2011), Troncoso et al. (2011), and Weintraub and Epstein (2002) for the Chilean forest industry and Broman et al. (2009) and Kong et al. (2012) for the Swedish forest industry. We now group the remaining literature into specific product-based value chains. Note that the origin of each product-based value chain is the same—the tree—but the processing, transportation, storage, and market prices are different for each part of the tree. For example, limbs and small logs could be used for pulp or fuel, while other logs could be sent to sawmills. Each log and the limbs from each tree are sent to divergent processes (mills) accordingly to the end product's requirements. This situation creates many product-based value chains from the same harvested tree.

12.3.1.1   Forest Fuel Value Chain

The earliest published work on forest value chains in general is Eriksson and Björheden (1989), in which the authors develop what is in effect a value chain model for the forest fuel supply of a subsidiary of the Swedish Cellulose Company: transportation, storage, and processing decisions are modeled over a long-term planning horizon consisting of 2-month time periods, and the results of the optimization then allowed managers to devise better rules for operational planning.

To quote the paper directly: "the result of the LP model shows that optimizing forest-fuel production essentially means minimizing transport costs."

In a later publication, Gunnarsson et al. (2004) present a model applied on data provided by Swedish biofuel supplier Sydved Energileveranser AB. Six scenarios are run, comprising a basic case and variations upon it, such as increased demand, changed chipping capacity, and so on. The authors analyze these scenarios and make a case for encouraging the use of models such as theirs within decision support systems.

12.3.1.2   Roundwood and Lumber Value Chain

As we will see below in Section 12.3.2, the main focus of Olsson (2004) is forest road investments. However, one of the papers comprising this PhD thesis takes a broader perspective by presenting a stochastic model used for scenario optimization in the roundwood value chain. Although not tested on real-world data, that paper's main contribution is to show that stochastic formulations can model the roundwood value chain more accurately than deterministic ones.

Vila et al. (2006) presents "a generic methodology to design the production–distribution network of divergent process industry companies in a multinational context." To illustrate their methodology, the authors applied it to a fictional, but realistic Canadian softwood lumber industry case.

12.3.1.3 Pulp Value Chain

Several contributions focusing on the pulp value chain have been published from a fruitful collaboration between Swedish academics and pulp producer Södra Cell AB (for a general overview, see Carlsson et al. [2009]). These publications report useful results that led to several recommendations implemented by Södra.

Bredström et al. (2004) describe the pulp value chain and two different integer programs devised to model it. The paper furthermore elaborates on the techniques developed to solve these models in a branch-and-cut-and-price framework, such as the branching rules and column generation algorithms used to optimize the models in a timely manner.

Gunnarsson et al. (2006) present an integrated strategic production planning model for Södra, which is more general than previous models from Bredström et al. (2004), and it turned out, more accurate as well. The model was deemed suitable for practical use as a decision support system and was thus used to make the annual budgeting for 2005–2007.

Carlsson and Rönnqvist (2005) present a model used to analyze wood flows (and by extension, transportation costs) both from the forest to the pulp mills (road transportation) and from the pulp mills to the company's international clients (maritime transportation). Gunnarsson and Rönnqvist (2008) concentrate on the heuristic solution methods applied to solve the model in Carlsson and Rönnqvist (2005).

12.3.2 FOREST ROAD NETWORKS

Before moving on to cover tactical problems, we devote this section to OR methods applied to the management of the forest road networks. This includes both tactical and strategic planning issues, depending on the size and scope of the network. Indeed, a short road accessing a harvest compartment might only see use for a season, whereas a road accessing a forest region will see much more traffic spread over several decades.

This body of work is only a small subset of the vast amount of publications pertaining to road design in general. However, forest roads have particular characteristics that warrant special consideration, the most prominent particularity being that most forest roads see relatively little traffic. Quoting Church et al. (1998): "Roads represent an investment in the transportation system and an improvement in the asset value of the forest. However, they also consume productive forest land and can become environmental (and economic) liabilities. An important challenge for professional foresters is to design road networks that minimize the amount of road constructed and maintained in an active state while still meeting access demands for multiple-entry silviculture systems, stand-tending operations, fire suppression, salvage operations, and motorized-vehicle recreation. Achieving this balance requires thoughtful planning to design appropriate road construction and maintenance strategies."

12.3.2.1 Network Design

Most OR work in forest road network design appears to be in helping the forest engineer plan the road network, rather than precisely planning the exact layout of each road. While earlier work could only rely on crude models of the terrain elevation (Liu and Sessions, 1993), recent advances in computing power have enabled researchers to use models based on geographic information systems. Indeed, the solutions generated by the method presented in Dean (1997) closely match existing road networks and are realistic, according to the author. His conclusion is that such decision-helping tools warrant further consideration. NETWORK 2000 (Chung and Sessions, 2000) and NETWORK 2001 (Chung and Sessions, 2001) are evolutions of an earlier algorithm, NETWORK II (also by John Sessions). NETWORK 2001 aims to help transportation planners to design a suitable network, while minimizing road construction and transportation costs, as well as minimizing total open road length, also taking into account road deactivation costs. Anderson and Nelson (2004) describe a method to generate a forest road network covering a whole island off the Canadian West Coast. Incidentally, the article also gives a thorough overview of the literature in this field. Rather than planning whole road networks, Ghaffariyan et al. (2010) focus on optimizing road density for stand access, for the purpose of minimizing the cost of logging.

12.3.2.2 Profile Design

In forestry as in general, following the decision to build a road to link up several points on a map, the actual road layout must be engineered in detail. Specifically, the horizontal and vertical road profiles must meet certain constraints, in order to be safely passable by the vehicular traffic, as well as to account for its effect on soil erosion. Considering the cost and extent of labor involved, it is therefore hardly surprising that OR methods have been studied in order to minimize these profiling costs subject to these constraints, and the particular case of forest roads is no exception.

Apart from Kanzaki (1973) and Douglas and Henderson (1988), most OR publications on designing forest road profiles have been relatively recent. Ichihara et al. (1996) propose a genetic algorithm for optimizing forest road profiles, and Tan (1999) a heuristic for planning the location of new forest roads. The remaining recent literature can be centered on two principal authors: Abdulla Akay and Kazuhiro Aruga.

Akay's work revolves around his development of TRACER, his forest road alignment decision support system. This development work, including the optimization algorithms at the heart of TRACER, was first described in his Ph.D. dissertation (Akay, 2003). Akay and Sessions (2005) give an overview of the use of TRACER in real-world applications, while Akay et al. (2005) present an exhaustive review of computer-aided forest road design systems.

Aruga's work comprises several recent publications that explore the application of metaheuristics in designing road profiles (Aruga, 2005; Aruga et al., 2005a, 2005b, 2006). Aruga et al. (2007) also consider the impact of erosion of the road surface in their profile design algorithm.

12.3.2.3   Road Upgrade and Deactivation

Increasingly, environmental costs need to be taken into account, in addition to operational maintenance costs of road networks. Indeed, forest roads contribute to soil erosion, and furthermore, the aggregate used to surface the roads has turned out to be the cause of accelerated sediment delivery to streams, itself a factor in the degradation of salmonid habitats in the western United States. In order to manage the environmental costs arising from the existence of forest road networks, Thompson et al. (2010) propose a multiobjective mathematical programming model to study the impact of upgrading certain road segments with culverts and the like. Pulkki (1996), on the other hand, shows that constructing loops in forest access road networks results in coverage inefficiencies and excessive water-crossings, most of which can be avoided with minimal impact on transportation cost and also results in reduced road construction costs and environmental impacts. The scarcity of adequate aggregate quarries has further encouraged the forest industry to reclaim the aggregate from the surface of decommissioned forest roads. Sessions et al. (2006) and Thompson and Sessions (2008) have proposed mathematical models to determine optimal aggregate recycling policies. Coulter et al. (2006) propose several heuristics for the scheduling aspect of road maintenance operations.

Long-term forest road management has also proven to be a field ripe for study. In a similar perspective as the previously cited works, Eschenbach et al. (2004) develop a model for determining road removal plans within a watershed. Olsson (2004) is a doctoral dissertation that comprises a number of papers proposing several optimization methods for forest road investments, in the context of forestry in northern Sweden. The models developed in Henningsson et al. (2006) also focus on the trade-off of road investments versus transportation costs in Sweden, with a time horizon of around 10 years. Anderson et al. (2006) follow up on Anderson and Nelson (2004) and propose a model for determining optimal road class within a similar time horizon, but with a focus on forestry on Hardwicke Island, British Columbia.

12.3.3   INTEGRATING TRANSPORTATION IN HARVEST PLANNING

Long-term forest road maintenance is closely related to harvest planning. It is, therefore, necessary to take a holistic approach in the mathematical modeling of tactical problems such as harvest planning (Weintraub and Navon, 1976). This is evident in the RoadOpt decision support system (Karlsson et al., 2006), as well as in other decision support systems, such as the one presented in Stückelberger et al. (2006), for instance. Note that Epstein et al. (1999) provide a more recent follow-up on these efforts in the Chilean forest industry, and Church et al. (1998) give a good general overview of locational issues in forest management.

Nelson and Finn (1991) present an analysis of the effect on harvest yields of road construction schedules and of forest management techniques such as harvest exclusion periods. Sessions (1992) discuss the access of timber harvest sites, as does Murray (1998) in a more abstract manner, considering routes rather than specific roads. Clark et al. (2000) propose a method for harvest scheduling and the construction of roads to access harvest stands, as do Richards and Gunn (2000) and Richards and Gunn (2003). Weintraub et al. (1995) present a 0–1 mixed-integer programming (MIP) formulation for the problem of planning simultaneously harvest and transportation, but owing to the size of practical models, the authors propose a heuristic approach to solve it approximately. Guignard et al. (1998) and Andalaft et al. (2003) subsequently show how to strengthen this MIP with cuts to help reduce the optimality gap during optimization.

12.3.4 TACTICAL TRANSPORTATION PROBLEMS

This section is dedicated to transportation planning problems with a tactical planning horizon. It is divided into two subsections covering wood flow cost analysis, route planning, and backhauling.

12.3.4.1 Wood Flow Cost Analysis

OR methods have been applied in cost analyses of various aspects of forestry operations. Weintraub and Navon (1976) is an early publication of this type, with the following publications appearing in recent years. Murphy and Stander (2007) describe how to simulate some scenarios in order to show that an optimal solution of a forest transportation problem with uncertain parameters can be very unstable with respect to these uncertainties and that a robust approach should be preferred. Frisk et al. (2010) present an application of game theory techniques to devise an allocation of the costs and savings resulting from the collaboration of two or more companies in forest transportation. Han and Murphy (2011, 2012a) propose a sensitivity analysis of the cost effectiveness of hauling woody biomass, which is a waste product of the forest industry. McDonald et al. (2001a) show how to simulate a logistic network in order to gauge the potential gains from collaborative log hauling.

12.3.4.2 Route Planning and Backhauling

In the last decade, a number of publications have shown how OR methods could improve transportation planning in the forest industry. The main focus has been on the study of the impact of backhauling, which consists of routing log trucks in such a manner as to minimize the amount of time that trucks run empty, which would otherwise be as high as 50% of the time (this is possible because paper mills and sawmills require different types of wood).

Palander et al. (2004) present an economic argument for backhauling, using a fictional Scandinavian energy-wood transportation network. Puodziunas et al. (2004) make a similar case in the context of roundwood transport in Lithuania. Palander and Väätäinen (2005) also make a case for interenterprise collaboration in Finland using backhauling. Forsburg et al. (2005) and Carlsson and Rönnqvist (2007) report on the development of FlowOpt, a decision support system for transportation planning, which implements these ideas and takes backhauling into account.

On a related note, Carlgren et al. (2006) discuss log sharing combined with transportation planning, while Mendell et al. (2006) evaluate the potential gains from shared log-trucking resources in the southern United States.

12.3.5   Log-Truck Scheduling

Log-truck scheduling has been the focus of a lot of work, chiefly because of the significant cost savings obtainable simply by improving the dispatching of log-trucks to collect timber at the harvest sites, a rationalization that incurs little to no downside to the parties involved. These cost savings are a consequence of the relative bulk of the commodities transported and of the relative length of the distances over which the commodities are transported, in contrast with other vehicle routing problems. Rönnqvist (2003) gives an overview of these OR efforts, a perspective which is all the more interesting considering how the author collaborated on many of these. Epstein et al. (1999) also give an overview and follow-up of various OR systems, but specifically in the Chilean forest industry. Audy et al. (2012) provide an in-depth and up-to-date summary and comparison of all the algorithms and decision support systems proposed to dispatch log-trucks, and we refer the reader to that report for a detailed analysis of the features and capabilities of existing methods. This section is divided into two subsections; first, we cover decision support systems used in an actual production setting, following which we cover log-truck dispatch methods, usually illustrated by case studies.

12.3.5.1   Decision Support Systems

An early form of decision support system for log-truck scheduling can be found in Shen and Sessions (1989) and has reportedly found application in the Chinese forest industry. To the best of our knowledge, ASICAM is the first well-known system to have been developed as such and to have been successfully applied. The ASICAM system is described first-hand in Weintraub et al. (1996) and has been subsequently reviewed a few years later in Epstein et al. (1999). Cossens (1993) has studied its potential application in New Zealand. Internally, ASICAM relies on various heuristics and simulation techniques so as to reliably optimize daily operations planning for about 200 trucks within a couple of minutes on what was for the time a reasonably powerful desktop PC (Intel 486 CPU with coprocessor).

Linnainmaa et al. (1995) provide an overview of the EPO system developed for a pulp and paper company operating in Finland, Enso-Gutzeit (later Stora Enso). At its heart, EPO is a dispatching system similar to ASICAM, but which also features some strategic planning capabilities. The EPO system features a complete graphical interface and was designed to run on HP9000 workstations, with access to the company mainframe and full road, truck, and timber databases, the latter being regularly updated by the foremen at the harvest sites. Rummukainen et al. (2009) describe a subsequent version of the dispatching module that uses tabu search.

Eriksson and Rönnqvist (2003) present Akarweb, the first web-based dispatch system to implement backhauling, which dispatches log-trucks on routes designed to reduce the time during which they run unloaded. In a sense, it is related to the FlowOpt system, in that it is concerned with the operational aspect of backhauling, while the latter is tactical in scope, analyzing wood flows, devising routes, and integrating these with harvest operations (Carlsson and Rönnqvist, 2007; Forsburg et al., 2005).

Also incorporating backhauling, the Virtual Transportation Manager (VTM) is a Canadian project developed and described in Audy et al. (2007b). In the words of the authors, "the VTM is a web-based system developed to allow collaborative route planning among many transportation managers, from different companies or business units forming a coalition." Audy et al. (2007a) and Marier et al. (2007) describe the underlying algorithms in further detail.

Andersson et al. (2008) present RuttOpt, a modular dispatching system designed for the Swedish forest industry and relying on the NVDB, a national road database. RuttOpt applies linear programming and tabu search optimizations to generate schedules for a set of trucks, which are stored in a database and can be presented in the form of Gantt schedules, maps, reports, and so on. Flisberg et al. (2009) describe this hybrid optimization approach in greater detail. The linear programming optimization in RuttOpt applies a column generation scheme proposed in Palmgren et al. (2003, 2004), in which each column is a possible truck route and is generated heuristically. Note that these last two papers have also been published in the form of a Ph.D. thesis (Palmgren, 2005).

Finally, Acuna et al. (2011) present FastTRUCK, a recently developed dispatching system for the Australian forest industry.

12.3.5.2 Further Case Studies

The contributions presented in this subsection describe additional optimization algorithms for the log-truck scheduling problem. In contrast to some of the systems presented earlier, these methods have not been tested in a production setting. Nevertheless, they have usually been validated using realistic case studies and may have influenced subsequent development efforts.

Indeed, in most cases, some kind of proof-of-concept is required before delving into more substantial, practical work, and the optimization proper constitutes but a small fraction of the tasks of a decision support system (data retrieval, presentation, etc.). We have chosen to group these case studies by region, because although log-truck scheduling operations are superficially similar throughout the world, local geography, traditions, legislation, and organizational cultures can still make these case studies distinct enough to warrant specific consideration.

During the last few years, there has been a growing body of work published in Canada on the subject, approaching it variously as a vehicle routing problem (Gingras et al., 2007), as a constraint satisfaction problem (El Hachemi et al., 2008, 2009, 2010, 2013), or as a pickup and delivery problem (Audy et al., 2011). In the United States, McDonald et al. (2001b, 2010) approach the log-truck scheduling problem with the intent of studying, via a simulation model, the potential cost savings obtainable from collaboration between companies. Han and Murphy (2012b) similarly study the possible cost savings in hauling woody biomass in Oregon. In the southern hemisphere, Rönnqvist and Ryan (1995) propose a real-time truck dispatch system capable of handling around 100 trucks for the New Zealand forest industry. Murphy (2003) presents a 0–1 MIP model for log-truck scheduling in New Zealand, with motivations similar to those presented in McDonald et al. (2001b). Rey et al. (2009) describe a column generation scheme similar to that of Palmgren (2005), but for solving an MIP, and apply it to a case study in the Chilean forest industry. Finally, in central Europe, Hirsch (2011) applies tabu search to minimize empty truckloads in round timber transport, and Gronalt and Hirsch (2007) propose a tabu search strategy for scheduling trucks in Austria.

12.4   RESEARCH CHALLENGES IN WOOD TRANSPORTATION PLANNING

Within the forest products sector, most transportation research deals with the forest-to-mill part of the chain, and very little is being done on the mill-to-market part of the chain and on the integration between the two. In particular, there is a need to understand the flow of information back up the chain from the markets. There is also the necessity to better understand and model the operations and the capacity issues at the mills, and to go beyond the current black box optimization approach. Most research projects deal with tactical/operational planning, but the industry is in need of real-time control tools that deal with information flows to provide better decisions on transportation. Such tools should of course handle the uncertainty inherent to real-time decision processes, but uncertainty should also be taken into account at the planning level. There are several sources of uncertainty, such as forest fires, extreme weather conditions, inaccuracies in the information collected relative to the availability of products and the inventory levels, unexpected variations in the demands at the mill, and so on. In addition, it is important to quantify the value of information to reduce the uncertainty and enable better, more robust, decision-making that can adapt in case of unforeseen events and still remain efficient. Techniques need to be developed to assess the benefits of merchandizing yards, and to develop models and methods for their location and their optimization. Multimodal and intermodal transportation of wood products, through truck, train, water, and barge/ship, has also been identified as an important research topic. Because they have an impact on all forest value chain activities, including transportation, environmental concerns, in particular the development of green products and carbon sequestration effects, can become an area of potentially fruitful research.

The challenges in wood transportation planning in Canada are the same, but some characteristics are worth emphasizing. As mentioned in Section 12.2.3, distances between forests and mills often exceed 150 km and forest road networks are expensive to build and maintain. Rail transportation is rarely applicable, water transportation is mainly concentrated in the West Coast region, and dispersion of transformation sites complicates the hauling process and reduces its efficiency. Trucking legislation is provincial, which leads to a large variety of trailer configurations used in each province.

With transportation-related research in the forest sector, it is important to link researchers to the industry and build research partnerships. In particular, researchers and industry representatives in Canada have identified with high priority four broad research areas:

- Real-time planning and management of materials and information flow (quality, yields, costs) starting from the markets/demand, and tracing back up the value chain through all distribution channels, manufacturing/processing, and raw material procurement (i.e., from market to forest)
- Merchandizing yards—benefits, location, planning, inventories, and control
- Real-time transportation and logistic systems planning and control
- Multimodal and intermodal transportation benefits, planning, and control

These research areas address global value chain issues, which are highly relevant to the forest products sector, as we have seen in Section 12.3.1. To the best of our knowledge, studying the benefits of multimodal wood transportation from an OR perspective has been seldom addressed in the literature (one notable exception is the work of Forsburg et al., 2005, which considers railroad transportation). Real-time planning and control of forest operations, in particular, real-time wood transportation, has been the object of very few studies (in Section 12.3, we mention only one contribution in this area: Rönnqvist and Ryan, 1995). This topic is highly relevant to the industry because it has the potential to overcome some of the limitations of existing dispatching systems by allowing a better treatment of the uncertainty inherent to forest operations and wood transportation planning.

REFERENCES

Ackerman, P. A. and R. E. Pulkki. 2004. Shorthaul pulpwood transport in South Africa: A network analysis case study. *The Southern African Forestry Journal* 201:43–51.

Acuna, M., M. Brown, and L. Mirowski. 2011. Improving forestry transport efficiency through truck schedule optimization: A case study and software tool for the Australian industry. In Proceedings of the FORMEC 2011, October 9–13, 2011, Graz, Austria.

Akay, A. E. 2003. Minimizing total cost of construction, maintenance, and transportation costs with computer-aided forest road design. Ph.D. Thesis, Oregon State University, OR.

Akay, A. E., K. Boston, and J. Sessions. 2005. The evolution of computer-aided road design systems. *International Journal of Forest Engineering* 16:73–79.

Akay, A. E. and J. Sessions. 2005. Applying the decision support system, TRACER, to forest road design. *Western Journal of Applied Forestry* 20:184–191.

Andalaft, N., P. Andalaft, M. Guignard, A. Magendzo, A. Wainer, and A. Weintraub. 2003. A problem of forest harvesting and road building solved through model strengthening and Lagrangean relaxation. *Operations Research* 51:613–628.

Andersson, G., P. Flisberg, B. Lidén, and M. Rönnqvist. 2008. RuttOpt: A decision support system for routing of logging trucks. *Canadian Journal of Forest Research* 38:1784–1796.

Anderson, A. E. and J. Nelson. 2004. Projecting vector-based road networks with a shortest path algorithm. *Canadian Journal of Forest Research* 34:1444–1457.

Anderson, A. E., J. D. Nelson, and R. G. D'Eon. 2006. Determining optimal road class and road deactivation strategies using dynamic programming. *Canadian Journal of Forest Research* 36:1509–1518.

Aruga, K. 2005. Tabu search optimization of horizontal and vertical alignments of forest roads. *Journal of Forest Research* 10:275–284.

Aruga, K., W. Chung, A. E. Akay, J. Sessions, and E. S. Miyata. 2007. Incorporating soil surface erosion prediction into forest road alignment optimization. *International Journal of Forest Engineering* 18:24–32.

Aruga, K., J. Sessions, and A. E. Akay. 2005a. Heuristic planning techniques applied to forest road profiles. *Journal of Forest Research* 10:83–92.

Aruga, K., J. Sessions, A. E. Akay, and W. Chung. 2005b. Simultaneous optimization of horizontal and vertical alignments of forest roads using tabu search. *International Journal of Forest Engineering* 16:137.

Aruga, K., T. Tasaka, J. Sessions, and S. Miyata. 2006. Tabu search optimization of forest road alignments combined with shortest paths and cubic splines. *Croatian Journal of Forest Engineering* 27:37–47.

Audy, J. F., S. D'Amours, and L.-M. Rousseau. 2007a. Collaborative planning in a log truck pickup and delivery problem. In Proceedings of the Sixth Triennial Symposium on Transportation Analysis, June 10–15, Phuket, Thailand.

Audy, J.-F., S. D'Amours, L.-M. Rousseau, J. Favreau, and P. Marier. 2007b. Virtual transportation manager: A web-based system for transportation optimization in a network of business units. In Proceedings of the Third Forest Engineering Conference, October 1–4, Mont-Tremblant, Canada.

Audy, J.-F., N. El Hachemi, L. Michel, and L.-M. Rousseau. 2011. Solving a combined routing and scheduling problem in forestry. In Proceedings of the Industrial Engineering and Systems Management (IESM), 1–10, Metz, France, May 25–27.

Audy, J.-F., S. D'Amours, and M. Rönnqvist. 2012. Planning methods and decision support systems in vehicle routing problem for timber transportation: A review. Technical Report CIRRELT-2012-38, CIRRELT, Canada.

Bredström, D., J. T. Lundgren, M. Rönnqvist, D. Carlsson, and A. Mason. 2004. Supply chain optimization in the pulp mill industry: IP models, column generation and novel constraint branches. *European Journal of Operational Research* 156:2–22.

Broman, H., M. Frisk, and M. Rönnqvist. 2009. Supply chain planning of harvest and transportation operations after the storm Gudrun. *INFOR: Information Systems and Operational Research* 47:235–245.

Carlgren, C.-G., D. Carlsson, and M. Rönnqvist. 2006. Log sorting in forest harvest areas integrated with transportation planning using backhauling. *Scandinavian Journal of Forest Research* 21:260–271.

Carlsson, D., S. D'Amours, A. Martel, and M. Rönnqvist. 2008. Decisions and methodology for planning the wood fibre flow in the forest supply chain. In *Recent Developments in Supply Chain Management*, Helsinki University Press, Finland, 11–39.

Carlsson, D., S. D'Amours, A. Martel, and M. Rönnqvist. 2009. Supply chain planning models in the pulp and paper industry. *INFOR: Information Systems and Operational Research* 47:167–183.

Carlsson, D. and M. Rönnqvist. 2005. Supply chain management in forestry—Case studies at Södra Cell AB. *European Journal of Operational Research* 163:589–616.

Carlsson, D. and M. Rönnqvist. 2007. Backhauling in forest transportation: Models, methods, and practical usage. *Canadian Journal of Forest Research* 37:2612–2623.

Chung, W. and J. Sessions. 2000. NETWORK 2000: A program for optimizing large fixed and variable cost transportation problems. In Proceedings of the Eighth Symposium on Systems Analysis in Forest Resources, September 27–30, G. J. Arthaud and T. M. Barrett (eds.), Snowmass Village, CO, pp. 109–120.

Chung, W. and J. Sessions. 2001. NETWORK 2001—Transportation planning under multiple objectives. In Proceedings of the International Mountain Logging and 11th Pacific Northwest Skyline Symposium, 10–12, Seattle, WA, December 10–12.

Church, R. L., A. T. Murray, and A. Weintraub. 1998. Locational issues in forest management. *Location Science* 6:137–153.

Clark, M., D. Meller, and P. McDonald. 2000. A three-stage heuristic for harvest scheduling with access road network development. *Forest Science* 46:204–218.

Cochran, J. J., L. A. Cox, P. Keskinocak, J. P. Kharoufeh, J. C. Smith, S. D'Amours, R. Epstein, A. Weintraub, and M. Rönnqvist. 2011. Operations research in forestry and forest products industry. In *Wiley Encyclopedia of Operations Research and Management Science*, J. J. Cochran, L. A. Cox, P. Keskinocak, J. P. Kharoufeh and J. C. Smith (eds.). Hoboken, NJ: John Wiley & Sons.

Cossens, P. 1993. Evaluation of ASICAM for truck scheduling in New Zealand. Technical report no. 01435273, Logging Industry Research Organisation, Rotorua, New Zealand.

Coulter, E. D., J. Sessions, and M. G. Wing. 2006. Scheduling forest road maintenance using the analytic hierarchy process and heuristics. *Silva Fennica* 40:143.

D'Amours, S., J.-M. Frayret, and A. Rousseau. 2003. De la forêt au client: Pourquoi viser une gestion intégrée du réseau de création de valeur? Technical report, FOR@C, Université Laval, Quebec City, Canada.

D'Amours, S., M. Rönnqvist, and A. Weintraub. 2008. Using operational research for supply chain planning in the forest products industry. *INFOR: Information Systems and Operational Research* 46:265–281.

Dean, D. J. 1997. Finding optimal routes for networks of harvest site access roads using GIS-based techniques. *Canadian Journal of Forest Research* 27:11–22.

Desautels, R., R. Després, F. Dufresne, G. Gilbert, S. Leblanc, L. Méthot, Y. Provencher, G. Rochette, B. Senécal, and C. Warren. 2009. Voirie forestière. In *Manuel de foresterie*, Chapter 30, Éditions MultiMondes, Québec, Canada, pp. 1187–1244.

Douglas, R. A. and B. S. Henderson. 1988. Computer-assisted resource access road route location. *Canadian Journal of Civil Engineering* 15:299–305.

El Hachemi, N., M. Gendreau, and L.-M. Rousseau. 2008. Solving a log-truck scheduling problem with constraint programming. In *Integration of AI and OR Techniques in Constraint Programming for Combinatorial Optimization Problems*, L. Perron and M. A. Trick (eds.), *Lecture Notes in Computer Science*, Vol. 5015, pp. 293–297. New York: Springer.

El Hachemi, N., M. Gendreau, and L.-M. Rousseau. 2009. A heuristic to solve the weekly log-truck scheduling problem. In Proceedings of the International Conference on Industrial Engineering and Systems Management, May 13–15, Montreal, Canada.

El Hachemi, N., M. Gendreau, and L.-M. Rousseau. 2010. A hybrid constraint programming approach to the log-truck scheduling problem. *Annals of Operations Research* 184:163–178.

El Hachemi, N., M. Gendreau, and L.-M. Rousseau. 2013. A heuristic to solve the synchronized log-truck scheduling problem. *Computers and Operations Research* 40:666–673.

Epstein, R., R. Morales, J. Seron, and A. Weintraub. 1999. Use of OR systems in the Chilean forest industries. *Interfaces* 29:7–29.

Epstein, R., M. Rönnqvist, and A. Weintraub. 2007. Forest transportation. In *Handbook of Operations Research in Natural Resources*, A. Weintraub, C. Romero, T. Bjorndal, R. Epstein and J. Miranda (eds.), Vol. 99, pp. 391–403. Boston, MA: Springer.

Eriksson, L. O. and R. Björheden. 1989. Optimal storing, transport and processing for a forest–fuel supplier. *European Journal of Operational Research* 43:26–33.

Eriksson, J. and M. Rönnqvist. 2003. Decision support system/tools: Transportation and route planning: Akarweb—A web based planning system. In Proceedings of the Second Forest Engineering Conference, May 12–15, Växjö, Sweden, pp. 48–57.

Eschenbach, E. A., R. Teasley, C. Diaz, and M. A. Madej. 2004. Decision support for road decommissioning and restoration by using genetic algorithms and dynamic programming. In Proceedings of the Redwood Forest Science Symposium: What Does the Future Hold, Rohnert Park, CA, March 15–17.

Flisberg, P., B. Lidén, and M. Rönnqvist. 2009. A hybrid method based on linear programming and tabu search for routing of logging trucks. *Computers and Operations Research* 36:1122–1144.

Forsburg, M., M. Frisk, and M. Rönnqvist. 2005. FlowOpt—A decision support tool for strategic and tactical transportation planning in forestry. *International Journal of Forest Engineering* 16:101–114.

Frisk, M., M. Göthe-Lundgren, K. Jörnsten, and M. Rönnqvist. 2010. Cost allocation in collaborative forest transportation. *European Journal of Operational Research* 205:448–458.

Ghaffariyan, M. R., K. Stampfer, J. Sessions, T. Durston, M. Kuehmaier, and C. H. Kanzian. 2010. Road network optimization using heuristic and linear programming. *Journal of Forest Science* 56:137–145.

Gingras, C., J.-F. Cordeau, and G. Laporte. 2007. Un algorithme de minimisation du transport à vide appliqué à l'industrie forestière. *INFOR: Information Systems and Operational Research* 45:41–47.

Gronalt, M. and P. Hirsch. 2007. Log-truck scheduling with a tabu search strategy. In *Metaheuristics, Operations Research/Computer Science Interfaces Series*, K. F. Doerner, M. Gendreau, P. Greistorfer, W. Gutjahr, R. F. Hartl and M. Reimann (eds.), pp. 65–88. Boston, MA: Springer.

Guignard, M., C. Ryu, and K. Spielberg. 1998. Model tightening for integrated timber harvest and transportation planning. *European Journal of Operational Research* 111:448–460.

Gunnarsson, H. and M. Rönnqvist. 2008. Solving a multi-period supply chain problem for a pulp company using heuristics: An application to Södra Cell AB. *International Journal of Production Economics* 116:75–94.

Gunnarsson, H., M. Rönnqvist, and D. Carlsson. 2006. Integrated production and distribution planning for Södra Cell AB. *Journal of Mathematical Modelling and Algorithms* 6:25–45.

Gunnarsson, H., M. Rönnqvist, and J. T. Lundgren. 2004. Supply chain modelling of forest fuel. *European Journal of Operational Research* 158:103–123.

Han, S.-K. and G. E. Murphy. 2011. Trucking productivity and costing model for transportation of recovered wood waste in Oregon. *Forest Products Journal* 61:552–560.

Han, S.-K. and G. Murphy. 2012a. Predicting loaded on-highway travel times of trucks hauling woody raw material for improved forest biomass utilization in Oregon. *Western Journal of Applied Forestry* 27:92–99.

Han, S.-K. and G. E. Murphy. 2012b. Solving a woody biomass truck scheduling problem for a transport company in western Oregon, USA. *Biomass and Bioenergy* 44:47–55.

Henningsson, M., J. Karlsson, and M. Rönnqvist. 2006. Optimization models for forest road upgrade planning. *Journal of Mathematical Modelling and Algorithms* 6:3–23.

Hirsch, P. 2011. Minimizing empty truck loads in round timber transport with tabu search strategies. *International Journal of Information Systems and Supply Chain Management* 4:15–41.

Ichihara, K., T. Tanaka, I. Sawaguchi, S. Umeda, and K. Toyokawa. 1996. The method for designing the profile of forest roads supported by genetic algorithm. *Journal of Forest Research* 1:45–49.

Kanzaki, K. 1973. On the decision of profile line of forest road by dynamic programming. *Journal of the Japanese Forestry Society* 55:144–148.

Karlsson, J., M. Rönnqvist, and M. Frisk. 2006. RoadOpt: A decision support system for road upgrading in forestry. *Scandinavian Journal of Forest Research* 21:5–15.

Kong, J., M. Rönnqvist, and M. Frisk. 2012. Modeling an integrated market for sawlogs, pulpwood, and forest bioenergy. *Canadian Journal of Forest Research* 42:315–332.

LeBel, L., D. Cormier, L. Desrochers, D. Dubeau, J. Dunnigan, J. Favreau, J.-F. Gingras, M. Hamel, P. Meek, J. Michaelsen, C. Sarthou, and N. Thiffault. 2009. Opérations forestières et transport des bois. In *Manuel de foresterie*, Chapter 31, Éditions MultiMondes, Québec, Canada, pp. 1245–1304.

Linnainmaa, S., J. Savola, and O. Jokinen. 1995. EPO: A knowledge based system for wood procurement management. In Proceedings of the Seventh Annual Conference on Artificial Intelligence, August 20–23, Montreal, Canada, pp. 107–113.

Liu, K. and J. Sessions. 1993. Preliminary planning of road systems using digital terrain models. *International Journal of Forest Engineering* 4:27–32.

MacKay, D. 1978. *The Lumberjacks*. Toronto and New York: McGraw-Hill Ryerson.

Marier, P., J.-F. Audy, C. Gingras, and S. D'Amours. 2007. Collaborative wood transportation with the Virtual Transportation Manager. In Proceedings of the International Scientific Conference on Hardwood Processing, Quebec City, Canada, September 24–26.

McDonald, T., S. Taylor, and J. Valenzuala. 2001a. Potential for shared log transport services. In Proceedings of the 24th Annual Meeting of the Council on Forest Engineering, Snowshoe, WV.

McDonald, T. P., K. Haridass, and J. Valenzuala. 2010. Mileage savings from optimization of coordinated trucking. In Proceedings of the 33rd Annual Meeting of the Council on Forest Engineering: Fueling the Future, Auburn, AL, June 6–9.

McDonald, T. P., S. E. Taylor, R. B. Rummer, and J. Valenzuela. 2001b. Information needs for increasing log transport efficiency. In Proceedings of the First International Precision Forestry Cooperative Symposium, Seattle, WA, pp. 181–184.

Mendell, B. C., J. A. Haber, and T. Sydor. 2006. Evaluating the potential for shared log truck resources in middle Georgia. *Southern Journal of Applied Forestry* 30:86–91.

Murphy, G. E. 2003. Reducing trucks on the road through optimal route scheduling and shared log transport services. *Southern Journal of Applied Forestry* 27:198–205.

Murphy, G. E. and H. Stander. 2007. Robust optimisation of forest transportation networks: A case study. *Southern Hemisphere Forestry Journal* 69:117–123.

Murray, A. T. 1998. Route planning for harvest site access. *Canadian Journal of Forest Research* 28:1084–1087.

Nelson, J. D. and S. T. Finn. 1991. The influence of cut-block size and adjacency rules on harvest levels and road networks. *Canadian Journal of Forest Research* 21:595–600.

Olsson, L. 2004. Optimisation of forest road investments and the roundwood supply chain. Ph.D. Thesis, Department of Forest Economics, Swedish University of Agricultural Sciences, Umea.

Palander, T. and J. Väätäinen. 2005. Impacts of interenterprise collaboration and backhauling on wood procurement in Finland. *Scandinavian Journal of Forest Research* 20:177–183.

Palander, T., J. Väätäinen, S. Laukkanen, and J. Malinen. 2004. Modeling backhauling on Finnish energy–wood network using minimizing of empty routes. *International Journal of Forest Engineering* 15:79–84.

Palmgren, M. 2005. Optimal truck scheduling: Mathematical modeling and solution by the column generation principle. Ph.D. Thesis, Linköpings Universitet, Sweden.

Palmgren, M., M. Rönnqvist, and P. Varbrand. 2003. A solution approach for log truck scheduling based on composite pricing and branch and bound. *International Transactions in Operational Research* 10:433–447.

Palmgren, M., M. Rönnqvist, and P. Varbrand. 2004. A near-exact method for solving the log-truck scheduling problem. *International Transactions in Operational Research* 11:447–464.

Pulkki, R. E. 1982. Introduction to long-distance transport of wood. In Proceedings of the Seminar on Water Transport of Wood, R. E. Pulkki (Ed.), National Board of Vocational Education, Forestry Training Programme for Developing Countries, Helsinki, Finland, pp. 179–206, May 31–June 18.

Pulkki, R. E. 1996. Water crossings versus transport cost: A network analysis case study. *Journal of Forest Engineering* 7:59–64.

Rey, P. A., J. A. Munoz, and A. Weintraub. 2009. A column generation model for truck routing in the Chilean forest industry. *INFOR: Information Systems and Operational Research* 47:215–221.

Richards, E. W. and E. A. Gunn. 2000. A model and tabu search method to optimize stand harvest and road construction schedules. *Forest Science* 46:188–203.

Richards, E. W. and E. A. Gunn. 2003. Tabu search design for difficult forest management optimization problems. *Canadian Journal of Forest Research* 33:1126–1133.

Rönnqvist, M. 2003. Optimization in forestry. *Mathematical Programming* 97:267–284.

Rönnqvist, M. and D. Ryan. 1995. Solving truck despatch problems in real time. In Proceedings of the 31th Annual Conference of the Operational Research Society of New Zealand, Wellington, New Zealand, August 31–September 1, pp. 165–172.

Rummukainen, H., T. Kinnari, and M. Laakso. 2009. Optimization of wood transportation. In Papermaking Research Symposium, Kuopio, Finland.

Sessions, J. 1992. Using network analysis for road and harvest planning. In Proceedings: Workshop on Computer Supported Planning of Roads and Harvesting, J. Sessions (Ed.), 3. Feldafing, Germany. Published by Oregon State University, OR, pp. 26–35.

Sessions, J., K. Boston, R. Thoreson, and K. Mills. 2006. Optimal policies for managing aggregate resources on temporary forest roads. *Western Journal of Applied Forestry* 21:207–216.

Shen, Z. and J. Sessions. 1989. Log truck scheduling by network programming. *Forest Products Journal* 39:47–50.

Statistics Canada. 2012. Rail in Canada 2009, Table 4, Rail transportation, Length of track operated at December 31, all carriers. http://www.statcan.gc.ca/pub/52-216-x/2009000 /t001-eng.htm (accessed: May 6, 2016).

Stückelberger, J., H. Heinimann, W. Chung, and M. Ulber. 2006. Automatic road network planning for multiple objectives. In Proceedings of the 29th Annual Meeting of the Council on Forest Engineering, Coeur d'Alene, Idaho July 30–August 2, pp. 233–248.

Tan, J. 1999. Locating forest roads by a spatial and heuristic procedure using microcomputers. *International Journal of Forest Engineering* 10:91–100.

Thompson, M. and J. Sessions. 2008. Optimal policies for aggregate recycling from decommissioned forest roads. *Environmental Management* 42:297–309.

Thompson, M., J. Sessions, K. Boston, A. Skaugset, and D. Tomberlin. 2010. Forest road erosion control using multiobjective optimization. *Journal of the American Water Resources Association* 46:712–723.

Toivonen, T. 1979. Waterway standards for floating, bundle towing and the construction of special channels. In Proceedings of IUFRO S3.01-04 Symposium on Water Transport of Wood, R. E. Pulkki and P.-J. Kuitto (eds.), Res. Notes 40, Department of Logging and Utilization of Forest Products, University of Helsinki, Finland, June 29–July 4, pp. 93–108.

Troncoso, J., S. D'Amours, P. Flisberg, M. Rönnqvist, and A. Weintraub. 2011. A mixed integer programming model to evaluate integrating strategies in the forest value chain: A case study in the Chilean forest industry. Technical report CIRRELT-2011-28, CIRRELT, Canada.

Uusitalo, J. 2010. *Introduction to Forest Operations and Technology*. Tampere, Finland: JVP Forest Systems Oy.

Vila, D., A. Martel, and R. Beauregard. 2006. Designing logistics networks in divergent process industries: A methodology and its application to the lumber industry. *International Journal of Production Economics* 102:358–378.

Weintraub, A., A. Magendzo, A. Magendzo, D. Malchuk, G. Jones, and M. Meacham. 1995. Heuristic procedures for solving mixed-integer harvest scheduling-transportation planning models. *Canadian Journal of Forest Research* 25:1618–1626.

Weintraub, A. and D. Navon. 1976. A forest management planning model integrating silvicultural and transportation activities. *Management Science* 22:1299–1309.

Weintraub, A. and R. Epstein. 2002. The supply chain in the forest industry: Models and linkages. In *Supply Chain Management: Models, Applications, and Research Directions*, J. Geunes, P. M. Pardalos and H. E. Romeijn (eds.), Vol. 62, pp. 343–362. Boston, MA: Kluwer Academic Publishers.

Weintraub, A., R. Epstein, R. Morales, J. Seron, and P. Traverso. 1996. A truck scheduling system improves efficiency in the forest industries. *Interfaces* 26:1–12.

Wibstad, K. 1979. Water transport of wood: regional report from Europe. In Proceedings of IUFRO S3.01-04 Symposium on Water Transport of Wood, R. E. Pulkki and P.-J. Kuitto (eds.), Res. Notes 40, Department of Logging and Utilization of Forest Products, University of Helsinki, Finland, June 29-July 4, pp. 9–18.

Index